Model Theory and the Philosophy of Mathematical Practice

Formalization without Foundationalism

JOHN T. BALDWIN

University of Illinois, Chicago

CAMBRIDGE
UNIVERSITY PRESS

CAMBRIDGE
UNIVERSITY PRESS

University Printing House, Cambridge CB2 8BS, United Kingdom

One Liberty Plaza, 20th Floor, New York, NY 10006, USA

477 Williamstown Road, Port Melbourne, VIC 3207, Australia

314-321, 3rd Floor, Plot 3, Splendor Forum, Jasola District Centre, New Delhi - 110025, India

79 Anson Road, #06-04/06, Singapore 079906

Cambridge University Press is part of the University of Cambridge.

It furthers the University's mission by disseminating knowledge in the pursuit of education, learning and research at the highest international levels of excellence.

www.cambridge.org
Information on this title: www.cambridge.org/9781316638835
DOI: 10.1017/9781316987216

© John T. Baldwin 2018

First published 2018
First paperback edition 2019

A catalogue record for this publication is available from the British Library

ISBN 978-1-107-18921-8 Hardback
ISBN 978-1-316-63883-5 Paperback

Contents

Figures

Acknowledgments

My interest and ability to tell the story of modern model theory to philosophers were immensely stimulated by my frequent attendance, since 2009, at Midwest PhilMath Workshop (MWPMW), which is led by Mic Detlefsen. Conversations with the participants provided further impetus. But this book would not have been written without the constant encouragement and advice of Juliette Kennedy and Andrès Villaveces in developing the plan and then the details of the text. On a less personal basis, the work of Ken Manders in philosophy and Robin Hartshorne's exposition of Euclid and Hilbert were both provocative and inspiring. Michael Hallett John Mumma, and Robert Thomas were particularly helpful with comments, translations, and advice on the geometry chapters. The support of the University of Illinois at Chicago and the Simons Foundation is gratefully acknowledged.

I gratefully acknowledge the comments, conversations, references, and/or texpertise of the following mathematicians and philosophers and others who slip my mind at the moment: Aida Alibek, William D'Allesandro, Sharon Baldwin, Will Boney, Alexandre Borovik, Greg Cherlin, Philip Ehrlich, Paul Eklof, Sol Feferman, Curtis Franks, James Freitag, Isaac Goldbring, Bradd Hart, Robin Hartshorne, William Howard, Maciej Kłeczek, Chris Laskowski, Michael Lieberman, Kenneth Manders, David Marker, Julien Narboux, Victor Pambuccian, David Pierce, Marcus Rossberg, Dana Scott, Saharon Shelah, Wilfrid Sieg, Kostya Slutsky, Craig Smorynski, Daniel Sutherland, Jouko Väänänen, Sebastien Vasey, Carol Wood, Fernando Zalamea. In particular, Rami Grossberg and Dimitris Tsementzis each read most of it, while Neil Barton read it all, and they offered invaluable corrections and comments. The errors that remain are mine.

The articles [Baldwin 2013a, Baldwin 2017a, Baldwin 2017b, Baldwin 2014] have been expanded and reshaped in this volume. The first three overlap primarily with Chapters 9–12; the last with Chapters 1–3 and 14. One diagram was taken from [Baldwin 1988a] and several from David Joyce's edition of *Euclid's Elements* on the Internet. I thank Cambridge University Press, Oxford University Press, the Association of Symbolic Logic, and Professor Joyce for permission to use this material.

Introduction

The announcement[1] for a conference on Philosophy and Model Theory in 2010 began:

Model theory seems to have reached its zenith in the sixties and the seventies, when it was seen by many as virtually identical to mathematical logic. The works of Gödel and Cohen on the continuum hypothesis, though falling only indirectly within the domain of model theory, did bring to it some reflected glory. The works of Montague or Putnam bear witness to the profound impact of model theory, both on analytical philosophy and on the foundations of scientific linguistics.

My astonished reply to the organizers[2] began:

It seems that I have a very different notion of the history of model theory. As the paper at [Baldwin 2010] points out, I would say that modern model theory begins around 1970 and the most profound mathematical results including applications in many other areas of mathematics have occurred since then, using various aspects of Shelah's paradigm shift. I must agree that, while in my view there are significant philosophical implications of the new paradigm, they have not been conveyed to philosophers.

This book is an extended version of that reply to what I will call *the provocation*.[3] I hope to convince the reader that the more technically sophisticated model theory of the last half century introduces new philosophical insights about mathematical practice[4] that reveal how this recent model theory resonates philosophically, impacting in particular such basic notions as syntax and semantics,[5] structure, completeness, categoricity, and

---

[1] This delightful Paris conference was organized by Brice Halimi, Jean-Michel Salanskis, and Denis Bonnay. The full announcement is here. http://dep-philo.u-paris10.fr/dpt-ufr-phillia-philosophie/la-recherche/les-colloques/philosophy-and-model-theory-314434.kjsp
[2] Letter to Halimi, September 20, 2009.
[3] The announcement contains a number of astute observations that we will comment on in due course.
[4] See page 5.
[5] We use semantics for 'semantics of formal language' (page 300). Among the accomplishments of model theory alluded to in the provocation are contributions to the philosophy of language as in [Button & Walsh 2016]; our topic here is the philosophy of mathematics.

axiomatization. Thus, large parts of the book are devoted to introducing and describing, for those not familiar with model theory, such topics as the stability theoretic classification of first order theories, its applications across mathematics, and that its interaction with classical algebra is inevitable (Chapter 5.6). Much of this exposition will be in the context of discussing the *paradigm shift*.[6] In short, the paradigm around 1950 concerned the study of *logics*; the principal results were completeness, compactness, interpolation, and joint consistency theorems. Various semantic properties of theories were given syntactic characterizations but there was no notion of partitioning all *theories* by a family of properties. After the paradigm shift there is a systematic search for a finite set of syntactic conditions which divide first order theories into disjoint classes such that models of different theories in the same class have similar mathematical properties. In this framework one can compare different areas of mathematics by checking where theories formalizing them lie in the classification.

Framework for Formalization: We *always* speak of formalizing a *particular* mathematical topic. A *formalization* of a mathematical area specifies a vocabulary (representing the primitive notions of the topic), a logic, and the axioms[7] (postulates in the technical sense of Euclid) for that topic. By a topic we might mean group theory or algebraic geometry, or perhaps set theory. We argue that comparing (usually first order) formalizations of different mathematical topics is a better tool for investigating the connections between their methods and results than a common coding of them into set theory. Avoiding a global foundation allows us to evade the Gödel phenomena and study instead different 'tame'[8] areas of mathematics: e.g., any stable or o-minimal theory, real algebraic geometry, differentially closed fields, etc.

---

[6] See further detail on page 8. While I invoke Kuhn's term, I don't want to take on all of its connotations. My meaning is attuned with Harris ([Harris 2015], 23) 'I soon found myself caught up in the thrill of the first encounter between two research programs, each of a scope and precision that would have been inconceivable to previous generations, each based on radically new heuristics, each experienced by my teachers' generation as a paradigm shift.' He later gives Weil, Grothendieck, and Langlands as examples of creators of paradigm shifts. The authors in the collection *Revolutions in Mathematics* [Gillies 2008a] generally argue that a tenable notion of revolution in mathematics must be much more restrictive than Kuhn's.

[7] A *theory* is the set of consequences of such an axiom set. Thus, except for the trivial case of all validities, a theory is always about some specific topic.

[8] Model theorists loosely call a theory tame if it does not exhibit the *Gödel phenomena* – self-reference, undecidability, and a pairing function. One source of the word is Grothendieck's notion of 'tame topology'. See Chapter 6.3 for Wilkie's explanation and [Teissier 1997] and [Dries 1999].

In Chapter 1 we elaborate on this notion of formalization and explain the importance of each of the choices: vocabulary, logic, and axioms.

We develop some uses of first order formalization in studying the organization of mathematical practice, and some consequences of this new organization for traditional mathematics, and explore how the analysis of formalization here affects some standard topics in the philosophy of mathematics. This book supports four main theses.

Theses:

(1) Contemporary model theory makes formalization of *specific mathematical areas* a powerful tool to investigate both mathematical problems and issues in the philosophy of mathematics (e.g. methodology, axiomatization, purity, categoricity, and completeness).
(2) Contemporary model theory enables systematic comparison of local formalizations for distinct mathematical areas in order to organize and do mathematics, and to analyze mathematical practice.
(3) The choice of vocabulary and logic appropriate to the particular topic are central to the success of a formalization. The technical developments of first order logic have been more important in other areas of modern mathematics than such developments for other logics.
(4) The study of geometry is not only the source of the idea of axiomatization and many of the fundamental concepts of model theory, but geometry itself (through the medium of geometric stability theory) plays a fundamental role in analyzing the models of tame theories and solving problems in other areas of mathematics.

At first glance the first thesis may seem banal. Isn't this just the justification for the study of symbolic logic? Isn't this claim just a rehash of positivistic themes of the 1930s? Not at all. The examples illustrating the first aspect of Thesis 1, mathematical problems, concern specific[9] mathematical topics and we address them using modern model theoretic techniques (e.g. Chapter 6 and Theorem 10.3.3). For the second aspect one might ask, 'What is the philosophy of mathematics?' Avigad [Avigad 2007] answers as follows, 'Traditionally, the two central questions for the philosophy of mathematics are: What are mathematical objects? How do we (or can we) have knowledge of them?' The traditional ontological issues are not in the scope of this book. Rather we address the second question by studying epistemological

---

[9] We indicate similarities and distinctions from the original foundational project on page 9.

issues concerning the organization and understanding of mathematics as it is practiced. By a local foundation for a mathematical area, we mean a specification of the area by a set of axioms. Hilbert (in the *Grundlagen*) and Bourbaki use informal axioms. A key goal[10] of this book is to show the mathematical advantages of stating these axioms in (usually first order) formal logic.

Ken Manders clarifies the contrast between the traditional focus on reliability and our focus on the clarity and interaction of mathematical concepts. He outlines the distinction between the foundational[11] and the model theoretic approach. He begins colloquially[12] by stipulating that traditional epistemology concerns the correctness of mathematical assertions taken as knowledge claims.

Is it **all right?**, traditional epistemology asks about knowledge claims. All schools in 'logical foundations of mathematics' share this concern for reliability. But a long-term look at achievements in mathematics shows that *genuine mathematical accomplishment consists primarily in making* **clear** *by using new concepts*: …Representations and methods from the reliability programs are not always appropriate. We need to be able to emphasize special features of a given mathematical area and its relationships to others, rather than how it fits into an absolutely general pattern. Model theoretic algebra works in just this way. A model theoretic approach may be able to bring out the point of algebraic methods in number theory and geometry. [Manders 1987]

Manders argues that a crucial aspect of mathematical progress is the introduction of new concepts to clarify a particular area. We illustrate this insight in various contexts, e.g., in our discussion of purity in Chapter 12. While in [Manders 1987] Manders discusses only model theoretic algebra, we argue that not only the standpoint of model theoretic algebra (Chapter 4.4), but to an even greater degree the standpoint of the 'one model theory' (Chapter 5.6) that was obtained by integrating the methods of classification theory (Chapters 5 and 6) with model theoretic algebra, clarify and unify concepts in various areas of mathematics by finding unexpected similarities across fields of mathematics. Thus we don't abandon the epistemological enterprise but we focus on clarification rather than verification. How does one shape mathematical theories to best represent

---

[10] The details of formalization are laid out in Chapter 1. In that chapter we consider the preformal mathematics as a set of concepts; in Chapter 9 we take Detlefsen's view of a data set of accepted propositions. Our view certainly fits within the notion of the *hypothetical conception of mathematics* described in [Ferreirós & Gray 2008].

[11] That is, a proposed global axiomatic system to include all (or almost all) of mathematics.

[12] The italics in the quotation are mine, but the boldface is original.

the inherent logic of the material? What similar patterns of reasoning or combinatorial features appear in various areas of mathematics?

While investigating how mathematicians arrive at clear concepts, we examine how the understanding of mathematical concepts changes (say the notion of number for the Greeks or for us now), and how formalization forces a clarifying analysis of concepts. For example, the first order axiomatization of geometry (Chapters 9–10) provides much finer information than the second order axiomatization.

One might incorrectly suspect the book is a defense of the 'formalism' leg of the foundational triumvirate. Rather, we deal with 'formalization' as a scheme for organizing mathematics without addressing any of the ontological concerns of 'formalists'.

The second thesis addresses the deep interactions between model theory and traditional mathematics. We describe how the new paradigm, by focusing attention on the content of particular fields rather than on a reduction to a global theory of all mathematics, provides connections across fields of mathematics that leads to mathematical advances (Chapter 6). This fact is base data for the study of mathematical practice. Thus the epistemological focus extends from reliability to more general concerns of clarity and coherence. Our approach is aligned with those grouped as *philosophy of mathematical practice* on the web page of the APMP:[13] 'Such approaches include the study of a wide variety of issues concerned with the way mathematics is done, evaluated, and applied, and in addition, or in connection therewith, with historical episodes or traditions, applications, educational problems, cognitive questions, etc.' These approaches are exemplified, though in many different ways, by such authors as Arana, Ehrlich, Hallett, McLarty, Maddy, Mancosu, Manders, Schlimm, and Tappenden.

This movement[14] is more fully described in the introduction to *The Philosophy of Mathematical Practice* [Mancosu 2008b] in which Mancosu makes the philosophical aims more precise. He notes that Benaceraff's 'rightly influential' articles set the guiding question as 'how, if there are

---

[13] Association for the Philosophy of Mathematical Practice. http://institucional.us.es/apmp/index_about.htm

[14] Various terms have been used to describe this well-known distinction between various approaches to the philosophy of mathematics. The side I refer to as 'traditional philosophy of mathematics' is dubbed 'philosophy of Mathematics' (Harris, page 30 of [Harris 2015]) or 'Foundations of Mathematics' (Simpson in clarifying his view on the Foundations of Mathematics Listserv). While what I call 'philosophy of mathematical practice' becomes 'philosophy of mathematics' (Harris) or 'foundations of mathematics' (Simpson) and philosophy of 'real mathematics' for Corfield [Corfield 2003]. Unlike Corfield, as a mathematician and model theorist for forty odd years, I regard model theory as 'real mathematics'.

abstract objects, could we have access to them?' Then he goes on to describe the positive goals of his book.

> The authors in this collection ... believe that the epistemology of mathematics has to be extended well beyond its present confines to address epistemological issues that have to do with fruitfulness, evidence, visualization, diagrammatic reasoning, understanding, explanation, which are orthogonal to the problem of access to 'abstract objects'. ([Mancosu 2008b], 1–2)

In this spirit the goal in this book is to study not just the logical foundations of mathematics, but to understand the role of logic in contemporary mathematics. This discussion invokes not only some existing philosophical literature, but programmatic pronouncements by such authors as Bourbaki, Hilbert, Hrushovski, Kazhdan, Macintyre, Pillay, Shelah, Tarski, and Zilber that often have influenced, if not determined, mathematical practice. We introduce Shelah's methodological command 'find dividing lines' on the next page; it recurs often in the text when illustrating the organization of model theory. In *The Statesman*, a dialog devoted in part to analyzing 'good definition', Plato advises to 'cut through the middle.' Chapter 13 develops the analogy between this dictum of Plato and Shelah's principle of seeking 'dividing lines' to understand the relations among mathematical theories.

The third thesis has several aspects. Both mathematical and philosophical questions may have different answers depending on the choice of logic. Chapter 8 expounds the vast variation in the amount of entanglement with axiomatic set theory among first order, infinitary, and second order logic. Metatheoretic investigation of first order logic gives finer information than second order logic about categoricity, definability, and axiomatization. This finer information, particularly about definability, provides not only spectacular pure model theoretic results[15] but also new tools for the study of traditional mathematics.

Shapiro's *Foundations without Foundationalism: A Case for Second Order Logic* inspired the title of this work. He writes,[16] 'One of the main themes of this [Shapiro's] book is a thorough antifoundationalism ... The view under attack is the thesis that there is a unique best foundation of mathematics and a concomitant view that there is a unique best logic – one size fits all.'

---

[15] One goal of this book is to explain the mathematical, philosophical, and methodological significance of the *main gap theorem* (Theorem 5.5.1). It established a dichotomy between two kinds of theories: ones where the models are decomposed in a uniform manner in terms of countable models and those which there are the maximal number of models in $\aleph_\alpha$ for every $\alpha > 0$. See page 64.

[16] ([Shapiro 1991], 220)

We have gone to some lengths to identify inadequacies of first order logic, and we have shown how second order logic, with standard semantics, overcomes many of these shortcomings.' Indicating his alternative on page 29, he quotes Skolem's characterization of an 'opportunistic' view of foundations: to have a foundation which makes it possible to develop present day mathematics, and which is consistent so far as known yet. Shapiro proclaims, 'We might say that it is foundations without foundationalism.' So he still seeks reliability but without the high standard, maximally immune to rational doubt, and reductionist nature of traditional programs.

Shapiro lays out a positive argument for founding mathematics using second order logic. He argues that basic analysis is comfortably axiomatized in second order logic and that notions such as closure of a subgroup, well-order, and infinite are all naturally defined in terms of second order logic. He identifies first order logic as deficient because it is 'subject to the compactness and Löwenheim theorem' and thus allows nonstandard models (page 112).

The view here is somewhat orthogonal. Our position is not anti-foundationalist; we just focus on clarity and organization rather than reliability. On the one hand, we argue that one devise local foundations for particular areas of mathematics rather than seeking a single foundation. And we agree that normal informal mathematical reasoning would most easily be formalized in second order logic. But in contrast to discussing the foundations of arithmetic and analysis, our focus is on the role of formalization in solving problems of modern mathematics. For us, compactness and categoricity in power for first order logic are not deficits but powerful tools for understanding mathematics.

We could restate the fourth thesis as: geometry is the missing link that must be added to Bourbaki's three 'great mother-structures' (group, order, topology) (page 62) that are intended to organize mathematics. Geometry is a broad term; we will focus on three examples in this book: Euclidean axiomatic geometry and its descendants, real and complex algebraic geometry, and combinatorial geometry. The interaction among these subjects unites what might appear to be disparate facets of this book. In studying the axiomatization of elementary geometry we highlight Hilbert's use of only first order axioms to prove this geometry admits a system of coordinates (as in high school geometry) over some field. Thus, it satisfies the crucial property of geometry – a clear concept of dimension. This notion is distilled into defining *combinatorial geometry.*[17] This general notion applies equally

---

[17] See Chapter 5.4.

well to finite as well as 'continuous' geometries. The stability classification of theories allows one to determine those theories T whose models admit (at least locally) combinatorial geometries. For those that do, it is possible to develop a structure theory for the models of T, where the building blocks of the models are geometries. The methods involved in this result have many applications in traditional topics such as algebraic geometry. Conversely, Hrushovski's field configuration, ultimately inspired by Hilbert's proof of the existence of a field in a Euclidean geometry, constructs classical groups and geometries from general model theoretic hypotheses with no algebraic or geometric hypotheses. Thus classical mathematics and model theory are inextricably intertwined.

After explaining in the next few paragraphs what I mean by the paradigm shift, I will try to clarify the purpose of this book by comparing approaches to the methodology of mathematics articulated by Franks, Maddy, Manders, and Tarski. Then I will move to a more detailed discussion of the contents of the book. I make reference in this discussion to many concepts of model theory and some deep mathematics. Familiarity with at least an upper-division undergraduate logic course is assumed in many places. More advanced model theoretic notions are introduced, I hope gently, and indexed. Some details in the introduction are intended for those with more background, but they should become clearer by examining the treatment in the text. There is little attempt to explain in any depth the concepts and results in other areas of mathematics. I do attempt to give a broad picture to show interactions both between these areas and with model theory.

Features of the Paradigm Shift: The paradigm shift that swept model theory in the 1970s really occurred in two stages.

The initial impetus is a subtle and gradual switch from emphasizing pure logic to applied logic. While these terms are related to the connotation evident in the names chosen, the actual distinction is technical. An *applied logic* has constant symbols, a *pure logic* does not (page 32). As we discuss in Chapter 3.1, work of Parsons and Väänänen on *internal categoricity* emphasizes the technical importance of this distinction for second order logic and its methodological significance for first order logic permeates this book. Henkin's defense of his proving completeness for *uncountable vocabularies*[18] clarifies his motivation and the shift, 'While this is not of especial interest when formal systems are considered as logics – i.e., as means for analyzing the structure of languages – it leads to interesting applications

---

[18] We discuss this quotation from [Henkin 1949] further on page 96.

in the field of abstract algebra.' Although this pure/applied distinction was standard in the 1940s, its absence from modern mathematical logic texts [Shoenfield 1967, Chang & Keisler 1973, Marker 2002], witnesses an aspect of the paradigm shift.

During the first stage in the 1950s and 1960s the focus switched from the study of properties of logics[19] to the study of particular (primarily first order) theories (the logical consequences of a set of axioms) and properties of theories and their impact on the models of the theories.[20] Robinson's identification of *model complete* theories is an early example of studying a *class* of theories; another is Morley's analysis of $\aleph_1$-*categorical* theories. The Ax–Kochen–Ershov proof of the Lang conjecture proceeds by identifying *complete* theories of Henselian valued fields.[21]

In the second stage, Shelah's decisive step was to move from merely identifying some fruitful properties (e.g. complete, model complete, $\aleph_1$-categorical) that might hold of a theory to a *systematic* classification of theories. As described in Chapter 5.3, he divides complete first order theories into four categories, each characterized by a syntactic property. The aim is to determine the class of theories whose models have a structure theory in a precise sense: each model is determined by a *system* of cardinal

---

[19] Thus typical theorems involved such notions as decidability and interpolation and made assertions that were true for *any* theory. See Chapters 1.3 and 8.1.

[20] Contrast this study of particular fields (theories) with the goals enunciated by Russell in the preface to [Russell & Whitehead 1910]. 'We have however avoided both controversy and general philosophy and made our statements dogmatic in form. The justification for this is that the chief reason in favour of any theory on the principles of mathematics must always be inductive, i.e., it must lie in the fact that the theory in question enables us to deduce ordinary mathematics. In mathematics the greatest degree of self-evidence is usually not to be found quite at the beginning but at some later point; hence the early deductions, until they reach this point, give reasons rather for believing the premises because true consequences follow from them, than for believing the consequences because they follow from the premises.' We agree with this analysis of the process of axiomatization. We argue however that experience has shown that the formulation of such premises for the principles of mathematics (in the *Principia* formulation or from a contemporary perspective ZFC) stray too far from the actual practice of mathematics to properly illuminate it. Thus we pass from global to local foundations. In Chapter 3 we celebrate the success of some second order axiomatizations that meet Russell's criterion in that they imply the results known in the area. Most of the text describes various first order theories whose axiomatizations are based on the same principle. We illustrate in Chapter 9 for the case of Hilbert's geometry the aptness of the end of Russell's paragraph, 'All that is affirmed is that the ideas and axioms with which we start are sufficient, not that they are necessary. Our notion of 'immodest descriptive axiomatization' (Chapter 9.1) identifies axiom systems that overreach their goal.

[21] Ax and Kochen won the 1967 American Mathematical Society Cole prize in number theory for their solution of Lang's conjecture that for each positive integer d there is a finite set Y_d of prime numbers, such that if p is any prime not in Y_d then every homogeneous polynomial of degree d over the p-adic numbers in at least $d^2 + 1$ variables has a nontrivial zero. Ershov obtained the result independently in the Soviet Union.

invariants. Shelah introduces the methodological precept of a 'dividing line' (Chapters 2.4, 8.4, and 13). He formulates each dividing line property so that theories that fall on one side (e.g. unstable) are creative; their models cannot be systematically analyzed as composed of small models, essentially new models are increased as the cardinality increases, and there is the maximal number of models. Models of theories on the other side (stable) have a 'structure theory' (i.e. they admit a local dimension theory). In the classifiable[22] case, each model is determined by a (well-founded) tree of cardinal invariants. This book investigates some consequences for mathematics and philosophy of mathematics of that paradigm change from: study the properties of logics (compactness, interpolation theorems, etc.) to: study virtuous properties of theories. A property of theory T is virtuous (Chapter 2.3) if it impacts the understanding of the models of T. A property is a 'dividing line' if both it and its complement are virtuous.

While the stability classification provides precise mathematically formulated dividing lines, 'tame vs. wild' is a less formal notion. The first order theory of arithmetic, $\text{Th}(N, +, \cdot)$, is seen to be wild[23] because it admits a pairing function and so loses the essential geometric distinction between a structure A and the 'plane over it', A^2; thus there can be no notion of dimension. And so, there is no geometry on $(\mathbb{N}, +, \cdot)$. Since humans can only comprehend a small number of alterations of quantifiers, the existence of definable sets of arbitrary quantifier rank prevents a clear intuition[24] of the structure $(N, +, \cdot)$. Shelah's taxonomy further shows that arithmetic has both of the strongest non-structure properties (the strict order property and the independence property[25]). Together these conditions help to explain why little of modern *algebraic* number theory takes place directly in first

---

[22] A theory is classifiable if it satisfies each of the three dividing lines superstable, NDOP, and NOTOP and is shallow; see page 139 for these acronyms.

[23] Undecidablity is often taken as a sign of wildness. But it is easy find non-recursively axiomatized theories that are extremely tame (strongly minimal) and decidable theories can be wild (atomless Boolean algebras).

[24] By intuition, I mean the usual usage of mathematicians, a rough understanding of a concept or mathematical object and not any of the technical philosophical meanings. My intent is in the spirit of the first page of the *Grundlagen* [Hilbert 1971] where Hilbert writes, 'This problem [axiomatizing geometry] is tantamount to the logical analysis of our intuition of space'. Shapiro ([Shapiro 1991], 39) suggests 'tentative preformal beliefs'. Although one often thinks of the natural numbers as a clearly given structure, this assertion rests on a confusion. As Roman Kossak has pointed out, a clear intuition or vision of the natural numbers with successor is often confused with a clear intuition of arithmetic, the natural numbers with both addition and multiplication; few, if any, actually have the second intuition. See Chapter 1.2 for the role of the vocabulary and page 79 for further development of 'intuition of a structure'.

[25] These are syntactical conditions, which each imply the maximal number of models in each power; technical definitions are on page 130.

order Peano arithmetic (as opposed to logical analysis showing a particular result is provable in Peano). Rather, auxiliary more tame structures such as algebraically closed or valued fields provide the framework for proofs of number theoretic results.[26]

We will explore the origins in the 1950s of this shift, its fruition in the 1970s, and the more mature pattern that developed in the 1980s. We consider some consequences of modern model theory for mathematics and for the philosophy of mathematics. Consequences for the latter arise in two ways: using the insights of modern model theory to develop existing research lines in philosophy (purity, categoricity, etc.); initiating the study of issues (e.g. the exceptional nature of $\aleph_0$, the role of model theory in organizing mathematics) in the philosophy of mathematics that are first seen from this new perspective. These issues all contribute to the emerging study of the philosophy of mathematical practice (e.g. [Mancosu 2008a]).

The Context of This Work: Why should this change of mathematical goal from studying logics to studying the classification of theories have any impact on the study of the philosophy of mathematics? In contrast to the traditional program that studies the *foundations of mathematics* by constructing a single formal theory supporting all of mathematics, this shift empowers a strategy of defining many different formal theories to describe particular areas of mathematics.

Tarski's phrase, the methodology of deductive systems, is at the heart of the discussion. We proceed less ambitiously than Tarski, whose *Introduction to Logic and to the Methodology of the Deductive Sciences* aimed

to present to the educated layman ...that powerful trend ...modern logic ... [which] seeks to create a common basis for the whole human knowledge. ([Tarski 1965], xi)

Thus, unlike the analytic philosophy of the 1930s or even the work of Putnam and Montague mentioned in *the provocation*, there is no claim that the methods considered here are broadly applicable to the foundations of science. Rather than the broad program espoused by Tarski, our more modest goal is to expound for philosophers and mathematicians how the formal methods, initially springing from Tarski, Robinson, and Malcev but greatly extended in the wake of the paradigm shift, enhance the pursuit and organization of mathematics and the ability to address certain philosophical issues in mathematics.

---

[26] The epistemological significance of such a reduction is explored in [Manders 1987].

Our approach is closer to that of two works of Maddy: *Second Philosophy: A Naturalistic Method* [Maddy 2007] and *Defending the Axioms* [Maddy 2011]. In the latter, she analyzes the methodology and the justification of the axioms of one particular first order theory, set theory. To see the analogy with the current book, we modify her[27] Second Philosopher's account of issues arising in the study of set theory, replacing each occurrence of 'set' in Maddy's text by '*model*'.

When our Second Philosopher is confronted with contemporary *model* theory, we've seen that questions of two types arise. The first group is methodological: what are the proper grounds on which to introduce *models*, to justify *model*-theoretic practice, to adopt *model*-theoretic axioms? The second group is more traditionally philosophical: what sort of activity is *model* theory? how does *model*-theoretic language function? what are *models* and how do we come to know about them? ([Maddy 2011], 41)

Some of these questions resonate immediately in their new context. Our discussion of formalization[28] in Chapter 1 considers the grounds for introducing models. On the other hand, we are not trying to formalize model theory, so the explicit question 'What are the proper grounds to adopt model theoretic axioms?' is not in view. We seek rather to analyze the fundamental techniques and principles of model theory. Indeed, one might wonder whether a potential formalization is remotely possible. The 'grounds for adopting axioms' question raised by Maddy is one level too abstract for our study of model theory. Finding grounds for accepting the axioms is a task for a model theorist in formalizing any particular theory, including set theory, not about the field of model theory itself. As *the provocation* indicated, thinking of set theory in the same way as any other first order theory has proved its worth as a methodological standpoint in the last 50 years.[29] For us, the issue of intrinsic and extrinsic justification for axioms, a central concern for Maddy [Maddy 2011], arises for any field of mathematics, from Euclid/Hilbert geometry (Chapter 9) to Hrushovski's contemporary theory of algebraically closed fields with an automorphism (Chapter 2.4).

The specific questions Maddy labels as 'traditionally philosophical' in the quotation at hand are among those this book intends to address. Much of the book describes 'the activity of model theory' and tries to explicate some

---

[27] In [Maddy 2007], Maddy begins to describe the method of inquiry of a *Second Philosopher*, by distinguishing it from the Cartesian method.

[28] See page 33.

[29] See [Shelah 1999] for Shelah's pragmatic approach to the choice of axioms for set theory.

of the directions of research. Thus, by *model theory* we almost always[30] refer to the study of the interaction between a collection of sentences in a formal language[31] and structures that satisfy those sentences. Much of our analysis concerns the function of model theoretic language. In seeking to understand how we come to know about models, we will study their properties and their relation with theories and classes of theories. While we focus on first order theories, the properties of models will be second order: e.g. prime, saturated, and universal.

While in [Maddy 2011] Maddy discusses the relation between ontological issues and the choice of axioms for set theory, here we deal more with her earlier methodological concerns:

In sum, then, the Second Philosopher sees fit to adjudicate the methodological questions of mathematics – what makes for a good definition, an acceptable axiom, a dependable proof technique – by assessing the effectiveness of the method at issue as means towards the goal of the particular stretch of mathematics involved. Straightforward examination of the historical record suggests that theories about the nature of mathematical existence and truth don't play an instrumental role in these determinations, but this is not to say that such metaphysical questions evaporate completely from the second-philosophical point of view. ([Maddy 2007], 359)

Here are some pertinent examples of methodological issues: Shelah's program of setting 'dividing lines' as normative assertions about the notion of 'good definition' (Chapter 13); the appropriate axioms for geometry (Chapter 9); and the nature of proof techniques in model theory (Chapters 4, 5, and 8). Two further methodological schemes are the role of test questions (Chapter 13) and the use of strong hypotheses (e.g. extensions of set theory) to obtain conclusions that one hopes to later establish with weaker hypotheses (Chapter 8.6). Thus our approach falls between those of Tarski and Maddy. Our scope is much narrower than Tarski's logic of deductive systems but also wider (from one standpoint) than Maddy's. The aim of Maddy's *Defending the Axioms* is to justify one first order theory for all of mathematics; in contrast, we are trying to understand what the goals of justification should be for different theories of various areas of mathematics. We approach global mathematical issues not by seeking a common foundation but by finding common themes and tools

---

[30] Chapter 14 allows more generality.
[31] Thus a 'formal theory', abbreviated in context to 'theory', reflects a mathematical theory in the usual informal sense such as field theory or matrix theory.

for various areas, not in terms of the topic studied, but *in terms of common combinatorial and geometrical features isolated by formalizations of each area.*

Another inspiration for this work is Franks' study of the Hilbert program in *The Autonomy of Mathematical Knowledge.* In his introduction, Franks strikes a chord that will resonate in the current book: 'The first theme is that questions about mathematics that arise in philosophical reflection – questions about how and why its methods work – might be best addressed mathematically.' But then he continues:

> The second theme arises out of the first. Once one sees mathematics potentially providing its own foundations, one faces questions about the available ways for it to do so. The two most poignant issues are how a formal theory should refer to itself and how properties about a theory should be represented within that theory. [Franks 2010a]

Here we part company. Since I seek no global theory of mathematics, there is no self-reference problem. Indeed, two key insights of modern model theory are that (i) large amounts of modern mathematics can be better understood by formal systems which are tame (page 148), so do *not* support self-reference, and (ii) this tameness is actually constitutive of the fertility[32] of these theories in mathematical practice.

As is common in model theory, we adopt a rather strong metatheory, ZFC.[33] However, we are interested (Chapter 8.6) in the possibility of weakening or strengthening model theoretic results within the general framework of axiomatic set theory. Such investigations can clarify the distinctions between making hypotheses about a specific topic and postulating general combinatorial principles in set theory. We do deny that the reduction of mathematics to set theory, designed for reliability purposes, is adequate for the understanding of mathematical practice. This is not to reject the question of justifying the ZFC axioms but to table it while discussing the role of formalization in clarifying mathematical discourse. In fact, the local foundations can provide greater reliability. Remark 10.2.3 notes that Tarski's autonomous foundation for geometry is finitistically consistent so we observe a weaker metatheory in that case.

The book is arranged as a web as well as a narrative. That is, we try to expound the basics of various model theoretic notions in terms of their methodological significance. A notion often has more than one such

---

[32] That is, we show (Chapters 2.4 and 6) how the fact of the tameness can be exploited to prove mathematical theorems.

[33] Zermelo–Fraenkel set theory with the axiom of choice.

significance, so we have extensive cross-referencing in the text. A further goal is to attack the idea that after the foundational crises of the early twentieth century mathematicians stopped engaging with philosophical issues. To that end, we frequently quote from expository articles, International Congress of Mathematics talks, and other sources in which mathematicians have laid out programs that not only raise specific mathematical problems but proclaim norms for 'good mathematics' and fruitful directions for research. We now summarize in more detail the contents of the book to fill out the description we have just given of the overall goals.

This book is *not* a text in model theory. We give some basic definitions, state some milestone results, and give a feel for the methods involved in establishing them. But we rely on such expositions as the Stanford Encyclopedia of Philosophy for basic notions, [Marker 2002], [Baldwin 1988a], [Baldwin 2009], and [Hodges 1987] for more technical concepts, and refer the interested to advanced texts and papers in stability theory for further details. We do however sketch major ideas and results of model theory to illustrate our Theses 1–4. The exposition is organized around methodological and historical themes. References play different roles that we try to distinguish; in the history sections we are trying to identify the original source; but references to texts aim for accessible accounts. Shelah's theorem II.2.13 of [Shelah 1978] reports nine equivalent definitions of a stable theory. These diverse statements are methodologically crucial. One of the key points of his 'dividing lines' program (Chapter 13.3) is that equivalent versions of the same definition play entirely different roles. We will in fact discuss three or four equivalents to stability in different sections of the book (Chapter 2.4, Theorem 5.3.5, Theorem 8.2.4).

We now describe the four parts of the book and connect them with our general theme.

1. Rethinking Categoricity: Michael Detlefsen raised a number of questions about the role of categoricity. The attempt to answer these questions has shaped a good portion of this book.

Question I:[34] (A) Which view is the more plausible – that theories are the better the more nearly they are categorical, or that theories are the better the more they give rise to significant non-isomorphic interpretations?

(B) Is there a single answer to the preceding question? Or is it rather the case that categoricity is a virtue in some theories but not in others? If so, how do we tell these

---

[34] Question I was questions III.A and III.B in a 2008 letter. Question II was question IV in the Detlefsen letter. I thank Professor Detlefsen for permission to quote this correspondence.

apart, and how do we justify the claim that categoricity is or would be a virtue just in the former?

Question II: Given that categoricity can rarely be achieved, are there alternative conditions that are more widely achievable and that give at least a substantial part of the benefit that categoricity would? Can completeness be shown to be such a condition? If so, can we give a relatively precise statement and demonstration of the part of the value of categoricity that it preserves?

Discussion revealed different understandings of some basic terminology. Does categoricity mean 'exactly one model', full stop? Or does it mean exactly one model in a given cardinal? Since Morley's ground-breaking categoricity theorem[35] the actual meaning among model theorists for the colloquial 'categorical' is 'categorical in an uncountable cardinal'. In the usual first order model theoretic situation, the one model interpretation is trivial (it means finite). Is a theory automatically closed under (deductive/semantic) consequence? Is the topic 'theory' or 'axiomatization'? Detlefsen's concerns were primarily about first or second order axiomatizations to provide descriptive completeness[36] for a particular area or for all of mathematics. In contrast, model theorists consider primarily (complete) first order theories. These different perspectives yield two roles for formalization in mathematics: as a foundational tool and as a device in the mathematician's toolbox.

In Chapter 1, we carefully define the terms that caused confusion and describe the modus operandi of this book. To understand a particular field of mathematics one lays out a formal theory to describe it. Three crucial decisions must be made: what are the primitive concepts of the field? What logic is most appropriate for the formalization? What are the fundamental principles of the field that should be laid down as axioms? Chapter 1.1 explains the intent of formalization. The inextricable link between vocabulary and structure is explored in Chapter 1.2. In particular, the notion of isomorphism is meaningless unless a vocabulary is specified. The detailed definition of *full formalization* distinguishes logic, theory, and axiom and emphasizes their distinct roles, both technically and epistemologically. In Chapter 1.3, we follow Magidor [Magidor 2015] in discussing the

---

[35] Morley's theorem asserts that a countable first order theory is categorical in one uncountable cardinal κ (all models of that cardinality are isomorphic) if and only if it is categorical in all uncountable powers. Morley received the Steele prize from the American Mathematical Society in 2003 for the seminal influence of this work.

[36] In [Detlefsen 2014], Detlefsen distinguishes between 'descriptive completeness' and 'completeness for truth'. We address this notion at length in Chapter 9.1.

entanglement of various logics with axiomatic set theory. Chapter 1.4 moves from logics to theories, completing the basis for the discussion of the four theses.

In Chapter 2.1, we draw on the work of Lakatos and Manders to distinguish several degrees of formalization as a tool to clarify concepts. That is, we think of formalization and in particular modern classification theory as a tool that allows both the investigation of specific areas of mathematics and a systematic way to examine relations between fields. In this sense it is reminiscent of Wilson's theory [Wilson 2006] of facades as connecting devices between views of a topic. In Chapter 2.1, we explore the connections of Wilson's facades connecting various areas of study with our development of a systematic connection among areas of mathematics. We draw heavily on Manders' earlier study [Manders 1984, Manders 1987, Manders 1989] of epistemological aspects of model theory.

We then (Chapter 2.2) contrast the nuanced opinion of David Kazhdan, an expert in group representations who has used model theory to help formulate the notions of motivic integration, with Bourbaki's blunt proclamation, 'We emphasize that it [formalization[37]] is but one aspect of this [the axiomatic] method, indeed the least interesting one.'[38] In fact, our goal is to reverse Bourbaki's aphorism by illustrating the importance of formalization and in particular formal definition as a mathematical tool. In Chapter 2.2, we elaborate on the distinction (made for example by Bourbaki [Bourbaki 1950]) between the axiomatic method and formal methods.

We continue in Chapter 2.3 by analyzing Detlefsen's Question I.B from page 15. Before determining the virtue of categoricity, one must clarify what is meant by virtue. Chapter 2.3 considers what might constitute virtue *in this context* and provides a criterion for a 'good or virtuous property': *a virtuous property has significant mathematical consequences for theories or their models.* This criterion leads to a strong form of Thesis 1: *Formalization impacts mathematical practice most directly not because of its foundational aspect but by the direct application of formal methods to ordinary mathematical problems.* In particular, the notion of a (complete) first order theory is a significant mathematical tool (Chapter 4.3). And identifying specific virtuous properties of first order theories is a key to using that tool.

We return in Chapter 3 to Detlefsen's question IA to evaluate the role of categoricity based on this notion of virtuous properties. We build on the [Button & Walsh 2016] distinction between determining *reference* and

---

[37] They use the phrase 'logical formalism' but meaning what we have called formalization.
[38] Page 223 of [Bourbaki 1950].

truth value to analyze categoricity. We note that with a metatheory as strong as ZFC, reference is not really an issue.[39] If, even with ZFC foundations, the problem is taken as to find a description in the given vocabulary, the key idea of Thesis 3, *specifying the logic* comes into play. There is not a uniform answer to a question of the form: how does logic X affect notion Y? For fixed Y, the answer will differ widely depending on whether the logic X is first order, infinitary, or second order. Most of Chapter 3.1 employs the full second order semantics, where categoricity determines a reference for a description of a structure. We argue that categoricity is interesting for a few second order sentences describing particular structures. But this interest arises from the importance of *those* axiomatizations of *those* structures, not from any intrinsic consequence of second order categoricity for arbitrary theories. From this standpoint the categoricity of a particular axiom set is crucial. In contrast, contemporary model theory focuses either on the properties of (complete) first order theories and classes of those theories or on studying particular theories, usually those arising in mathematical practice. Here, the particular choice of axioms is not important for studying consequences but can be crucial for the process of verifying that a structure is a model of the theory. We conclude the section with a discussion of *internal categoricity* and the distinction between 'determinate reference' and 'determining reference'.

In Chapter 3.2, we discuss three goals Meadows [Meadows 2013] proposed for a categoricity theorem and argue they can be better realized by $L_{\omega_1,\omega}$-sentences or theories than by second order sentences or theories. In contrast to the ad hoc nature of second order categoricity, *categoricity in power* has immense structural consequences (Chapter 3.3, Chapter 5.4). Zilber conjectures (page 83) that that virtue holds in the strongest way: any first order theory categorical in an uncountable power arises from a canonical structure[40]. In Chapter 3.4 we elaborate our views on the role of formalization in mathematics and begin the generalization from categoricity in power to classification theory. Models of $\aleph_1$-categorical theories are the foundation blocks of models of many stable first order theories.

2. The Paradigm Shift: In exploring the progression from studying logics, to theories, to classes of theories (and their models), we see the beginnings of the paradigm shift in the 1950s. In Chapters 4.1 and 4.2 we analyze some

---

[39] Often structuralists seek a distinct foundation; such considerations are too far from our main line to be developed here.

[40] His canonical structures include the complex field, even when equipped with exponentiation.

of the commonplaces of today that were thorny issues around 1930. Partly because of the failure to understand the vast distinctions among the various logics, the notions of semantic completeness, syntactic completeness, and categoricity were often confused in the 1920s. We focus on the Löwenheim–Skolem–Tarski and completeness theorems to see how early work of Carnap, Gödel, Tarski, and Malcev illustrate the importance for model theory of distinguishing the precise logic of the investigation. At present, there is essentially no model theory of second order logic, a richly developed model theory of first order logic, and a comparatively rudimentary model theory for infinitary logic.

In the remainder of Chapter 4, we fix our logic as first order; we study theories in general. Specific theories, given by appropriate choice of primitive concepts and axioms, apply to various areas of mathematics. Our study in Chapter 4.3 of why *complete first order theories* are a good unit for specifying a mathematical topic is deepened in Chapter 5. Robinson's emphasis on complete theories of significant algebraic structures [Robinson 1956] and his important criteria of quantifier elimination and model completeness (Chapter 4.4) begin the paradigm shift. Chapter 4.5 explains the notion of interpretability between theories as a tool to connect various areas preparing for a later study of several philosophical issues. Bi-interpretablity allows us to realize the extent to which extension by *first order explicit definability* is harmful for some purposes (it does not preserve purity, Chapter 9) or helpful for others (clarifying what a structure is, Chapter 4.6). We conclude this chapter by exploring the role of vocabulary and logic in studying the notion of structure in relation to two recent philosophical investigations. The first (Chapter 4.6) is the Button–Walsh notion of informal isomorphism expounded in their forthcoming book [Button & Walsh 2017] and the second is a discussion of one aspect of the development of homotopy type theory (Chapter 4.7) as a proposed foundation of mathematics. That is, we show that while the slogan 'isomorphism is equivalent to identity' is anathema to a model theorist, a more careful reading of this 'univalence axiom' eliminates the conflict.

In Chapter 5 we lay out the second stage of the paradigm shift, from the ground-breaking work of Morley and Vaught to the seminal stability hierarchy (Chapter 5.5) and then the developing use of that hierarchy in mathematics (Chapter 6). The interaction between the syntactically definable properties of theories and the Boolean algebra of definable sets in their models is crucial. Chapter 5.1 elucidates the somewhat broader notion of 'syntactic' entailed by this move. We examine in Chapters 5.2 and 5.3 two important components of the paradigm shift. We study the notion of a universal domain, both (a) its epistemological virtues in terms of Manders' concept of 'domain extension' and (b) the mathematical objectives foreshadowed in Weil and Jónsson. Then, we turn to Morley's and subsequently Shelah's analysis of the type space of models of a theory and the introduction of the stability hierarchy with its simplifying effect on the saturation spectrum.

Much of the historical approach in Chapter 5 reinforces Thesis 4 on the role of geometry by discussing two instances of the long-term impact of Hilbert on model theory. In using models to demonstrate the independence of various axioms of geometry, Hilbert builds on Bolyai, Klein, Beltrami, Poincaré, and others, but he moves from examples to a method. He introduces two uses of the notion of interpretation; in one, the models he constructs to illustrate independence of axioms are interpretations of geometry into field theory; in the other, his definition of multiplication and addition of line segments is an interpretation of field theory into geometry. The notion of interpretation (Chapter 4.5) provides a uniform account of both directions. The second impact of his interpretation of the field and resulting coordinatization was the intimate contact between field theory and geometry yielding the subject of *combinatorial geometry*. In Chapters 5.5–5.6 we discuss the use of combinatorial geometry by model theorists, which in practice develops Bourbaki's goal of finding an 'organizing hierarchy of structures.' The generalized notion of independence created by Shelah allows the assignment of a dimension by finding geometries in a wide variety of mathematical objects. *Geometric stability theory* applies the classification of these geometries via the Zilber trichotomy to obtain deep results in pure model theory with application in many areas.

The great advance in developing classification theory arose from the following natural problem with an unnatural or rather an unexpected method of solution. Morley conjectured that except for the obvious exception[41] a countable, first order theory should have at least as many models in κ^+ as in κ. Shelah's strategy to solve the Morley conjecture was pivotal in the paradigm shift. Let $I(T, \kappa)$, the spectrum function[42], denote the number of non-isomorphic models of T with cardinality κ. We sketch the strategy. Step 1: calculate $I(T, \kappa)$ for each T. Step 2: observe that for each T the function is increasing. Step 1 seems much harder than the original problem. But in fact, the strategy turns 'Is the spectrum function

---

[41] There are $\aleph_1$-categorical theories with infinitely many countable models. See Chapter 3.3 and Chapter 8.3.

[42] The spectrum function assigns to κ the number of models with cardinality κ. See page 138.

increasing?' into two problems. (i) Partition[43] the countable theories into a finite number, say N, of kinds and each kind into classes, indexed by $i < N$ and countable ordinals, such that all theories in the same class $X_{i,\alpha}$ have the same spectrum function $f_i(\kappa, \alpha)$. (ii) Determine the spectrum function for each class of theories. Thus Shelah sees a much greater regularity in the universe. There are not $2^{\aleph_0}$ different functions, indexed by $2^{\aleph_0}$ complete theories, that are each $I(T)$ for some T but rather a finite number of functions $f_i(\kappa, \_)$, such that each $I(T, \kappa)$ is $f_i(\kappa, \alpha)$ for i and α determined by its class. The main gap is a dichotomy between those theories whose models are controlled (classifiable) and those which are wild. The rough idea was to count the number of models. But the result is much stronger. The models of a classifiable theory with cardinality λ are determined by a tree with countable height and width λ of cardinal invariants (each at most λ). After opening Chapter 5.5 with an analysis of the meaning of classification in mathematics, we expound Shelah's main gap classification in more detail, stressing the connections with geometry.

The discovery (Chapter 5.6) by Zilber and Hrushovski that any structure satisfying certain specific purely model theoretic conditions interprets one of three classical groups is a key step both technically and methodologically. It supports Thesis 4 by showing geometry is fundamental to understanding the notion of model theoretic classification. Hilbert's coordinatization of the Euclidean field is the ultimate inspiration for Hrushovski's result. Thus the study of Euclidean geometry in Chapters 9–12 provides insight into one of the most technical and fruitful results in model theory. The interaction of model theory with real and complex algebraic geometry springs from two sources. On one hand these subjects can be seen as investigating definable sets and, in that sense, as a special case of model theoretic investigation. On the other, Hrushovski's Trichotomy Theorem 5.6.4 not only solved purely model theoretic problems but is a crucial step to prove important results in Diophantine geometry.[44]

The argument for Thesis 2 has two parts. Chapter 5 begins with a gradual switch from a study of logics to a study of theories, spearheaded by the work of Henkin, Robinson, Tarski, and Vaught. The remainder explores the

---

[43] A partition of a collection is a division of the collection into non-overlapping subgroups such that each member of the collection is in one of the subcollections. The ordinal α is the depth of T (page 140).

[44] Diophantine geometry is a modern approach in number theory. Rather than solving equations by working only with integers (i.e. in 1st or 2nd order arithmetic), the problem is posed as: find the integral (rational) solutions among those in $\mathbb{C}$ of an equation with coefficients in say $\mathbb{Q}$. This allows the application of powerful tools in the tame world of algebraic geometry to the wild world of arithmetic. This technique underlies e.g. the proof of Fermat's last theorem.

impact of Morley's result, the broad outline of stability theory, the significance of geometric stability, and the development of general model theoretic concepts which might be applied in different areas of mathematics. Chapter 6 explores the use of model theoretic concepts to solve problems that mathematicians regard as emblematic of tameness from important mathematical theories (e.g. algebraically closed fields, ACF) that happen to be $\aleph_1$-categorical, through careful selection of axioms and vocabularies that are both mathematically fruitful and model theoretically accessible.

The examination of groups of finite Morley rank[45] in Chapter 6.1 is a template for possible future case studies of the role of formalization in mathematical practice. This more detailed section suggests how future work on the philosophy of mathematical practice could proceed, by examining the particular epistemological issues that arise in a specific field of mathematics. Even these brief descriptions make the general point clear: these tools are used by model theorists and by mathematicians in other areas. These applications are usually thought of as belonging to algebra but significant advances in analysis are made using first order model theoretic methods we describe in Chapter 6.3 as *axiomatic* and *definable* analysis. The first applies mainly stability theoretic tools while the second draws primarily on o-minimality. Examples include arithmetic algebraic geometry, differential algebra, and asymptotic analysis. All of the model theoretic analysis depends crucially on the notion of *definability* (Definition 1.3.4) which arises from the distinction of *syntax* from *semantics*.

Pillay [Pillay 2000] and others have argued that only first order or at least only compact logics are fruitful for the study of mathematics. And it is certainly true that the overwhelming majority of applications of model theory in mathematics have been made using compact logics[46] and almost all of that for first order logic. Some rejoinders to this claim are explored in Chapters 7 and 14. Model theory is itself mathematics and the study of infinitary logics is important in its own right. In particular, several crucial first order concepts were reverse engineered from infinitary notions. And there are applications, notably Zilber's study of the complex field with exponentiation, where the infinitary viewpoint has inspired major advances in number theory. Thus, our Thesis 3 does not restrict to first order logic but only points to its current success.

We begin Chapter 7 by considering the mathematical need for infinitary logic to express a basic insight in algebraic geometry from the 1940s:

---

[45] This subject is a hybrid of model theory and algebraic groups inspired by the classification of finite simple groups. A group is *simple* if it has no proper normal subgroup.

[46] A logic is said to be compact if it satisfies the compactness theorem; see the discussion of continuous logic in Chapter 4.2.

the Lefschetz principle. Infinitary logic is slightly behind the state of first order logic in 1965. There is a categoricity theorem for sentences of $L_{\omega_1,\omega}$ (Chapter 7.1), which approximates Morley's result for first order logic. The result is weaker in two ways: it relies on an extension of ZFC, the very weak generalized continuum hypothesis (VWGCH[47]) and the hypothesis is on models with cardinality less than $\aleph_\omega$, not on a single cardinal. We then sketch Zilber's use of $L_{\omega_1,\omega}(Q)$ to ground the theory of complex exponentiation. The study of infinitary logic relies in places on extensions of set theory. Some, such as VWGCH, are mild. On the other hand, large cardinals have recently had a major impact – raising serious set theoretic issues. There are consequences of large cardinal axioms that are algebraic in nature, not just combinatorial. We explore these results in Chapter 14 along with our general discussion of the common generalization of many infinitary logics as abstract elementary classes. In Chapter 7.2 on Vaught's conjecture, we return to the question of the proper choice of logic by examining the interaction of first order and infinitary logic as well as the use of descriptive set theory in the 50-year study of whether a theory can have exactly $\aleph_1$ countable models.

Chapter 8 expounds the interaction of model theory and set theory and its connection with such philosophical puzzles as indiscernability, identity, and isomorphism. The set theoretic role in Tarski's formulation of the definition of truth is fundamental. But is there a deeper connection? Around 1970, model theory seemed to be deeply entangled with both cardinality and cardinal arithmetic and thus with axiomatic set theory. Surprisingly, understanding the entanglement with cardinality resulted in largely eliminating the ostensible entanglement of first order logic with cardinal arithmetic. Chapter 8.1 describes how the advent of classification theory and the switch of focus from *logics* to *theories* obviated this issue. Nevertheless a few important principles from combinatorial set theory remain essential tools. They are used to construct *sequences of indiscernibles*; is it only a pun to think of the Leibniz principle (Chapter 8.2)?

Both Morley's categoricity theorem and his conjecture that the number of models of a theory in an *uncountable* κ increases with κ depend essentially on the restriction to uncountable κ. The same example illustrates both: vector spaces over $\mathbb{Q}$ have infinitely many countable models, one in each finite dimension, but only one model in every uncountable power. So Morley's theorem is not about transfer among *all* infinite cardinals. What is exceptional about $\aleph_0$? In Chapter 8.3 we explore how a second combinatorial

---

[47] For every $n \in \omega$, $2^{\aleph_n} < 2^{\aleph_{n+1}}$.

technique, the method of stationary sets, provides a set theoretic explanation for the model theoretic fact that in many ways $\aleph_0$ is an exceptional cardinal. We proceed then to discuss three ways in which model theory is entangled with set theory. Chapter 8.4 suggests that model theory has an inherent but quite subtle *entanglement with cardinality*[48]. We argue that the interpretation of the equality symbol as identity determines the role of counting problems as a test question in model theory. The subtlety is that such cardinal dependent properties as categoricity *in power* and stability in κ are proved equivalent to *syntactic properties* of countable (for countable vocabularies) sets of formulas (Chapter 5.3). And this discovery removes the dependence of first order model theory on *cardinal arithmetic* and thus on extensions of ZFC (Chapter 8.1). The combinatorial principles described in Chapter 8.2 require uncountable sets and even $\beth_{\omega_1}$. Does this mean that at least the Axiom of Replacement is crucial for model theory (Chapter 8.5)? Chapter 8.6 exhibits three ways that axiomatic set theory can engage model theory: oracular, metatheoretic, and actual entanglement. The last engagement occurs mostly in infinitary logic and is detailed in Chapter 14. We close in Chapter 8.7 with the far-sighted remarks of Sacks on the foundation of model theory.

3. Axiomatizing Geometry: Part III addresses Thesis 4 and the philosophical aspect of Thesis 1. We address such longstanding issues in the philosophy of mathematics as completeness, categoricity, syntax and semantics, axiomatization, and purity. With one major exception, the modern model theory involved is simply the importance of specifying vocabulary and logic when formalizing a topic (Part I). These specifications allow us to clarify the role of Hilbert's groups of axioms in terms of the logic in which each of them is formulated. Hilbert forged in the *Grundlagen* several model theoretic techniques that are fundamental to modern logic. The great step, culminating in Hilbert, of replacing the notion of axiom as 'obviously true' by a set of axioms as an 'implicit definition' of a collection of primitive concepts is the fundamental idea of model theory. Some readers may prefer a more chronological approach and read Part III before Part II; notions such as interpretability arise in both and are cross-referenced.

We investigate the role of *axiomatization* in mathematics from several perspectives. We extend Detlefsen's notion [Detlefsen 2014] of a descriptive axiomatization in Chapter 9, by introducing the notion of a *modest*

---

[48] Kennedy suggested the term in analogy with the continuing discussion of the entanglement of second order logic and set theory (e.g. [Kennedy 2015]).

descriptive axiomatization: the axioms should not be significantly stronger than the theorems they are trying to ground. We contrast *modesty* and *purity* (continued in Chapter 12.4); formalization is helpful for deciding one but not the other. We argue that the second order axiomatization of Euclidean geometry is immodest. To justify this claim, we correlate (following [Hartshorne 2000]) various axiom sets with specific collections of theorems from Euclid. Then, we expound Hilbert's great achievement in the *Grundlagen*; he provides a *first order* axiomatization for all the essentials of Euclidean geometry. The technical heart of this result is proving the bi-interpretability of (appropriate) geometries with appropriate ordered fields (Chapter 9.4). We support Thesis 4 by finding in Euclid and Hilbert the roots of the fundamental correspondence between algebra and geometry that underlies geometric stability theory (Chapter 5.4). The fundamental shift from the Greek dichotomy between magnitude and number to the modern conception of numbers as measures of magnitudes is studied in Chapters 9.2 and 9.5. In Chapter 10, we consider the role of transcendental numbers from the viewpoints of Euclid, Archimedes, Descartes, Tarski, and Hilbert. In Chapter 10.2, we consider the relations between the geometries of Descartes, arguably axiomatized by Tarski, and the Hilbert geometry. Using a basic result about o-minimal theories, we extend Hilbert's (Tarski's) first order axiomatization to provide a first order theory that justifies the use of π in formulas for the area and circumference of a circle in Euclidean (Chapter 10.1) and 'Cartesian' (Chapter 10.3) geometry.

Chapter 11 examines various overloadings of the word complete. We first discuss Kreisel's analysis (in an unpublished paper) of Poincaré's use of 'complete' in his review of the *Grundlagen*. Then we report the distinction among various formulations of the 'completeness of geometry': syntactic, semantic, order. We contrast the effect of first and second order axiomatizations to emphasize that first order axioms are sufficient for grounding geometry while the second order axioms are needed to ground mathematical analysis. And finally we return to the role of Henkin's completeness theorem for second order logic in establishing internal categoricity for ZFC^2. We dispute, on this ground, Feferman's argument that the CH is an indefinite mathematical proposition. Returning to [Button & Walsh 2016], we distinguish between showing a theory refers to a unique object and identifying that object.

4. Methodology: The last part of the book considers three aspects of the methodology of model theory. In Chapter 12 we attempt and fail to make the notion of 'mathematical purity' a formal notion. That is, 'purity' cannot be explained simply by asserting that 'the content of the conclusion is explicit in the vocabulary of the axioms of an appropriate formal theory'. We agree with Arana [Arana 2008a] that the purity or impurity of a particular argument depends essentially on the particular content. The crux of our argument is:[49] if a proposition X is expressible in a first order theory, and X is shown to hold in each model of that theory, even by introducing radically new methods and concepts that are not formalizable in first order logic, then the completeness theorem ensures there is a formal first order proof of X. We contend that Hilbert's argument for the impurity of a proof of the Desargues theorem from assumptions about 3-space is itself even more impure, because it depends on algebra. A fully rigorous account of his proof (Chapter 12.5) relies on the *completeness theorem* and the notion of *interpretability*. Hilbert's argument was less formal. Indeed, how could it have been otherwise in 1899?

Chapter 13 explores Shelah's dicta on the introduction of metamathematical definitions in model theory. We place his procedure in the context of existing philosophical work (Lakatos, Tappendon, Werndl) analyzing the justification of definitions. We explore the strength of Shelah's particular set of dividing lines recounting examples where they solve problems (e.g. finite axiomatizability) quite different from the original intent. Following Coffa [Coffa 1991], we see the historical accounts of nineteenth century mathematics as overemphasizing rigor and reliability as opposed to a search for clearly defining mathematical notions and their relations. We close Chapter 13 by turning to one of the earliest studies of classification, Plato's *Statesman*. We distinguish the study of definition, classification, and taxonomy, concluding that Shelah actually articulates the building of a taxonomy.

Much of this book developed the role of formalization as a mathematical tool. Kennedy [Kennedy 2013] has advanced the notion of formalism-freeness or formalism-independence as a description of concepts that can be developed in distinct formalisms. In Chapter 14, after exploring this notion, we look at model theoretic examples, investigating the effect of dropping the syntax side of the syntax–semantics distinction and just studying models. Two principal avatars of this approach are Shelah's notion of abstract elementary class (AEC) and Zilber's concept of a Zariski-geometry. The development of abstract elementary classes proceeded for 40 years almost exclusively in a formalism-free manner. This chapter surveys recent interactions of AEC with category theory and large cardinals and a renewed

---

[49] This fact was already pointed out by Hilbert and Ackermann [Hilbert & Ackermann 1938].

use of syntax that has produced significant advances. We summarize the arguments for our four theses in Chapter 15.

Summary: Having surveyed the contents, I reiterate the main theme. André Weil [Weil 1950] wrote, 'If logic is the hygiene of the mathematician, it is not his source of food; the great problems furnish the daily bread on which he lives.'

In this book we draw on the writings of philosophers and mathematicians to show that while there may have been some sense in this view when Weil was young, logic plays a much more significant role in mathematics today.[50] The view of logic as solely hygiene was a reasonable one in the 1920s; that was its first great task. But as logic has developed so has its epistemological scope. The early concerns with reliability engendered the analysis of computability in the 1930s. The concern with the clarification of concepts, stemming from both Frege and Hilbert, leads to the work of Malcev, Robinson, and Tarski. Model theorists think of symbolic logic as a tool for investigating mathematics, not just the foundations of mathematics; modern model theory builds on this insight with both some spectacular success in traditional mathematics and a systematic way to examine relations between fields.

---

[50] The focus on model theory in this book just reflects my background; similar examples of concrete applications of formal methods can be found in other areas of logic: proof mining (better bounds in algebra and functional analysis); set theoretic resolution of the Whitehead problem and issues in general topology: applications of descriptive set theory to dynamical systems, combinatorics, and analysis; use of computability hierarchies in computer science, in differential geometry, and in studying randomness; etc.

Refining the Notion of Categoricity

We reported in the introduction (page 15) several questions raised by Mic Detlefsen. In less precise fashion, they were: Is it better for a theory to be categorical or not? How do we justify whether categoricity is a *virtue* for a theory? Is completeness a good approximation to categoricity? This part has several themes that together respond to these questions.

(1) The exact modern meaning of such terms as vocabulary, theory, and logic significantly influences the answer to these questions. These meanings were developed to address epistemological concerns.

(2) The answers to such questions as these are highly dependent on the logic in which the theory is formalized. While a strong logic makes it easier to find categorical theories, this may in fact be a disadvantage. Too many theories may be categorical. The axiomatization may obscure the fundamental ideas of the area.

(3) We precisely define our notion of a 'virtuous property'.

(4) We argue that 'categoricity in power' and 'completeness' are virtuous properties that spawned a family of others resulting in the role of modern model theory as both a mathematical tool and a schema for organizing mathematics.

Chapters 1 and 2 lay out the basic mathematical and philosophical (respectively) terminology of this book. Chapters 1.1 and 1.2 primarily address theme (1). Chapter 1.3 clarifies the role of various logics while properties of theories and axioms are examined in Chapter 1.4. Chapter 2 addresses philosophical issues about our notion of formalization. First we stress that it is a process and then we distinguish two possible goals: foundational and instrumental. Chapter 2.3 expounds the criterion of theme (3). Chapter 2.4 outlines how these virtuous properties can serve as organizing principles for mathematics and introduces the stability hierarchy, a set of virtuous properties that provide a specific method for such an organization that has powerful consequences for finding invariants for models.

These distinctions underlie the argument in Chapters 3.1 and 3.2, which deal with theme (2). Chapters 3.3 and 3.4 develop the notion of categoricity in power, whose powerful consequences in finding invariants for models signal the importance of studying classes of theories. Thus we initiate the study of the paradigm shift which occupies Part II.

1 | Formalization

Suppose we want to clarify the fundamental notions and methods of an area of mathematics and choose to formalize the topic. Our notion of a formalization of a mathematical topic[1] involves not only the usual components of a formal system, specification of ground vocabulary, well-formed formulas, and proof but also a semantics. From a model theoretic standpoint the semantic aspect has priority over the proof aspect. The topic could be all mathematics via e.g. a set theoretic formalization. But our interest is more in the *local foundations* of, say, plane geometry or differential fields. We set the stage for developing Thesis 1, by focusing on a specific vocabulary, designed for the topic,[2] rather than a global framework. In any case, a mathematical topic is a collection of concepts and the relations between them. There are course other less restrictive notions of axiomatization, and often such a 'formalism' deliberately omits the semantic aspect. But we want the wider notion here as it reflects the model theoretic perspective.

It is not accidental that 'formalization' rather than 'formal system' is being defined. The relation between intuitive conceptions about some area of mathematics (geometry, arithmetic, Diophantine equations, set theory) and a formal system describing this area is central to our concerns. The first step in a formalization is to list the intuitive concepts which are the subject of the formalization. The second is to list the key relations the investigator finds among them. In stipulating this view of formalization, we are not claiming to fix the only meaning of the term but only the meaning most suitable for the discussion here.

Definition 1.0.1 *A full formalization involves the following components.*

(1) *Vocabulary: specification of primitive notions.*
(2) *Logic:*

---

[1] In general, context, area, and topic are synonyms in this book. See Chapter 12.3.
[2] Thus, propositional logic and its variants, which analyze general reasoning as opposed to specific mathematical content, are not discussed here.

(a) *Specify a class[3] of well-formed formulas.*

(b) *Specify truth of a formula from this class in a structure.*

(c) *Specify the notion of a formal deduction for these sentences.[4]*

(3) *Axioms: specify the basic properties of the situation in question by sentences of the logic.*

We think of the specification in Definition 1.0.1 as given in informal set theory; the usual mathematicians' remark 'could formalize in ZFC' applies. In the critique of some problems from a calculus text (Example 1.2.3) we will see how crucial item (1) is to avoid confusions in a situation where there is no thought of a formal language as described in (2)(a). An even more basic example of the importance of fixing vocabulary is the confusion in elementary algebra between whether the minus sign represents binary subtraction or the unary operation of taking the additive inverse.

[Church 1956] defines a *pure functional calculus* to have *no constants*. If there are constants it is an *applied* logic (or calculus). Thus, in Church's notation the first step in the paradigm shift is switching from pure to applied logic. In one sense, they are interchangeable for first order sentences. The validity of a first order sentence $\phi(\overline{A})$ is equivalent to that of $\forall \overline{X}\phi(\overline{X})$ so the difference appears small. But the emphasis is significant. We discuss the greater distinction for second order logic on pages 72 and 257.

1.1   The Concept of Formalization

In Section 4.1 of [Arana & Mancosu 2012], Arana and Mancosu draw an important distinction between 'intuitive' and 'formal' content. We expand a bit on that distinction. First we stress that the 'intuitive content of a proposition' is often ill-defined; there is some background context of the proposition that needs to be clarified. As a simple example, consider 'Are there infinitely many primes?' At the most basic level, one might take this to mean 'prime natural numbers'. But this is still an ambiguous phrase. The meaning of 'prime' depends on the structure placed on the natural numbers. The question can (as in [Arana 2014]) be asked of the structures (Chapter 1.2) $(N, \times, 1)$, $(N, \times, 1, <)$, $(N, +, \times, 0, 1)$, or many other choices. Arana investigates how such variants affect the 'purity' of a proof of infinitely many

---

[3] Most logics have only a set of formulas, but some infinitary languages have a proper class of formulas. While we consider only classical logics, the framework doesn't make this restriction.

[4] Such a notion will generally involve 'logical axioms'; when we speak of axioms we generally mean those in item 3. Euclid more precisely calls them *postulates* but such a distinction now seems pedantic. We sometimes discuss logics that have no deduction system.

primes. A natural response from a mathematician might be, 'What ring[5] are you talking about?' So we speak not of formalizing a proposition but of formalizing a topic or context – those words are interchangeable. A first step towards formalization is to further specify the intuitive content. What concepts are involved? By the *intuitive content* of a *context* or a *topic* we mean an individual (or community) understanding of a complex of concepts; it is not a property of a proposition. The daily work of a mathematician is to take such situations and write down more precise descriptions. Formalization is a very stringent scheme for writing such a description. In this chapter, we lay out one schema for making such descriptions.[6]

In [Arana & Mancosu 2012], formal content is defined somewhat vaguely as the 'inferential role of a statement'. Here[7] we will say the *formal content* of a collection of statements Φ is the collection of their models. When the completeness theorem holds, we can equally well say the deductive consequences of Φ.

In the Introduction, page 12, we promised to address the modification for model theory of an admonition of Maddy to find the proper ground for 'introducing models'. This task is easy; first there are intuitions of mathematical objects. Our goal is to clarify our understanding by formalizing them as mathematical structures. Sometimes we can just define a structure: the natural numbers is the set of finite ordinals. More often we describe a collection of structures in a fixed vocabulary that satisfy certain axioms.

We distinguish two degrees of formalization. We call the first framework[8] Euclid–Hilbert and the second Hilbert–Gödel–Tarski. In the first we have the notions of axioms, definitions, proofs, and, with Hilbert, models. Proof is carried out in natural language with no explicit rules of deduction. This Hilbert is the author of the *Grundlagen*. We could treat formal content as consequences of the axioms or the collection of models of the axioms in this framework but the second description would be more by

---

[5] The ring of integers is only one example. A ring with identity is a structure with universe A and operations $+, \times, 0, 1$ such that $(A, +, 0)$ is an abelian group, $\times$ is associative and distributes over $+$. The ring $\mathbb{Z}(\sqrt{-5})$ illustrates issues with the general definition of prime.

[6] Note that our focus on formal languages is a definite choice; Turing's 'formalization' of the notion of computability goes in a very different direction.

[7] See also Chapter 12.1.

[8] We do not explore the various contributions of Boole, Frege [Gillies 2008b], Peano, Pieri, Peirce, Russell–Whitehead, Schroeder, etc. Hilbert is taken to symbolize the development of modern symbolic logic around the turn of the twentieth century. He stands out for his contributions to both the syntactic and semantic components. In some presentations, we have added Vaught in naming the second notion because the commitment to specific vocabulary and truth in a single model is definitively nailed down as a model theoretic precept in [Tarski & Vaught 1956].

convention than a theorem. For Euclid–Hilbert logic is a means of proof. For Hilbert–Gödel–Tarski, logic is a mathematical subject. This Hilbert is the founder of proof theory. There are now explicit rules of proof and, after Gödel and Tarski, proofs (page 95) within set theory of the completeness theorem for first order logic. So our identification of the two notions of *formal content* is a theorem. A full formalization requires the Hilbert–Gödel–Tarski understanding. In the *Grundlagen* Hilbert proves the *compatibility* of his first order axioms by noting informally that the affine plane of a specific Pythagorean field (Definition 9.4.3) that he constructs satisfies those axioms. But it is only ([Moore 1988], 114) in paragraph 34 of [Hilbert 1918b] that he claims[9] to have proved the *consistency* of the axioms of geometry 'by showing any contradiction in the consequences of the geometric axioms must necessarily appear in the arithmetic of the real numbers as well.'

The process of specifying intuition forces clarification of the relations between the concepts. Thus the notion that for any line and any point not on that line there is a *unique* line parallel to the given line is a natural intuition about geometry. One of Euclid's great contributions is to realize that this intuition might require a separate declaration; it took 2000 years to confirm his doubt.

We will elaborate on our definitions of 'formalization' and 'formal content' in Chapter 12.3. In Chapter 14, we contrast the syntax/semantics approach here with the quite different notion of 'formalism-freeness'.

1.2   Vocabulary and Structures

We establish some specific notations which emphasize some distinctions between the mindsets of logicians, in particular, model theorists, and 'normal' mathematicians.

Definition 1.2.1

(1) *A vocabulary τ is a list of function, constant, and relation symbols.*
(2) *A τ-structure[10] $\mathcal{A} = \langle A, R_1, \ldots R_n \rangle$ is a set A (the domain of $\mathcal{A}$) with an interpretation of each symbol in τ. That is, for each n-ary relation[11] symbol R in τ, R^A is a collection of n-tuples from A.*

---

[9] The metatheory for this consistency claim is unclear. We now know the second order theory of geometry or the real field cannot be proved consistent by 'finitistic' means as Hilbert apparently eventually conceived them.

[10] Philosophers frequently use the word system ([Shapiro 1997], 92).

[11] Do the same for constant and function symbols.

(3) *A structure*[12] *is* many-sorted *if there are a family of unary predicates T_i and each of the variable, function, and relation symbols assign sorts to their arguments and, in the case of functions, values.*

We employ in *this chapter* a useful, but in general archaic, convention of writing $\mathbb{N}$ for the domain, functions, and relations and N for the domain of a structure. Later, we use current model theoretic notation and write N for both; there is little chance of confusion. Then $|N|$ denotes the cardinality of the domain, as $|X|$ always denotes the cardinality of a set X.

Specifying a vocabulary (signature, similarity type)[13] is only one aspect of the notion of a formal system. But it is a crucial one[14] and one that is often overlooked by non-logicians. From the standpoint of formalization, fixing the vocabulary is a first step, singling out the 'primitive concepts'. Considerable reflection from both mathematical and philosophical standpoints may be involved in the choice. For example, suppose one wants to study 'Napoleon's theorem' that the lines joining the midpoints of any quadrilateral form a parallelogram. At first sight, one might think the key notions (and therefore primitive concepts) are quadrilateral and parallelogram. But experience[15] even before Euclid showed that the central basic notions for studying the properties of quadrilaterals are point, line, incidence, and parallel[16] and the delineation of types of quadrilaterals is by explicit definition.

---

[12] There are some minor technical issues around many-sorted structures for the development of T^{eq} that are discussed in Chapter 4.6.

[13] The actual general definition of an abstract algebra was made by Garrett Birkhoff in [Birkhoff 1935] ('as no vocabulary suitable for this purpose is current'.) Moreover he explicitly introduces the idea of considering all algebras of a fixed *species* (vocabulary) and proves his famous theorem that a class of algebras is defined by a set of equations if and only if it is closed under homomorphism, subalgebra, and product. Years later Tarski extended this notion to include relation symbols and used *similarity type* for essentially this notion; sometimes it is called the *signature*. These notions are implicit in [Tarski 1950], [Robinson 1952], [Tarski 1954], and [Tarski & Vaught 1956]: the concept of two systems having the same type is defined there rather than the emphasis I have placed on choosing a vocabulary as part of formalization. These notions (a specification of a sequence of arities) are one level of abstraction higher than vocabulary. But they have the same effect in distinguishing the syntactic from the semantic and we will use vocabulary. In 1953 [Tarski et al. 1968] Tarski just specifies predicate and function symbols of prescribed arity. But Henkin [Henkin 1953] is completely modern. Still another 'synonym' is language. We explain on page 44 why we try to avoid this word. Although Malcev opens his 1936 paper [Malcev 1971b] by specifying a finite set of functions (predicates) defined on a set M, he doesn't use (per the index of his collected works) the words (similarity) type or signature until around 1960.

[14] Demopoulos [Demopoulos 1994] describes this issue as the role of non-logical constants in an insightful article on the Frege–Hilbert correspondence.

[15] Experience includes determining which definitions and postulates best conform to intuition and which allow a smooth development of the subject. Well-known precursors of Euclid include Eudoxus, Theudius, and Hermotimus ([Euclid 1956], 116).

[16] Whitehead and Leśniewski proposed alternatives basing geometry on regions rather than point/line for epistemological reasons, but that is not relevant to this point.

The choice of primitive notions for a topic is by no means unique. For example, formulated in a vocabulary with only a binary function symbol, the theory of groups needs $\forall\exists$-axioms (page 106) and groups are not closed under subalgebra. Adding a constant for the identity, and a unary function for inverse, turns groups into a universally ($\forall$: only universal quantifiers) axiomatized class that is closed under substructure. Alternatively, groups can be formulated with one ternary relation as the only symbol in the vocabulary. The three resulting theories are pairwise bi-interpretable.

The notion of isomorphism is often abused;[17] the notion only makes sense with respect to a specified vocabulary.

Definition 1.2.2 *We say two τ-structures A and B are isomorphic ($A \approx B$) if there is a bijection f between their domains such that for each τ-relation (analogously for function, constant) symbol R, $R^A(\boldsymbol{a})$ if and only if $R^B(f(\boldsymbol{a}))$.*

Thus we might construct a structure (N, S) in different ways; we could take the universe as the finite von Neumann or Zermelo ordinals.[18] In the vocabulary $\{S\}$, the structures are isomorphic; in the vocabulary $\{S, \epsilon\}$, they are not. We explore this distinction further in Chapter 4.7.

It is a commonplace in model theory that just specifying a vocabulary means little. For example in the vocabulary with a single binary relation, I can elect to formalize either linear order or successor (by axioms asserting the relation is the graph of a unary function). Thus, while I here focus on the choice of relation symbols, their names mean nothing; the older usage of signature or similarity type might be more neutral. The actual collection of structures under consideration is determined in a formal theory – by sentences in the logic. In the formalism-free approaches discussed in Chapter 14 the specification is in normal mathematical language. Having fixed a vocabulary with one binary relation, we say, e.g., 'Let K be the class of well-orderings of order-type $< \lambda$ such that ...'

But while axioms are necessary to determine the meanings of the relations in a vocabulary, the mere specification of the vocabulary provides important information. David Pierce [Pierce 2011] has pointed out the following example of mathematicians' lack of attention to vocabulary specification.

---

[17] Structure could be defined purely axiomatically. A category is defined as a collection of 'objects' and 'arrows' (or morphisms); objects A and B are 'isomorphic' if there is a pair of inverse morphisms f, g between them. That is, $f \circ g = 1_A$ and $g \circ f = 1_B$. This notion is useful in abstract category theory but most applications are to *concrete categories* of structures in the set theoretic sense where the vocabulary is specified.

[18] In the first case the successor of $\emptyset$ is $\{\emptyset, \{\emptyset\}\}$ while in the second it is $\{\emptyset\}$.

Example 1.2.3 (Pierce) Spivak's Calculus book is one of the most highly regarded texts in late twentieth-century United States. It is more rigorous than the usual Calculus I textbooks. Problems 9–11 on page 30 of [Spivak 1980] ask the students to prove the following are equivalent conditions on N, the set of natural numbers. This assertion is made without specifying the vocabulary that is intended for a structure $\mathbb{N}$ with domain N. In fact, N is described as the counting numbers,

$$1, 2, 3, \ldots$$

(1) induction $(1 \in X$ and $k \in X$ implies $k + 1 \in X)$ implies $X = N$.
(2) well-ordered Every non-empty subset has a least element.
(3) strong induction $(1 \in X$ and for every $m < k$, $m \in X$ implies $k \in X)$ implies $X = N$.

As Pierce points out, this doesn't make sense: (1) is a property of a unary algebra[19]; (2) is a property of ordered sets[20] (and doesn't imply the others even as ordered unary algebras[21]); (3) is a property of ordered unary algebras. In particular, (2) is satisfied by any well-ordered set while the intent is that the model should have order type ω.

It is instructive to consider what proof might be intended for (1) implies (3). Here is one possibility. Let X be a non-empty subset of N. Since every non-zero element of N is a successor (Look at the list!), the least element not in X must be $k + 1$ for some $k \in X$. But the existence of such a k contradicts property (1). There are two problems with this 'proof'. The first problem is that there is no linear order mentioned in the formulation of (1). The second is, 'what does it mean to "look at the list"?' These objections can be addressed. Assuming that $\mathbb{N}$ has a discrete linear order satisfying $(\forall x)(\forall y)[x \leq y \vee y + 1 \leq x]$ and that the least element is the only element which is not a successor resolves the first problem. This assertion follows informally (semantically) if one reads 'look at the list' as 'consider the natural numbers as a subset of the linearly ordered field of reals.'

As Pierce notes, a fundamental difficulty in Spivak's treatment is the failure to distinguish between the truth of each of these properties on the appropriate expansion of (N, S) and a purported equivalence of the properties – *an equivalence which can make sense only if the properties are expressed in the same vocabulary.*

---

[19] The vocabulary contains only the unary function S.
[20] The vocabulary contains only the binary function $<$.
[21] The vocabulary contains $<, S$.

But in another sense the problem is the distinction between Hilbert's axiomatic approach and the more naturalistic approach of Frege. I'll call Pierce's characterization of Spivak's situation, *Pierce's paradox*. It will recur;[22] Pierce writes:

Considered as axioms in the sense of Hilbert, the properties are not meaningfully described as equivalent. But if the properties are to be understood just as properties of the numbers that we grew up counting, then it is also meaningless to say that the properties are equivalent: they are just properties of those numbers. [Pierce 2011]

Note that this distinction about vocabulary is prior to distinctions between first and second order logic. We stated the difficulty in the purported equivalence of (1) and (2) in terms of second order logic. But the same anomaly would arise if Peano arithmetic (with a schema of first order induction) were compared with 'every *definable* set has a least element.'

Pierce's paradox is fundamentally a semantic remark. Two sentences are equivalent if they have the same models; this makes no sense if they do not have the same vocabularies or at least are viewed as sentences for a vocabulary that contains the symbols from each sentence. It might have been more precise for Pierce to say 'trivially false' rather than 'meaningless'. In the second and third cases enumerated by Pierce, it is clear that as sentences in the vocabulary with symbols $(S, <)$ they are simply not equivalent. And trivially they are both true in the structure $(N; S, <)$. It can be objected that it makes sense to prove one property of a given structure $\mathcal{A}$ implies the truth of another on $\mathcal{A}$ using properties of $\mathcal{A}$. That seems a normal enough mathematical strategy. But consider the case at hand: on $(N, <, S)$, well-order implies induction (i.e. order type ω). Why? Because it is a property of N that the order type is ω. But this seems to me to be just the type of argument I attribute to Spivak a few paragraphs up; it is hard to find a nontrivial phrasing of it.

We introduced vocabulary in Definition 1.0.1 as 'the specification of primitive notions.' Thus the choice of the vocabulary is the fundamental step in the formalization process. The vocabulary should focus attention on the concepts seen as most basic.[23] For example, in algebraic geometry, the crucial problems (solutions of systems of equations) are represented by conjunctions of systems of equations (perhaps of high degree in several variables), that is, conjunctions of atomic formulas in the vocabulary.

---

[22] See, in particular, just after Example 12.1.1.

[23] We don't attempt to analyze the meaning of 'basic' or 'natural' concept; we just rely on the usual understandings in mathematical practice. A more detailed analysis would distinguish between the criteria of 'reflecting intuition' and 'provides a clear framework'.

The following question is of real methodological importance; the two answers below are used in different contexts and with different results. What is the appropriate vocabulary and logic to study vector spaces over the reals?

Example 1.2.4 (Formalizing modules) Module is a generalization of vector space obtained by replacing the field of scalars by a ring R. The standard formalization is as a first order single sorted theory in a vocabulary with a symbol m_r for each $r \in R$ and $(+, 0)$. The axioms specify that the single sort is an abelian group and each unary function m_r behaves as the scalar multiplication by r on the vector space. The models of this theory are all modules over the ring R.

An alternative is a first order *many-sorted logic*.[24] One can approach real vector spaces in a 2-sorted logic with a sort F for field elements, a sort V for vectors with field operations on F, group operations on V, and scalar multiplication from $F \times V$ to V. The models of this theory are all pairs of a real closed field F and a vector space over F. If the intent is to study *real vector spaces* something must be done to restrict the F sort. For instance, one can add an $L_{(2^\omega)^+, \omega}$ axiom (Notation 1.3.5) insisting that every element of F is a unique realization of a cut in the subfield $\mathbb{Q}$ (each element of $\mathbb{Q}$ is definable over the empty set).

For different purposes each of these is a plausible approach. But since the 1970s at least, the first approach is almost universally adopted. Without some infinitary restriction on the real sort[25], the many-sorted theory is unstable (Definition 5.3.2); the single-sorted version is categorical in all powers.

An important notion that depends on the precise understanding of vocabulary is that of a *pseudo-elementary class*. A pseudo-elementary[26] class in a vocabulary τ is the collection of all *reducts* (forget the predicates and functions in $\tau_1 - \tau$) of models of a theory in a larger vocabulary τ_1. In earlier days, a universal vocabulary was often assumed, generally, containing infinitely many n-ary relations for each n. In contrast, we seek primitive terms which pick out the most basic concepts of the field in question and axioms which in Hilbert's sense give us an implicit definition of the area.

---

[24] The universe is a disjoint union of sorts (i.e. unary predicates). The relation and function symbols specify to which sorts they apply. By fiat, if there are infinitely many sorts one can ignore elements that are not in one of the given sorts. See Definition 4.6.2.

[25] For instance, one could write $\forall x[R(x) \rightarrow \bigvee_{r \in \mathbb{N}} x = r]$; but the advantages of first order logic are lost.

[26] Such a class is called *PC* if the τ_1 class is defined by a single first order sentence or *PC$_\delta$* if by a theory and *PC$_\delta$* in $L_{\omega_1, \omega}$ if the sentence is in $L_{\omega_1, \omega}(\tau_1)$.

Thus, we can formalize concepts such as real closed fields (RCF) or algebraic geometry[27] without reference to the construction in set theory of specific models.[28] This analysis is relevant to either traditional (global) or local foundations. For any particular area of mathematics, one can lay out the primitive concepts involved and choose a logic appropriate for expressing the important concepts and results in the field. While in the last quarter century model theory has primarily focused on first order logic as the tool, we discuss some alternatives in Chapters 7 and 14.

Still another example of the subtlety of choosing primitive terms is given in [Manders 1984]. Manders points out that the mutual interpretability between classical geometries and fields can only be treated as a transformation preserving model completeness by a very careful choice of the primitives for the geometry (particularly for geometries with an order on each line). He raises the general philosophical issue of obtaining a well-adapted logic for modern (i.e. scheme theoretic) algebraic geometry. As he put it (page 328), 'Why *must* even innovative attempts to use Tarski semantics, say with unobvious but geometrically intrinsic primitives, break down in describing modern algebraic geometry?' The stress on 'modern' here is essential; in Example 1.4.2 we discuss the naturality of the model theoretic formulation of affine schemes *over fields* and thus for Weil-style algebraic geometry; working over arbitrary rings in Grothendieck style is a different ball game.

Our entire discussion concerns what are sometimes called 'first order structures', which some distinguish from 'second order structures'. I find such notions[29] of '*nth* order structure' often conflate two distinct notions: (a) structure and (b) the semantics of a logic.

The notion of many-sorted (Chapter 4.6) structure allows one to treat uniformly structures of any (finite for simplicity) order. There is one sort symbol in the vocabulary for each order; there may be predicate symbols in the vocabulary relating elements from the same or different sorts.

A language (in the sense of [Shapiro 1991]) or a specification of well-formed formulas (2a) of Definition 1.0.1) or a grammar ([Manzano 1996], 6) describes the expressions of a logic based on such a vocabulary. There are countably many k-ary relation variables for each sort and for each k. As Section 3.3 of [Shapiro 1991] points out there can be both first order and

---

[27] See the discussion of Zariski geometries in Chapter 14.

[28] Geometry and analysis are presented in this way in e.g. [Spivak 1980], [Hilbert 1971], and [Heyting 1963].

[29] Corry's discussion [Corry 1992] of Bourbaki's awkward and unused attempt to provide such a formalization exemplifies the resulting confusion. Strangely, Corry makes no mention of the standard notion of structure discussed here.

higher order semantics for this language. But confusion is introduced when an unnecessary notion of a standard 'higher order structure,'[30] appears. The underlying structure is no different – only the scope of the quantifiers has changed. Valuable information is obtained by varying this scope as in the treatment of second order arithmetic [Simpson 2009], where one might quantify over the recursive sets or the π_1^1 sets.

Abraham Robinson's theory of *nonstandard analysis* involves not only a nonstandard (and importantly saturated) model of the real field but is also able to deal with higher order objects [Henson & Keisler 1986]. Because of its different techniques and formalization we do not include this important field in our analysis in this book.

The notion of fixing a vocabulary to study a family of structures is basic in modern model theory. But this convention is part of the paradigm shift. As noted on page 32, the study of logics in the early 1950s (e.g. [Church 1956]) speaks of the first order predicate calculus which has function and relation variables (which can't be bound) and might have function constants as well. Even the founding papers of model theory are ambiguous about this point, as discussed in footnote 13. But the modern notion is fixed, although [Chang & Keisler 1973] and [Shoenfield 1967] use 'language' rather than 'vocabulary'.

1.3 Logics

Most of this book studies theories. Nevertheless, the choice of a logic[31] in which to make a formalization is crucial. Choosing a logic which is too strong can (a) obscure the actual concept being formalized (see Chapter 11.2) and (b) as discussed later in this section, complicate the investigation because the logic itself is entangled with set theory[32] even though the issues being formalized are not. Definition 1.3.1 summarizes the notion of a *model theoretic logic*[33] $\mathcal{L}$ defined in [Barwise & Feferman 1985].

Definition 1.3.1 *A logic contains certain logical vocabulary: connectives, quantifiers, and a set of variables. For each (non-logical) vocabulary τ, the*

---

[30] A clear definition of this notion is given when ([Manzano 1996], 22) requires that 'all subsets of X' be the actual power set of X (in the ambient model of set theory).

[31] Since we are investigating the use of model theory as a tool to study mathematical structures as in standard mathematical practice and not general principles of reasoning, modal and propositional logics are not considered here.

[32] Recent studies of the entanglement of second order logic and set theory include entanglement of model theory with set theory [Kennedy 2015, Kennedy 2013, Parsons 2013, Väänänen 2012].

[33] In fact, the general notion is more abstract; it does not require the inductive definition of formulas we require in Definition 1.3.1.

collection of $\mathcal{L}(\tau)$-formulas that (for simplicity) is defined inductively in Tarski's natural way[34] *as written out in any beginning text. An $\mathcal{L}(\tau)$-formula with no free variables is called an $\mathcal{L}(\tau)$-sentence.*

Thus, a logic $\mathcal{L}$ is the pair $(\mathcal{L}, \models_\mathcal{L})$ such that $\mathcal{L}(\tau)$ is a collection of τ-sentences and for each $\phi \in \mathcal{L}(\tau)$ and each τ-structure $\mathcal{A}$, $\mathcal{A} \models_\mathcal{L} \phi$ is defined[35] *in the natural inductive way.*

All our logics will contain the = symbol and it will always be interpreted as identity. (See Chapters 4.7 and 8.4.)

We specified the existence of a deductive system in Definition 1.0.1 as one component of a full formalization. The study of such deductions was long the core of logic. Bernays stressed the importance of Hilbert's careful development of a logical calculus.

This procedure of the logical calculus supplements the method of axiomatic grounding of a science, to the extent that such a procedure makes possible, along with the exact laying down of the *presuppositions* as it is brought about by the axiomatic method, an exact pursuit of the *inference modes* with the aid of which one proceeds from the principles of a science to its conclusions. ([Bernays 1998], 195–196)

From a model theoretic standpoint, writing in 1922 Bernays missed a fundamental aspect of formalization. Guided by Tarski's notion of *semantic consequence* or *logical consequence*, there are two important notions of consequence for most of the logics studied in this book.

Definition 1.3.2 (Two notions of implication) *For any pair ϕ, ψ of $\mathcal{L}(\tau)$-sentences, we say ϕ logically (or semantically) implies ψ and write $\phi \models_\mathcal{L} \psi$ if for every τ-structure M, $M \models_\mathcal{L} \phi$ then $M \models_\mathcal{L} \psi$.*

A logic $\mathcal{L}$ with logical axioms and deduction rules has a natural notion of deduction. We write ψ can be deduced from ϕ as $\phi \vdash_\mathcal{L} \psi$.

We say a logic is complete if these notions agree.

Definition 1.3.3 (complete logic) *A logic $\mathcal{L}$ is deductively complete*[36] *if there is a deductive system*[37] *such that for every vocabulary τ and every sentence $\phi \in \mathcal{L}(\tau)$*

---

[34] Briefly the class of *atomic formulas* ($R(\mathbf{x})$ where R is a relation symbol in the vocabulary) is closed under the logical operations, disjunction, negation, and existential quantification in the first order case.

[35] We read this as ϕ is true in $\mathcal{A}$; we say ϕ is *valid* if it is true in every τ-structure.

[36] Apparently the first statement of the completeness theorem for first order logic in this form is in [Robinson 1951]; see Chapter 4.2.

[37] While there is a standard deductive system for second order logic, whether second order logic is complete depends on the choice of semantics (Henkin or 'full'). The use of $\models$ for semantic consequence was recommended in the foreword on terminology to [Addison et al. 1965]. Simon Kochen reports using the notation $M \models \phi$ in his 1959 Princeton PhD thesis and in [Kochen 1961]. Lyndon [Lyndon 1967] uses $\psi \models \phi$.

$$\vdash \phi \text{ if and only if } \models \phi.$$

A completeness theorem sets up a syntactic/semantic duality. The ability to translate results between syntax and semantic is the essence of model theory. If a logic $\mathcal{L}$ satisfies the completeness theorem then the deductive system provides additional (computability) information about the mathematical topic, but usually that is not central to the investigation. As we'll see, compactness[38] is a more used tool than completeness for model theory.

One meme to define model theory is 'the study of definable sets'.

Definition 1.3.4 *Let M be a structure and $X \subset M^n$. We call X $\mathcal{L}$-definable (with parameters $\mathbf{b}$) if there is an $\mathcal{L}$-formula $\phi(\mathbf{x}, \mathbf{y})$ and[39] $\mathbf{b} \in M$ such that*

$$X = \{\mathbf{m} \in M^n : M \models \phi(\mathbf{m}, \mathbf{b})\}.$$

Thus each natural number is definable in the real field,[40] but the set N of all natural numbers is *not* first order definable.[41] The particular logic $\mathcal{L}$ is usually clear from context and so omitted. A definable set which is definable without parameters is said to be *definable over the empty set*.

If our logic allows higher order variables, either $\mathbf{b}$ or $\mathbf{m}$ might name elements of sorts that ostensibly refer to higher order objects (page 41).

The usefulness of definability becomes clear with Cartesian geometry. The Greek notion of a conic section is transformed by Descartes so that, e.g., a parabola is the set of solutions of an equation $y = ax^2 + bx + c$.

Truth in a *specific* structure as in Definition 1.3.1 not only is a central tenet of model theory[42] but also reflects mathematical practice. For example, in his splendid popular account of the Langlands conjecture, Frenkel (my italics) writes,

The Shimura-Taniyama-Weil conjecture is a statement about certain equations. A large part of mathematics is in fact about solving equations. We want to know

---

[38] The *compactness theorem* asserts that if each finite subset Σ_0 of a set of sentences is satisfiable then so is Σ.

[39] We write $\mathbf{b}$ to denote a finite sequence $b_1, \ldots b_n$ without specifying the length of the sequence; $\bar{b}$ is also commonly used.

[40] Each $r \in \mathfrak{R}$ is definable *with parameters* by the formula $x = n^{\mathfrak{R}}$. Each $n \in N \subset \mathfrak{R}$ is defined *without parameters* by $x = 1^{\mathfrak{R}} \ldots + 1^{\mathfrak{R}}$ where 1 is added to itself $n - 1$ times. Pedantically we would insist on a new symbol $\hat{n}$ in the vocabulary such that $(\hat{n})^{\mathfrak{R}} = n$. Such a notation (a squiggle underneath the n is more common than a hat) is used for two weeks in an introductory model theory course. Then the instructor says, 'Without loss, we'll forget the hat.'

[41] See Tarski's theorem that every definable subset of the reals is a finite union of intervals (page 105).

[42] We have little to contribute to the important discussion (Etchmendy, Feferman, Hodges, Mancosu, et al.) as to when Tarski arrived at the notion of truth on a particular domain.

whether a given equation has a solution *in a given domain*; if so can we find one? If there are several solutions, then how many? Why do some equations have solutions and some don't? ([Frenkel 2013], 83)

Crucially, 'the natural inductive definition' of Definition 1.3.1 implies that the truth of a τ-sentence in a τ-structure M depends *only* on the isomorphism type of M. Thus the entire framework of 'model theoretic' logics depends on what Burgess [Burgess 2010] has called 'indifferentism to identity', ignoring a specific set theoretic construction of the model. This is one aspect of formalism-freeness (Chapter 14); but much more is entailed.[43]

Shapiro ([Shapiro 1991], 10) phrases 'the question' as '*What is the [correct or best] [language or logic] in which to [...]?*' Slightly modified, this question is a major theme of this book. The essential rewording[44] is to replace 'language or logic' by 'vocabulary and logic'. Shapiro ([Shapiro 1991], 5.3.2) correctly argues that when [...] means 'work in second order logic on the canonical structures' (arithmetic and analysis) the language/logic distinction is obliterated. But we argue throughout that preserving this distinction has significant mathematical consequences. In particular, we note after Definition 4.6.1 the immense differences among the structures $(\mathbb{N}, S)$, $(\mathbb{N}, +)$, $(\mathbb{N}, +, \times)$ from the standpoint of ordinary mathematics. And we fill the brackets with 'study virtuous properties of formalizations', those that have significant mathematical consequences.

We try (without full success) to avoid the word *language* because of the ambiguity deriving from several closely related uses. One may speak of the 'language of rings' meaning $+, \times, 0, 1$. A different and more specific version is for this phrase to imply that these operations obey the axioms of ring theory; see the discussion of vocabulary preceding Example 1.2.3. In a third version, one speaks of the language of first order logic, meaning the collection of formulas generated from a vocabulary (equivalently signature) by the finitary propositional connectives and existential and universal quantification.

---

[43] It is ironic that although model theory is based on the notion of identifying structures up to isomorphism with respect to the given vocabulary, Shelah has proved major results by deliberately ignoring this convention. For example, see the Whitehead problem discussion in Chapter 4.7 or the construction of many non-isomorphic models in Chapter 8.3. Still another example is Ehrlich's [Ehrlich 2001] use of Conway's surreal numbers to investigate real closed fields. These examples illustrate the power that derives from being careful about vocabulary and in contrast to the usual indifferentism, paying careful attention to how a structure is built in an expanded language (essentially set theory in the first and third case, an appropriate Skolem theory (page 91) in the second).

[44] Shapiro (page v of [Shapiro 1991]) defines language to include what is called vocabulary here with prescriptions about first versus second order variable. We address the variable issue in Chapter 4.6.

We describe here three logics: first order, infinitary, and second order; we discuss their historical development in Chapters 4 and 7.

Notation 1.3.5 *A logic (see Definition 1.3.1) $\mathcal{L}$ maps a vocabulary τ to the collection of $\mathcal{L}(\tau)$-sentences.*

(1) *The formulas of $L_{\omega,\omega}(\tau)$, first order logic, for the vocabulary τ consist of finite Boolean combinations of atomic τ-formulas augmented by existential, $\exists x$, and universal, $\forall x$, quantification.*
(2) *For any cardinal[45] κ, the formulas of $L_{\kappa,\omega}(\tau)$ infinitary logic[46] are defined analogously but now conjunctions and disjunctions of cardinality less than κ are allowed. $L_{\infty,\omega}$ allows conjunctions and disjunctions of arbitrary length.*
(3) *In second order logic, $L^2(\tau)$, predicate symbols of any arity m are allowed in constructing formulas and quantification over them is permitted as well.*

We divided the notion of formalization into five components: (1) specification of primitive notions, (2a) specifications of formulas and (2b) their truth, (2c) proof and (3) axioms. The standard account of a formal system (e.g. [Smullyan 1961]) includes (2a), (2c), and (3) but not (2b). From our standpoint and that of [Barwise & Feferman 1985], (2a), and (2b) are basic; (2c) may or may not exist. Second order order logic has a deductive system and (several) versions of (2b). By default in this book, we use the *full or standard* semantics, which interprets $M \models \forall X \phi(X)$ as 'for every subset A of M, $M \models \phi(A)$'. The 'Henkin semantics'[47] is really first order; the interpretation is 'for every a in the sort denoting subsets of M, $M \models \phi(A)$'. Henkin proved the completeness of second order logic relative to his semantics.

The vocabulary of *pure* second order logic contains only an element relation between *tuples* and relations of appropriate arity, and equality among individuals, and among predicates. The upward and downward Löwenheim–Skolem theorem both fail in second order logic. Note in particular that most 'large' cardinals are defined by sentences in second order logic.[48]

---

[45] Still more generally, the ω here can be replaced by λ and quantification over sequences of length less than λ are allowed.

[46] Chang's trick: the models of any sentence of $L_{\omega_1,\omega}$ are also defined by a first order theory and the omission of a type (e.g. [Baldwin 2009, Marker 2016]).

[47] See Definition 4.6.2, [Enderton 2007] and [Shapiro 1991].

[48] The second order sentence defining the first measurable cardinal κ has (at most) one model. Similarly, each model of ZFC has at most one cardinal κ satisfying the analogous first order sentence in the vocabulary with predicate symbol ϵ. See [Väänänen 2012].

We follow Magidor [Magidor 2015] in listing several criteria by which a logic may be *entangled with set theory*.

(1) Is satisfaction absolute[49] across universes of set theory?
(2) Does the set of validities of a logic $\mathcal{L}$ depend on the universe of set theory under consideration?
(3) What set theoretical assumptions are needed for getting a compactness cardinal for the logic and, if a compactness cardinal exists, computing it?
(4) Does the logic satisfy the Löwenheim–Skolem theorem (LST)? If not, what set theoretical assumptions are needed in order to get an LST cardinal for the logic[50] and if it exists how large must it be?
(5) Are large cardinal assumptions needed to get a Hanf number[51] for the logic?

By all of these measures, first order logic is not entangled with set theory while second order logic is highly entangled. Some extensions of first order logic are closer to first order on some of the measures (e.g. validity is absolute for $L(Q), L_{\omega_1,\omega}$) than others. There is an extensive interaction between the calculation of the cardinals described above and large cardinal axioms [Magidor 2015]. For example the LST number of second order logic is at least the first supercompact cardinal.

Another measure of entanglement of a logic $\mathcal{L}$ with set theory is to study the effect on the inner model L of set theory, if the definable sets used in the construction of the constructible sets, L, are taken to be $\mathcal{L}$-definable rather than first order definable. The intricate dependence of L on such logics as Chang's quantifier, $L_{\omega_1,\omega}$, second order logic, and the cofinality ω quantifier as investigated by Kennedy, Magidor, and Väänänen are discussed in [Magidor 2013] and [Kennedy et al. 2016].

We listed these criteria to contrast a deep entanglement of some logics with set theory with the situation obtained by restricting to first order logic and studying *theories* rather than *logics*. Most of the text deals with the major part of first order model theory that is not entangled with set theory.

---

[49] A property P of sets is said to be 'absolute' between a transitive model A of set theory and a transitive extension B if A and B both think all the objects satisfying P are in B and they agree on what they are. The most standard method for proving such a fact is the Shoenfield absoluteness lemma which states that a Π^1_2 predicate of the integers is absolute between models of set theory with the same ordinals. See, e.g. [Jech 1978] or [Kunen 1980].

[50] The LST cardinal for a logic $\mathcal{L}$ is the least cardinal κ such that if M is a structure for a vocabulary of cardinality $< \kappa$ then M has an $\mathcal{L}$-elementary submodel of cardinality less than κ. This notion weakens the Löwenheim–Skolem theorem by not insisting that all models have smaller elementary submodels but only that very big models do.

[51] The Hanf number for a logic $\mathcal{L}$ is κ if for every $\mathcal{L}$-theory T in a vocabulary of cardinality $< \kappa$, if T has a model of cardinality κ, it has arbitrarily large models.

Chapter 8 explores how an apparent entanglement was broken and the more subtle connections with infinitary logic.

1.4 Theories and Axioms

In this section we distinguish between axioms (a set of sentences) and a theory (a set of sentences closed under semantic consequence). Depending on the context this may or may not be an important distinction. In Chapter 3.1 we argue that the most important aspect of second order categoricity is to verify that a certain *axiomatization* of an intuition captures the concept completely.

Definition 1.4.1 *A theory T is a collection of sentences in some logic $\mathcal{L}$, which* is closed under semantic consequence.[52] *We say T is* axiomatized *by Σ if Σ is a subset of T and T is the set of logical consequences of Σ.*

For simplicity, we will usually assume that T is consistent (has at least one model), has only infinite models, and is implicitly in a countable vocabulary. We may emphasize this by writing T is *countable theory*.[53] I assume the existence of a semantics for each logic is defined in ZFC. The distinction between an axiom set and its consequences will be significant in Chapter 3.1 for second order logic and in considerations of finite axiomatizability of first order theories in Chapter 13.2. For first and second order logic, the question is whether a theory (which can only contain countably many sentences) is finitely axiomatized. For $L_{\omega_1,\omega}$ (which is closed under countable conjunction) the question is whether a theory is axiomatized by a single sentence or uncountably many.

The following example shows the value of formalization in a specific but abstract mathematical situation. Schemes were introduced by Grothendieck around 1960; the most basic kind is easily understood from a model theoretic standpoint.

Example 1.4.2 (Affine Schemes) The notion of a theory provides a general method for studying 'families of mathematical structures'. First order

---

[52] This is by no means a universal convention even in model theory. We adopt it to avoid a cumbersome use of the fairly trivial equivalence relation that theories, defined as sets of sentences, are equivalent if they have the same consequences. Even as late as Shoenfield's classic graduate text [Shoenfield 1967], the word theory is used for the entire syntactical apparatus: formal language, axioms and rules of inference for the logic, and specific axioms.

[53] Similarly, a theory with an uncountable vocabulary is called an *uncountable theory*.

axiomatizations may pick out salient features of a structure without determining the structure up to isomorphism. One obvious example is that algebraic geometers want to study 'the same' variety over different choices of a field k. This is crisply described as the solutions in the field k of the equations defining the variety. The Chevalley groups[54] can be seen as the matrix groups given by a specific definition interpreted with solutions in each field. 'Most' of the non-exceptional finite simple groups are Chevalley groups. Each family is an example of an affine scheme. In introducing the notion of an affine group scheme, Waterhouse begins with a page and a half of examples and defines a group functor. He then writes,

> The crucial *additional*[55] property of our functors is that elements of $G(R)$ are given by finding solutions in R of some family of equations with coefficients in k. ... Affine group schemes are exactly the group functors constructed by solutions of equations. But such a definition would be technically awkward, since quite different collections of equations can have essentially the same solutions. [Waterhouse 1979]

He follows with another page of proof and then a paragraph with a slightly imprecise (the source of the coefficients is not specified) definition of affine group scheme over a field k. Note that (for Waterhouse) a k-algebra is a commutative ring R with unit that extends k. Now, for someone with a basic understanding of definability in first order logic, an affine group scheme over k is a collection of equations ϕ over k that define a group under some binary operations defined by some equations ψ. For any k-algebra R, the group functor F sends R to the subgroup of R defined by those equations.[56] The key point is that, not only is the formal version more perspicuous, it underlines the fundamental notion. Definability is not an 'addition'; the group functor aspect is a *consequence* of the *equational* definition.

The meanings of such words as categoricity and theory have varied over time and among different groups of logicians during the twentieth

---

[54] A *classical group* over a field F is, roughly speaking, one of a specified family of matrix groups (special linear, orthogonal, symplectic, or unitary group) with coefficients from F. Thus the *special linear* group of dimension n over F is the group of invertible $n \times n$ matrices with determinant 1. The Chevalley groups are a subclass of these: simple finite groups of Lie-type.

[55] My emphasis. See text following quotation.

[56] We are working with the incomplete theory T in the vocabulary $(+, \cdot, 0)$ for rings with names for the elements of k. Two finite systems of equations over k, $\sigma(\mathbf{x})$ and $\tau(\mathbf{x})$, are equivalent if $T \vdash (\forall \mathbf{x})\sigma(\mathbf{x}) = \tau(\mathbf{x})$. The functorial aspects of F are immediate from the trivial observation that positive formulas (roughly, no negation signs) are preserved under homomorphism. Note that I am taking full advantage of k being a field by being able to embed k in each k-algebra. If we were to study modules over $\mathbb{Z}$, complications about finite characteristic would ensue. To be fair to Waterhouse's longer exposition, he is introducing further terminology that is useful in his development.

century. Without giving a serious historical account I want to settle on particular meanings commonly taken by contemporary model theorists. This important early twentieth-century concept, formulated in *applied logic*, had three names.

Definition 1.4.3 *T is* categorical[57] *or* monomorphic *or* univalent *if it has exactly one model (up to isomorphism). T is* categorical in power κ *if it has exactly one model in cardinality κ. T is* totally categorical *if it is categorical in every infinite power.*

A structure M is $\mathcal{L}$-categorical *if* $\text{Th}_{\mathcal{L}}(M) = \{\phi \in \mathcal{L}(\tau): M \models \phi\}$ *is categorical.*[58]

We defined the notion of a complete logic in Definition 1.3.3. For such a logic, conditions (1) and (2) below are equivalent conditions on a theory T.

Definition 1.4.4

(1) *A τ-theory T in a logic $\mathcal{L}$ is* semantically complete *if for every sentence $\phi \in \mathcal{L}(\tau)$*

$$T \models \phi \text{ or } T \models \neg\phi.$$

(2) *A τ-theory T in a logic $\mathcal{L}$ is* deductively complete or Post[59] complete *if for every sentence $\phi \in \mathcal{L}(\tau)$*

$$T \vdash \phi \text{ or } T \vdash \neg\phi.$$

Under these definitions, every categorical theory is semantically complete. Further, in a logic which admits upward and downward Löwenheim–Skolem theorem for *theories*, every theory that is categorical in some infinite cardinality is semantically complete. While 'categoricity in power implies completeness' holds for first order theories it fails for L^{II} and for theories.[60] Note that for any τ, any τ-structure M and any logic $\mathcal{L}$, $\text{Th}_{\mathcal{L}}(M) = \{\phi \in \mathcal{L}(\tau) : M \models \phi\}$ is a complete theory.

Of the logics in Notation 1.3.5, only first order has a fully satisfactory completeness theorem. Keisler [Keisler 1971] proves completeness [Karp 1964] of $L_{\omega_1,\omega}$ and provides a proof system; but it has an inference

---

[57] Model theorists scarcely know the word since in first order logic only finite structures can be categorical. The term 'totally categorical' came into use in the 1980s, as a strengthening of 'categorical in uncountable cardinals'.

[58] This ascription of a property originally defined for a theory as a property of a structure whose complete theory has that property is a common sight in model theory.

[59] Note that propositional logic, that is the empty theory in propositional logic is both Post and semantically complete. Pursuing a misleading analogy and thinking this equivalence held for stronger logics complicated the problem of isolating the notion of complete in the 1920s.

[60] E.g., for L^{II} take the disjunction of sentences characterizing distinct cardinals.

rule from infinitely many hypotheses. Both the logics $L_{\omega,\omega}$ and the logics $L_{\kappa,\omega}$ are sometimes called 'first order' because the quantification is only over individuals; usually for us, first order refers only to $L_{\omega,\omega}$.

As we will see, minor differences in terminology, apparently introduced only for convenience, such as demanding a theory be complete or closed under semantic consequence, or invoking Morley's quantifier elimination procedure (page 112), can signal major changes in viewpoint.

The model theoretic analyses discussed below depend on the description of some area of mathematics as the models of a formal first order theory. Actual formulas in the formal system appear rarely in the technical papers; the fact that particular concepts admit such a description is endemic and essential.

Note that one is not permitted to quantify over the *standard* natural numbers in a first order sentence.[61] We may say some property holds for every natural number, but only when the property is actually given by a family of first order formulas ϕ_n.

In Definition 1.0.1, the logic does not depend on the particular area of mathematics. Rather, particular vocabularies and axiomatizations/theories are chosen to represent particular topics.[62] And a logic is chosen that is suitable to express the concepts to be studied. Thus, while most applications of model theory use first order logic, Zilber (page 170) found $L_{\omega_1,\omega}$ essential for the study of complex exponentiation in order to prevent the Gödel phenomena[63] from preventing any useful model theory.

The importance for foundational issues of distinguishing logics, which really became clear only in digesting the Gödel theorems, is central to this book. As noted in Chapter 3.1, there are huge differences between the *sterile notion of second order categoricity* (Chapter 3.1) and the *fertile notion of categoricity in power* in first order logic (Chapter 3.3) and $L_{\omega_1,\omega}$ (Chapter 7 and 7.3). Turning to a more concrete example we point out in Chapter 9 that while Euclid's geometry can be formalized in $L_{\omega_1,\omega}$ (to include the axiom of Archimedes), Hilbert shows that that geometry could in fact be given a first order axiomatization since he avoids the Archimedean axiom in establishing the results from Euclid and in particular the area of polygons.

---

[61] Thus, in Definition 10.1.1, we need infinitely many sentences $i_n < \pi < c_n$ to specify π; the string of symbols $\forall n(i_n < \pi < c_n)$ is not a well-formed formula.

[62] This 'fact' is one step in the paradigm shift. Gödel's proof of the completeness theorem makes heavy use of relations symbols of all arities. Henkin's proof works in the given vocabulary.

[63] We use this phrase as a shorthand for incompleteness, undefinability of truth, existence of a pairing function, etc.

2 | The Context of Formalization

Bourbaki wrote 'We emphasize that it [logical formalism] is but one aspect of this [the axiomatic] method, indeed the least interesting one.'[1] In this chapter, we begin the argument that Bourbaki have it backwards. Formal methods are not a pedantic puddle on the path to find the methods and principles that underlie a mathematical result but rather a stepping stone to understanding.

We explore the epistemological purposes of formalization along the lines of Manders (Chapter 2.1) and contrast Bourbaki's views with the contemporary mathematician, Kazhdan (Chapter 2.2). In Chapter 2.3 we take up Detlefsen's questions (page 15), and provide a criterion for the virtue of a property of a theory. In Chapter 2.4 we begin the explanation of how virtuous formal properties can provide a more fruitful way of organizing mathematics than that proposed by Bourbaki.

2.1 The Process of Formalization

In Chapter 1.1 we gave a static description of a formalization because such full formalizations of various topics are taken as the data for the organization of mathematics via classification theory. Here, we take a somewhat broader view. Lakatos writes:[2]

For more than two thousand years there has been an argument between *dogmatists* and *skeptics* ... The core of this case-study will challenge mathematical formalism but will not challenge directly the ultimate positions of mathematical dogmatism. Its modest aim is to elaborate the point that informal, quasi-empirical, mathematics does not grow through a monotonous increase of the number of indubitably established theorems but through the incessant improvement of guesses by speculation and criticism, by the logic of proofs and refutations. ([Lakatos 1976], 4–5)

In contrast, I take formalization as part of that 'incessant improvement' and one that aims toward 'indubitably established theorems'. Our goal here

---

[1] Parenthetical references for pronouns added. ([Bourbaki 1950], 223).
[2] The omitted material in the quote, like much of Lakatos' introduction, is a diatribe against formalism, which is irrelevant to the project here.

is to move beyond case studies, to show how formalization can be a tool for systematizing the changing definitions of mathematical notions. In this section, I expand more on how this tool will be developed in the book and where it is applied.

This kind of systematic analysis is developed by Mark Wilson for physical science. My approach is less ambitious than Wilson's; I deal only with mathematics, not science writ large. Thus while I agree that it is folly to try to fix the 'meaning' (in whatever technical sense one wants) of a 'concept' once and for all across all contexts ([Wilson 2006], 87–96), my remedy, within mathematics, is a formalization that specifies both a formal language and a set theoretic semantics that fixes context more precisely than one can ever expect for physical science. Although Hilbert is often paraphrased by 'the axioms implicitly define the geometry', for him, the axioms are inspired by previous intuitions[3] (analogous to Wilson's real world). Nevertheless, in a mathematical investigation within a fixed formalization we can and must ignore these original intuitions and simply carry out a deduction in the system.[4] Thus Hilbert is not a *strict formalist*; his axioms are not arbitrary rules for a game. But having chosen the rules (logic and axioms), we must play by those rules. However, one must show that those intuitions are reflected by the axioms to justify the assertion that we have in fact formalized topic X. Thus, Hilbert's axiomatization of Euclidean geometry is justified by his derivation of the results of Euclid. After Tarski's definition of truth, we have an exact notion of what the Hilbert geometries are (those structures which satisfy Hilbert's axioms).

Wilson considers, as an example, the examination of a physical phenomenon at various scales and considers the physical theory at each scale as a 'patch'. As the size of an object approaches the boundary of two patches, there are local connections between the patches. The atlas (or facade) is the collection of these multiple patches together, in analogy to the mathematical notion of a manifold,[5] with connecting maps ([Wilson 2006], 377–378). Thus, Wilson is correlating many views of the same object.

My standpoint is rather different, but analogous. Take some parts of 'mathematics' as the subject area (as opposed to, say, Wilson's continuum mechanics) and the patches become various complete first order

---

[3] See footnote 24 in the Introduction.

[4] Of course, if we deduce a result that contradicts the intuition, we may decide that we had chosen the wrong formalization.

[5] An atlas consists of a set of charts; each chart (patch) looks like a piece of flat Euclidean space, connected by transition maps that account for the curvature of a more complicated space.

theories that describe different areas of mathematics. The links are made by interpretations or, more weakly but importantly, by satisfying the same virtuous property. We identify common characteristics of different areas with respect to our scheme.

The fact that different mathematical notions such as dimension, compact, etc. will have different denotations in the various fields is not anomalous but such occurrences become phenomena to be investigated. Is this replication of a name more than a coincidence? Thus I am proposing an atlas of mathematics as opposed to a single foundation. The goal is to study various properties that hold of different areas rather than manifold-like transitions between maps of the same area.

In particular, this approach can be seen as an attempt to implement the following project of Manders (see page 4):

We need to be able to emphasize *special* features of a given mathematical area and its relationship to others, rather than how it fits into an absolutely general pattern. ([Manders 1987], 196, emphasis added)

I posit three levels for this analysis:

(1) 'normal mathematical' exposition
(2) a formalization of a topic as in Definition 1.0.1
(3) the study of relations among various areas, mediated by the formalization.

Thesis (2) asserts advantages of full formalization for both mathematics and studies of mathematical practice. One of these advantages is the ability to precisely define various properties of a formalization so they can be compared. I dub the good ones virtuous properties in Chapter 2.3 and expand on them in Chapters 2.4 and 3. Among these properties are categoricity in power (categoricity (page 49) is not virtuous), completeness, model completeness, ω-stability etc. Many of the properties are virtuous because they allow the systematic comparison of theories about different areas.

There is no claim that such formalization is essential or even possible for all areas of mathematics. Rather, the claim is that such a formalization is useful for some mathematical purposes and for the epistemological project set by Manders. In [Manders 1984, Manders 1987], Manders stresses the role of interpretation (Chapter 4.5) in what corresponds to the third level of analysis here. However, interpretation between formal theories is only one such tool. In particular the relationships developed in the stability hierarchy (e.g. Chapter 5.5), while syntactic, are considerably looser than interpretation[6] while still preserving crucial features. The extensive but preliminary investigation in [Manders 1987] of local to global methods in number theory is an example of a connection that one would like to analyze using formal methods.

Let's consider how this analysis might play out in geometry. Which geometry? Geometry might mean the geometry of Euclid, Descartes, or Hilbert, or in modern times might be modified by tens of adjectives. We have no thought that these are differing aspects of the same 'concept'. Rather we think of these as some 15 or 20 or more different mathematical topics. What commonality leads to the many distinct uses of the word geometry?

In Chapters 9–12 we implement the first two levels of the analysis by considering various full formalizations of the plane geometries whose 'normal expositions' were by Euclid, Descartes, and Hilbert as well as the quite distinct area of metric geometry. In those chapters we study how different formalizations enhance (or not) the accessibility of certain concepts and illuminate (or not) the driving principles of the area.

Many, not all by any means, of these various parts of geometry can be fitted out with a suitable fully formalized theory and general tools from stability or o-minimality apply for example to algebraic geometry and real algebraic geometry respectively. In Chapter 5.4 we explain the fundamental role of combinatorial geometry in classification theory and in Chapter 6.2 observe how these tools including geometric stability theory have played significant roles in proving new results in Diophantine geometry and analysis.

2.2   Two Roles of Formalization

While the immense significance of formal methods in the traditional study of the foundations of mathematics cannot be denied, the effect of such methods on the normal conduct of mathematics is open to considerable doubt [Sacks 1975, Macintyre 2011]. Our goal here is not to join the anti-foundationalist parade but to describe a different contribution of formal methods to mathematics.

In Chapter 14 I discuss Kennedy's notion of 'formalism-free' approaches in logic and especially in model theory.[7] Here I proceed in the opposite

---

[6] While very informally one might say a theory is unstable if a linear ordering can be interpreted on n-tuple in a model of T, the domain of this ordering need not be definable so this is not an interpretation in the usual sense.

[7] See [Kennedy 2013], [Baldwin 2013a].

direction. As in Theses 1 and 2, the goal is to show how formal symbolic logic plays an increasingly important *instrumental role* in ordinary mathematical investigations and provides schemes for organizing mathematics aimed not at finding foundations but at identifying mathematical concepts that link apparently diverse areas of mathematics and often address specific mathematical problems. We will see in Chapter 3.3 that these ideas develop from appropriate weakenings of categoricity and that they provided an unexpected fulfillment of some hopes of Bourbaki. We specified in Definition 1.0.1 exactly what we mean by a formalization of an area of mathematics. What does this definition have to do with traditional mathematics? A prominent mathematician, David Kazhdan,[8] in the first chapter, entitled 'Logic', of his lecture notes on *motivic integration*, writes:

One difficulty facing one who is trying to learn Model theory is disappearance of the natural distinction between the formalism and the substance. For example the *fundamental existence theorem* says that the syntactic analysis of a theory [the existence or non-existence of a contradiction] is equivalent to the semantic analysis of a theory [the existence or non-existence of a model]. [Kazhdan 2006]

At first glance this statement struck me as a bit strange. The fundamental point of model theory is the distinction between the syntactic and the semantic. On reflection,[9] it seems that Kazhdan is making a crucial point which I elaborate as follows. The separation of syntax and semantics is a relatively recent development. Dedekind grasps it late in the nineteenth century.[10] But it is actually clearly formulated as a tool only in Hilbert's 1917–18 lectures [Sieg 1999]. And immediately Hilbert smudges the line in one direction by seeing the 'formal objects' as mathematical. As Sieg writes,

But it was only in his paper of 1904, that Hilbert proposed a radically new, although still vague, approach to the consistency problem for mathematical theories. During the early 1920's he turned the issue into an elementary arithmetical problem ... ([Sieg 1994], 75)

Hilbert began the study of metamathematics by considering the formal language and the deduction relation on its sentences as mathematical objects. Thus syntactical analysis is regarded as a study of mathematical

---

[8] Kazhdan is a leading scholar in the field of group representations. Motivic integration (page 155) is a new subject suggested by Fields medalist Kontsevich, in the 1990s. He originated it as a tool in geometry and physics [Denef & Loeser 2002]. Its development as a tool in number theory is a joint work of model theorists and traditional mathematicians (in particular Kazhdan and Hrushovski).

[9] I thank Ehud Hrushovski, Juliette Kennedy, and David Marker for illuminating correspondence on this issue.

[10] See footnote 11 of [Sieg 1994].

objects (substance).[11] *The first great role of formalization then, aimed to provide a global foundation for mathematics.* The Hilbert program treats the syntax as a mathematical 'substance'.

But Kazhdan is commenting on a smudge in the other direction; to prove the completeness theorem, Gödel constructs a model (a mathematical object) from the syntactical formulas.[12] When one views the completeness theorem solely from the standpoint of logic, the construction of 'models' from syntactic objects to make a statement about syntactic objects is less jarring. The surprise is when a real mathematical object arises from the syntactic paraphernalia. Marker emailed, 'I've found when lecturing that a similar stumbling block comes when giving the model theoretic proof of the Nullstellensatz ([Marker 2002], 88) or Hilbert's 17th Problem when the variables in the polynomial become the witnessing elements in a field extension.'

In [Kazhdan 2006], Kazhdan is not concerned with the global foundations of mathematics; he is concerned with laying a foundation for the study of motivic integration (Example 6.2.6). This exemplifies the following formulation of Thesis 1: *By specifying in a formal language the primitive concepts involved in a particular area of mathematics and postulating the crucial insights of that field (usually thought of as defining the concepts implicitly in the Hilbert sense), one can employ the resources of 'formalization' to the analysis of 'standard' mathematical problems.* The goal of this formalization is not foundational in the traditional sense but accords with Bourbaki's ideal of isolating the crucial constructions that appear in many places. The new feature is that one is able to identify certain common syntactic features in mathematical theories that discuss very different content. The bonus that Bourbaki did not foresee is the mathematical applicability of such resources as the completeness theorem, quantifier elimination, techniques of interpreting theories, and the entire apparatus of stability theory.

Bourbaki [Bourbaki 1950] distinguish between 'logical formalism' and the 'axiomatic method'; the second, as they are too modest to say, is best exemplified by the Bourbaki treatise.[13] 'We emphasize that it [logical

---

[11] This transition is seen even in the title of Post's 1920 thesis [Post 1967].

[12] In fact this blurring is frequent in the standpoint of the Schröder school of algebraic logic. Badesa makes this point in [Badesa 2004]; his argument is very clearly summarized in [Avigad 2006]. The combinatorics of the proofs of Löwenheim and Skolem are very close to those of Gödel. But Gödel makes the distinction between the syntactic and semantic clear since the warrant for the form of the syntactic configuration, which is interpreted as a model, is that it does not formally imply a contradiction.

[13] Mathias [Mathias 1992] has earlier made a more detailed and more emphatic (stressing the inadequacy of their formal system) but similar critique to ours of Bourbaki's foundations. In

formalism] is but one aspect of this [the axiomatic] method, indeed the least interesting one.'[14] In part, this remark is a reaction to the great pedantry of early twentieth century logic as the language of mathematics was made rigorous. It also is a reaction to the use of logic only for foundational purposes in the precise sense of finding a universal grounding for mathematics.[15] Bourbaki is reacting against a foundationalism[16] which sacrifices meaning for verifiability. The coding of mathematics into set theory performs a useful function of providing a basis; unfortunately, the ideas are often lost in the translation. In contrast, Thesis 2 asserts formalization is a means for analysis of ideas in different areas of mathematics.

In his remarks at the Vienna Gödel centenary symposium in 2006, Angus Macintyre wrote,

That the 1931 paper had a broad impact on popular culture is clear. In contrast, the impact on mathematics beyond mathematical logic has been so restricted that it is feasible to survey the areas of mathematics where ideas coming from Gödel have some relevance. [Macintyre 2011]

This sentence unintentionally makes a false identification.[17] Macintyre's paper surveys the areas of mathematics where he sees the ideas coming from 'the 1931 paper' have some relevance. But incompleteness is not the only contribution of Gödel. In Chapter 4.3 and Chapters 6.2 and 6.3 we barely touch the tip of the iceberg of results across mathematics that develop from the Gödel completeness theorem and the uses of formalization described in this book. It is perhaps not surprising that in 1939, Dieudonné [Dieudonné 1939] sees only minimal value in formalization,[18] 'le principal mérite de la méthode formaliste sera d'avoir dissipé les obscurités qui pesaient encore sur la pensée mathématique'; the first true application

---

[Mathias 2012], Mathias discusses of the interactions of Bourbaki (primarily Weil) with Rosser. He particularly emphasizes that Bourbaki employs a (somewhat jumbled) account of the 1920s version of the epsilon symbol which make their set theory especially clumsy. William Howard [Howard 2013] reports that Weil had the manuscript reviewed by Myhill and Tennenbaum (all at the University of Chicago in the early 1950s) and rejected Myhill's negative comments.

[14] Parenthetical words added. Page 223 of [Bourbaki 1950].

[15] I use the term global foundations for this study.

[16] In general, when I speak of 'foundationalism', I refer to the notion of obtaining a basis for certitude in an area (usually all) of mathematics. The discussion of formalization as a tool for organizing mathematics argues that comparing various formalizations of local areas of mathematics allows one to see important patterns across mathematics without attempting any global foundations.

[17] Macintyre has confirmed via email that he intended only to survey the influences of the incompleteness results, not the completeness theorem or work in set theory.

[18] 'The principal merit of the formalist method will be to have dissolved the obscurities which weigh down mathematical thought.' My translation.

of the compactness theorem in mathematics occurs only in Malcev's 1941 paper [Malcev 1971a].[19]

Bourbaki [Bourbaki 1950] hint at an 'architecture' of mathematics by describing three great 'types of structures': algebraic structures, order structures, and topological structures. As we describe below in Chapter 2.4, the methods of stability theory provide a much more detailed and useful taxonomy which provides links between areas that were not addressed by the Bourbaki standpoint.

2.3   A Criterion for Evaluating Properties of Theories

In this section we return to the questions (page 15) Detlefsen raised concerning the significance of categoricity and completeness. His questions rely on an unspecified notion of a *virtuous property* of a theory, 'Or is it rather the case that categoricity is a virtue in some theories but not in others? If so, how do we tell these apart, and how do we justify the claim that categoricity is or would be a virtue just in the former?' Here, we provide our criterion for what makes a property[20] of theories virtuous.

I leave the notion of a property of a theory undefined. But here are a number of examples of 'properties': categorical, complete, decidable, finitely axiomatizable, Π^0_2-axiomatizable, has at least one or only infinite models, interprets arithmetic, or satisfies the amalgamation property. Another family of properties refers to models of the theory: each model admits[21] a linear order of an infinite subset; each model admits a combinatorial geometry; each model admits a tree-decomposition into countable models; each model interprets a classical group; every definable subset of a model is finite or cofinite; every model is linearly ordered and every definable subset is a finite union of intervals. Each property P in the second group satisfies that for a complete theory T, 'each model of T satisfies P' is equivalent to 'some model of T satisfies P'. Further examples of the second sort and the significance of that kind of property for mathematics will be developed

---

[19] Malcev writes, 'The general approach to local theorems does not, of course, give the solutions to any difficult algebraic theorems. In many cases, however, it makes the algebraic proofs redundant.' Malcev goes on to point out that he significantly generalizes one earlier argument and gives a uniform proof for all cardinalities of an earlier result, whose proof by Baer held only for countable groups.

[20] We speak of mathematical properties that connect syntax and semantics, since we think Tarski's truth definition establishes the goals of giving mathematical (set theoretic) definitions of the relations between syntax and semantics. In Chapter 14 we follow Kennedy and Tarski in distinguishing 'mathematical properties' as referring solely to semantical properties.

[21] 'Admits' is jargon; it means the property described holds of or about the model.

below. Key to all of these notions is that they involve syntactic properties of the theory.

I take the word virtue[22] in Question I to mean: *the property of theories has significant mathematical consequences for* any *theory holding the property*. For categoricity to be virtuous it would have to be that some properties of theories are determined simply by them being categorical. One aspect of significance is that theories which have the property will display other notable similarities. Many examples of 'significant consequence' appear in Chapter 2.4. We will note in Chapter 3.1 that the mere fact of categoricity has few consequences for a second order theory. In contrast, as we'll see in Chapter 3.3, the property, categoricity in power, of a first order theory T gives rich structural information about the models of T.

Colin Rittberg asked in an email, 'Given that virtues are excellences, what is the connection between your definition of virtue and the good?' This is an excellent question in the context of *virtue epistemology*.[23] However, such connections are far from the themes of this book. We apply the term 'virtue' only in the narrow sense explained in the next paragraph and the following discussion on 'pragmatic virtue'.

In defining 'virtuous property', I am generalizing from my mathematical experience in an attempt to give a precise criterion describing those properties of formal theories which are useful in research. While I took the word virtue from Detlefsen's question, my answer conforms to mathematical practice in naming concepts. A concept is described by a natural language term with an intent of evoking a notion, and providing a technical meaning in a specific context; the precise natural language meaning is irrelevant. A property could be virtuous because it has useful equivalents. For example, a first order theory T is *model complete* if every submodel N of a model M is an elementary submodel $N \prec M$. This is equivalent to the statement: every formula $\psi(\mathbf{x})$ is proven equivalent in T to a formula $(\exists \mathbf{y})\phi(\mathbf{x}, \mathbf{y})$, where ϕ is quantifier free. The second version is of enormous help in analyzing the definable subsets of a model of T.

---

[22] In making this stipulation I am certainly not arguing that this the only way 'virtue of a property of a formal theory' could be defined. And certainly I am not attempting to define virtue in any wider context. Such notions as depth of theorem or beauty of a proof are notions that could be explored; but they are not properties of theories and so are not considered here.

[23] Virtue epistemology is a school that connects epistemology with moral philosophy and takes the platonic virtues seriously. It features an emphasis on agentive rather than theoretical virtues [Zagzebski 1996, DePaul & Zagzebski 2003]. Rittberg and Tanswell are integrating this viewpoint into the philosophy of mathematics. The discussion of virtue and the good in [Zagzebski 1996] raises for me a different question. 'Must a "mathematical platonist" accept the other platonic ideals such as the good?'

Rittberg had earlier raised a more specific objection, which I find more troubling since it is specifically mathematical, that 'inconsistency' has significant mathematical consequences and thus would be 'virtuous'. He objects that intuitively inconsistency is far from virtuous. Here are several responses to that objection. On the one hand consistency/inconsistency is the entry point. Any property under consideration will specify the theory is consistent; in that sense we could stipulate away the objection. On the other hand, inconsistency is a virtue in the sense described. Many arguments in mathematics use the method of contradiction. Success is reached precisely when one finds an inconsistency. Further, finding a putatively useful property is inconsistent may itself be fruitful. The notion of a Reinhardt cardinal was proved inconsistent in the 1970s; it serves as guidepost in the formulation of large cardinal axioms.

Model completeness is called by Manders ([Manders 1987], 203) an 'accessibility property.' Model completeness reduces the definable relations in a theory to those defined using $\exists_1$ formulas (page 105), which are more comprehensible. Thus our syntactic approach provides criteria to evaluate the accessibility of formalizations of various areas of mathematics. Manders [Manders 1984] argues for the functorality[24] of model completion as an illuminating commonality among real and complex algebraic geometry. We pointed in Chapter 1.2 (page 40) to Manders' insight that the specific vocabulary of the formalization is crucial in establishing the functorality of the transformation that presents the projective plane as a model completion of the affine plane. This suggests that one can find epistemological criteria for evaluating different formalizations of the same topic (i.e. choose the one that enables the functorality). Another example is demanding o-minimality to produce easily surveyable definable sets via cell decomposition (page 159). Such considerations are implicit in Chapter 9 on the axiomatization of geometry.

We have taken virtue for what might more precisely be called 'pragmatic virtue' – the property has important effects. We might consider accessibility as a separate virtue. However, it has significant mathematical consequences so we eschew at the moment such finer distinctions. Many of the properties we discuss lead either to 'structure' or 'non-structure'. One can of

---

[24] In Chapter 4.5, we discuss 'interpretation' as a functor between categories of models of two theories. By the functorality of model completion, Manders compares two model completions. One can first find the algebraic (real) closure field as the model completion of the theory of (ordered) fields. Now the bi-interpretation between the affine (projective) geometry over a field commutes with two model completions: the field to its closure, the affine geometry to the projective.

course refine the notion of non-structure (e.g. Theorem 7.2.4). Many model theorists (not Shelah) might regard structure as 'better than' non-structure. Taking this pragmatic definition is not to deny the possibility of such finer gradations or narrower definitions of virtue but simply to fix on a common denominator – having a significant mathematical effect. To determine the virtue of a property A ask, 'What consequences for the theory T or for models of the theory follow from A holding of T?' We will be particularly interested in information about how the models of the theory are constructed from simpler structures. In this context Shelah's notion of a *dividing line* lays out those properties which are 'most virtuous'.

A dividing line is not just a good property, it is one for which we have some things to say on both sides: the classes having the property and the ones failing it. In our context normally all the classes on one side, the 'high' one, will be provably 'chaotic' by the non-structure side of our theory, and all the classes on the other side, the 'low' one, will have a positive theory. The class of models of true arithmetic is a prototypical example for a class in the 'high' side and the class of algebraically closed field the prototypical non-trivial example in the 'low' side.[25]

Thus a property is a dividing line if both it and its complement are virtuous.

We will argue in general that *categoricity* is not very virtuous (Chapter 3.1). Its importance is as a signal that a theory (or rather an axiomatization) attempting to describe a particular structure has succeeded. But for first order theories, *categoricity in power* is a highly virtuous property of a theory.[26] Further we will argue that a small number of models in a cardinal is a sign that the models of the theory have a strong structure theory.

2.4 Virtuous Properties as an Organizing Principle

There are two rather different approaches to choosing virtuous properties. One is to emphasize that relations definable in a model of the theory are easy to describe and visualize; theories in which each formula is provably equivalent to one with no or at least few alternations of quantifiers are virtuous.[27] However, for applications, as emphasized in the Robinson–

---

[25] See ([Shelah 2009a], 3). Shelah elaborates this theme in Section 2, 'For the logically challenged,' of the same chapter. Note that the virtue of a property in terms of the significance of its consequences is a different issue from which if either of 'high/chaotic' or 'low/structured' is virtuous under some conception of virtue.

[26] It does not however reach the status of dividing line; there are few specific consequences of a theory simply failing to be categorical in any power.

[27] We explore Manders' epistemological discussion of these developments in Chapter 5.2.

Macintyre school, it is crucial that the primitive relations be understandable mathematical notions; see Chapter 4.4. In contrast, Morley (page 112) introduced 'quantifier elimination by fiat'; extend the theory by definitions of the form $\phi(\mathbf{x}) \leftrightarrow R_\phi(\mathbf{x})$. This apparently technical shift lays the ground-work for Shelah's approach to finding a family of virtuous properties, which is our next topic.

The second approach is exemplified by the stability hierarchy: a collection of properties of theories as envisioned in Chapter 2.3 that organize complete first order theories[28] into families of theories with similar mathematically important properties. Bourbaki have some preliminary notions of combining the 'great mother-structures' (group, order, topology). They write,[29] 'the organizing principle will be the concept of a hierarchy of structures, going from the simple to complex, from the general to the particular.'

Kreisel writes,[30]

Though called 'basic', these structures are not at all natural in the sense of 'jumping out at us'; some are new-fangled. As a matter of experience, knowledge about this small batch of objects allows connections between remarkably many areas; in other words, we have a style of *reasoning by analogy* with the property of being precise thrown into the bargain – but without the fiction such transfer principles must take the literary form of metatheorems. Moreover, since a wide range is covered by these rigorously defined structures, we combine *flexibility and rigor* – correcting the wide-spread assumption that flexibility is somehow a prerogative of vagueness.

Despite Kreisel's optimism we view Bourbaki's remarks as vague; there is no real explanation of how group, order, and topology are to interrelate. And given the results discussed in this book, we see Kreisel as overly pessimistic about the prospects of metatheorems. The key to model theoretic organization is identifying a fourth mother-structure, geometry. In the first instance, geometry is taken in a weak sense. In Chapter 5.4 we define the notion of *combinatorial geometry*, which axiomatizes the ability for a structure to have a dimension and explore model theoretic conditions for finding these geometries in models of theories. Kreisel wrote a few years before the proof of Theorem 5.6.4, which, with more precision than this Bourbaki article, we can rephrase as, 'If two geometries interact then there is a classical group enforcing the interaction and it depends on at most three elements.'

---

[28] We could as well say structures since each structure M determines a complete first order theory, Th(M).

[29] ([Bourbaki 1950], 228).

[30] I thank Dana Scott for donating his Kreisel material to me and connecting me with the Archives of the University of Konstanz who graciously permitted the quotations ([Kreisel 1984], 6) from unpublished work of Kreisel.

The precise form of this statement is in fact formulated as a metatheorem.[31] We describe the general model theoretic framework in Chapter 5; in the remainder of this section we sketch some of the impact of the stability hierarchy. Of course, as Kreisel correctly foresaw, the application is not in the form of substituting particular mathematical objects of field X into a template and pulling out an important theorem of field X. Rather the template provides a specific lemma and a method of building analogies that combined with an understanding of the area produces useful results.

Of course, one should realize that there have been sweeping and profound developments of the Bourbaki dream more directly in their school. Zalamea summarizes:[32] 'In Grothendieck's way of doing things, in particular, we can observe, firstly, the introduction of a web of incessant *transfers, transcriptions, translations* of concepts between apparently distant regions of mathematics, and, secondly, an equally incessant search for *invariants, protoconcepts, and proto-objects* behind that web of movements.' Grothendieck aims at the solution of specific problems that had arisen in the first half of the twentieth century. In contrast, Shelah looks for organization in the abstract; applications are fine but not essential. His response on receiving the 2013 AMS Steele prize makes clear his fundamental aims.

I am grateful for this great honour. While it is great to find full understanding of that for which we have considerable knowledge, I have been attracted to trying to find some order in the darkness, more specifically, finding meaningful dividing lines among general families of structures. This means that there are meaningful things to be said on both sides of the divide: characteristically, understanding the tame ones and giving evidence of being complicated for the chaotic ones. It is expected that this will eventually help in understanding even specific classes and even specific structures. Some others see this as the aim of model theory, not so for me. Still I expect and welcome such applications and interactions. It is a happy day for me that this line of thought has received such honourable recognition. Thank you. [Shelah 2013b]

While the exact meaning of Zalamea's 'transfers, transcriptions and translations' is unclear to me, the astonishing fact is that the properties isolated by Shelah without concern for applications show up in many areas of mathematics (Chapters 6 and 13.2). The most important of Shelah's innovations is to consider *classes of theories defined by certain (syntactic)*

---

[31] Any model of a stable theory where such geometries occur and interact nontrivially embeds an algebraically closed field with certain specified group actions. A field is *algebraically closed* (ACF) if every polynomial with coefficients in the field has a solution.

[32] ([Zalamea 2012], 140). This is in one of his thirteen profiles of twentieth century mathematicians, from Atiyah to Zilber.

properties of the theory. Such classes include ω-stable, stable, o-minimal, and 1-based. We first sketch Shelah's development of a collection of dividing lines which solved the test problem of calculating the spectrum function of a first order theory. Then we examine how this classification of theories has been developed and extended by many authors to obtain results across mathematics.

There are two key components to Shelah's solution of the spectrum problem:

(1) Producing a sequence of dividing lines between chaos and structure such that

(2) each model of a theory which satisfies 'the structural side' of *each* dichotomy is determined by a system of invariants.

The fundamental tool of this organization is the study of properties of *definable* relations, that is, relations definable within a structure in a given *formal* language. Depending on the situation, there are several reasons why the subclass of definable sets is adequate to this task. In algebraic geometry (both real and complex) it turns out that mathematicians are, in fact, only studying (some) of the definable sets in the first place. In the other direction, the Wedderburn theory for non-commutative rings is on its face second order because it involves quantification over ideals. But, for stable rings, there are enough first order definable ideals to obtain the classical structure theorems for stable rings [Baldwin & Rose 1977, Baginski 2009].

The general idea of a structure theory as in (2) above is to isolate 'definable' subsets of models of a theory that admit a dimension theory analogous to that in vector spaces, and then to show that all models are controlled by a family (indexed by a tree) of such dimensions. Thus, the notion of 'invariant' is refined to reflect the different interaction of 'parts' of a structure that each have a dimension. Theories that are categorical in power are the simplest case. There is a single dimension and the control is very direct.

Shelah's stability theory [Shelah 1990] provides a method to categorize countable[33] theories into two major classes (the *main gap*): those that admit a structure theory (*classifiable*) and those which are *creative/chaotic*. If a theory admits a structure theory, then all models of any cardinality are controlled by countable submodels via a mechanism which is the same for all such theories. In particular, this implies that the number of models in cardinality[34] $\aleph_\alpha$ is bounded by $\beth_\beta(\alpha)$ (where $\beta < |T|^+$). If T is not

---

[33] Much is known but the problem is not resolved for uncountable vocabularies.

[34] See page 142 for definition of $\beth_\alpha$.

classifiable no such system of assigning invariants exists. The number of models in $\aleph_\alpha$ of a *chaotic* (or more positively, a *creative*) theory is $2^{\aleph_\alpha}$; rather than adding another copy of a countable constituent, an essentially new method of creating models is needed in each cardinality. While counting the number of models in each cardinality is the test question for this program, the greatest benefit of studying this question lies in the development of tools to provide invariants and give detailed descriptions of each model. In the last 25 years, tools in the same spirit of definability (but considering different syntactical properties) allow the investigation of the definable subsets of both classifiable and some creative theories.

The discovery in the 1970s that mathematically important structures were analyzable via the stability hierarchy emphasized the significance of the hierarchy. Any abelian group A is stable (i.e. Th(A) is stable); more strongly, all vector spaces over countable fields are ω-stable. Less obviously, matrix groups over algebraically closed fields are ω-stable.

Shelah's original characterization of stability established that every unstable theory exhibited one of two syntactic properties: the *strict order property* or the *independence property* (Definition 5.3.2). The real field is a prototypic example of the first and the *random graph* of the second.[35] The study of o-minimality, simple theories (Definition 8.5.6), and recent advances in studying NIP theories (page 130) show that 'tame' is a broader category than stable. The history of these concepts over the last 20 years shows the interplay of pure and applied model theory. In 1980, Shelah named a theory with a property from III.7.11 of [Shelah 1978] as a *simple theory* in [Shelah 1980] with prototype the random graph. But the idea languished till the mid 1990s. Then on the one hand Hrushovski proved that the theory of algebraically closed fields with a (generic) automorphism[36] is simple and on the other Kim and Pillay [Kim 1998, Kim & Pillay 1997] proved symmetry of forking in a simple theory and identified the lack of stationary types[37] as the key property holding in stable theories and missing in simple ones. This fueled a ten year study of these theories by a number of authors. Similarly the study of NIP theories was fueled

---

[35] The random graph was (re)discovered by Rado in the mid 1960s. (See Wikipedia for earlier work by Ackermann, Erdős, and Rényi.) It plays an important role in computer science since it is the model of those sentences in a binary vocabulary whose probability in models of size n tends to 1 as n tends to infinity [Fagin 1976, Glebski et al. 1969].

[36] Difference fields (fields with endomorphism) were first studied by Ritt in the 1930s. Hrushovski's proof inaugurated deep connections among stability theory, difference fields, number theory, and, as in Chatzidakis' 2014 address to the International Congress of Mathematicians [Chatzidakis 2000a], algebraic dynamics.

[37] A type is stationary if it has a unique 'non-forking' (generic) extension.

by new pure results by Shelah and the discovery by various authors that examples of NIP theories arose in many areas of mathematics: any theory with finite VC-dimension[38] database theory, algebraically closed valued fields. This spurred renewed work in pure model theory establishing the properties of NTP_2-theories as a common generalization of NIP and simple.[39] Among the canonical structures, the complex field is a prototype for good behavior and the real field (and even the real exponential field) are o-minimal and so admit a sophisticated analysis of the definable sets. Of the canonical structures only arithmetic has so far resisted these methods of understanding. This is witnessed by its both having the independence property and being linearly ordered. Of course set theory is equally unruly; a definable pairing function[40] is incompatible with a global dimension theory.

A common reading of Shelah is that the more unstable a theory is the more it lies on the side of chaos: 'chaos' means 'many models'. But Shelah has pointed out[41] that this reading puts the cart before the horse: 'The aim is classification, finding dividing lines and their consequences. This should come with test problems. The number of models is an excellent test problem and few models is the strongest form of non-chaos.' But Shelah has suggested, beginning in the late 1970s, test problems for the study of theories with many non-isomorphic models, in particular, of unstable theories without the strict order property (e.g., simple theories): existence of saturated extensions [Shelah 1980], the Keisler order ([Shelah 1978], Chapter 6), and the existence of universal models [Kojman & Shelah 1992]. [Shelah 1996] extends this investigation by introducing the SOP_n and establishing SOP_4 as a dividing line for some purposes.

In the last few paragraphs we have glimpsed the ways in which complete *formal* theories provide a framework for analysis. In Chapter 6.3, we discuss some of the profound implications of the stability hierarchy of complete theories for work in conventional mathematics. Thus I argue that the study of complete theories (a) focuses attention on the fundamental concepts of specific mathematical disciplines and (b) even provides techniques for solving problems in these disciplines. Neither (a) nor (b) is the

---

[38] The fundamental idea behind the independence property arose independently in work of Vapnik, Chervonenkis, Sauer, and Shelah in the early 1970s. In addition to forming a dividing line in model theory, the notion plays a role in learning theory, computational geometry, and probability theory.

[39] See Chapter 6.4.

[40] A *definable pairing function* is a definable one-to-one function from $X \times X$ onto X. Such a function destroys a notion of dimension since each n-tuple is coded by a singleton.

[41] See page 5 of [Shelah 2009a].

driving motivation for Shelah; his aim is to 'understand the white space on the map'.[42] But they are the motivation for much work in model theory. And in particular such applications have led to developing finer gradations in the stability hierarchy. Deep work in simple theories in the 1990s was stimulated by specific algebraic questions [Chatzidakis 2000b]. Most strikingly, model theoretic results of Chatzidakis, Hrushovski, and Peterzil were used by Hrushovski ([Hrushovski 1996]) to find explicit bounds in the Manin–Mumford conjecture (in number theory). Similarly, the great interest in dependent (i.e. NIP – see page 130) theories in the new millennium is driven both by classification type questions [Shelah 2015] and more applied problems as in [Hrushovski & Pillay 2011] and [Haskell et al. 2007].

[42] Page 33 of [Shelah 2009a]: 'This work certainly reflects the author's preference to find something in the *white part of our map*, the "terra incognita" rather than understand perfectly what we have reasonably understood to begin with (which is exemplified by looking at abstract elementary classes on which our maps reflect our having little to say on them, rather FMR [finite Morley rank, Chapter 6.1] theories or o-minimal theories, cases where we had considerable knowledge and would like to complete it).'

3 | Categoricity

Recall (page 15) Detlefsen's first question:

(IA) Which view is the more plausible – that theories are the better the more nearly they are categorical, or that theories are the better the more they give rise to significant non-isomorphic interpretations?

Button and Walsh distinguish two issues for philosophers investigating categoricity ([Button & Walsh 2016], 283): 'determinacy of reference of mathematical language and the determinacy of truth values of mathematical statements.' As in their paper we focus on the question of determining reference for theories that are intended to describe a single isomorphism type of a structure. With the normal background of ZFC, one can define the vocabulary τ, the domain A, and the interpretation of the τ-relations on A. This description unambiguously refers to a particular structure. The isomorphism type of A is the class of all τ-structures isomorphic to A. The number theorist Barry Mazur wrote,

the objects that we truly want enter the scene only defined as equivalence classes of explicitly presented objects. That is, as specifically presented objects with the specific presentation ignored, in the spirit of 'ham and eggs, but hold the ham.' [Mazur 2008]

Importantly, one can also 'take the ham'. For example, use linear algebra to study a matrix group and apply the result to any isomorphic group (Chapter 4.7).

There are several ways that a serious issue arises. Attempting to reify the notion of isomorphism type strikes the obstacle that the collection of structures isomorphic to A is not a set. This can be resolved by bounding the set theoretic rank of the structures or considering the isomorphism class as definable in ZFC; either of these is adequate from a model theoretic perspective. Secondly, one can demand that the description be formulated in the *original vocabulary* and then the problem becomes categoricity. In this context, the study of categoricity begins with the choice of logic. We argue, working from our pragmatic notion of virtue, that categoricity is interesting for a few second order sentences describing particular structures.

But this interest arises from the importance of *those* axiomatizations of *those* structures and not from any intrinsic consequence of second order categoricity for arbitrary theories. From this standpoint, categoricity of a *particular* axiom set is crucial. We study second order categoricity from the standpoint of the full semantics in Chapter 3.1 and *internal categoricity* or provable categoricity both later in Chapter 3.1 and on page 257.

Categoricity of a theory in $L_{\omega_1,\omega}$ deserves more attention than it has received. Unlike second order logic there are clear *mathematical* consequences from the categoricity of an $L_{\omega_1,\omega}$-sentence. An obvious one is that the model is countable; slightly more technically, the model must be minimal. Thus $L_{\omega_1,\omega}$-categoricity has some virtue but in both $L_{\omega,\omega}$ and $L_{\omega_1,\omega}$ categoricity in power is a virtuous property in our sense (Chapter 2.3).

3.1 Categoricity of Second Order Theories

In this section we argue against categoricity per se as a significant property of second order theories, while acknowledging the importance of noticing certain *axiomatizations* are categorical. Thus, we must distinguish between the categoricity of an axiomatization[1] and the categoricity of a theory. One aim of axiomatization is to describe a particular, *fundamental* or *canonical* structure. There are really very few such structures. In addition to the reals $\mathfrak{R}$ and natural numbers $\mathbb{N}$ one could add $\mathbb{Z}, \mathbb{C}, \mathbb{Q}$ and of course Euclidean geometry. In the twentieth century such structures as the p-adic numbers or the complex field with exponentiation enter the canon. Categoricity is a necessary condition for calling a second order axiomatization of a theory successful. In fact, Huntington's initial name for categoricity, sufficiency,[2] suggests he thought it sufficient. But, given the ease described below of obtaining complete second order theories, it is not sufficient. One must examine the axiomatization in detail. The goal of an axiomatization is to illuminate the central intuitions about the structure. The real linear order could be given a categorical axiomatization by describing the construction of the rationals from the natural numbers and then the reals as Cauchy sequences of rationals. As pointed out in [Väänänen 2012], this construction takes place in $V_{\omega+7}$. But, it is Dedekind's categorical axiomatization of the real order as a separable (Dedekind) complete linear order that is mathematically useful. This axiomatization highlights the properties needed for the foundations of calculus [Spivak 1980] as opposed to a tedious

---

[1] Following modern model theoretic practice, I say a class is $\mathcal{L}$-axiomatizable if it is the class of models of a set of $\mathcal{L}$-sentences. If I mean *recursively* axiomatized, I add this adjective.

[2] See page 16 of [Awodey & Reck 2002a]. Veblen [Veblen 1904] introduces the term categoricity.

construction of the reals from the natural numbers. From the perspective of providing a *unique* description of our intuitions, even a categorical second order axiomatization (of say the reals) is subject to attack from radically different perspectives (e.g. constructive mathematics or Ehrlich's absolute continuum [Ehrlich 2012]). Thus, *the interest in categoricity is not really that the theory[3] is categorical but in the categoricity of the particular axiomatization that expresses the intuitions about the target structure[4]* and allows the *development of the mathematics* of structure. Indeed, informal second order reasoning is a normal way to do analysis.[5]

This focus on categoricity of canonical structures begins with Peano and Dedekind. Blanchette provides an apt moniker for this view: model-centric [Blanchette 2014]. She contrasts it with the 'deductivist approach' by Hilbert and Peano. The early axiomatizations of geometry exhibit both aspects (Chapters 9–11). A definite model of geometry is intended, but Hilbert in particular is modeling his exposition on Euclid.

For any logic $\mathcal{L}$ it is evident that *categoricity* of $\mathrm{Th}_{\mathcal{L}}(M)$ implies $\mathrm{Th}_{\mathcal{L}}(M)$ is semantically complete (Definition 1.4.4); the converse fails if the logic has only a set of theories as there is a proper class of structures. Many *second order* theories are categorical. Consider the following little-known results.[6]

Theorem 3.1.1

*(1) **Marek–Magidor/Ajtai** (V=L) The second order theory of a countable structure is categorical.*

*(2) **H. Friedman** (V=L) The second order theory of a Borel structure is categorical.*

*(3) **Solovay** (V=L) A recursively axiomatizable complete second order theory is categorical.*

*(4) **Solovay/Ajtai** It is consistent with ZFC that there is a complete finitely axiomatizable second order theory with a finite vocabulary[7] that is not categorical.*

---

[3] That is, the collection of all true sentences about the structure.

[4] In Chapter 7.3, we exhibit a family of axiomatizations in infinitary second order logic, that arise systematically from knowing the structure of models in certain first order theories.

[5] See 5.2.2 of [Shapiro 1991].

[6] These results appeared in a paper by Marek [Marek 1973], its review by Magidor, and in a thread on FOM (http://cs.nyu.edu/pipermail/fom/2006-May/010544.html). They were summarized by Ali Enayat at http://mathoverflow.net/questions/72635/categoricity-in-second-order-logic/72659#72659. Solovay's forcing argument for independence is announced at http://cs.nyu.edu/pipermail/fom/2006-May/010561.html. Corcoran reports in his review of [Fraïssé 1985], where Fraïssé gives another proof of (1), that Fraïssé had conjectured (1) in 1950. Jouko Väänänen has relayed some further history. Ajtai in [Ajtai 1979] both proves (1) from $V = L$ and shows it is consistently false. The equation 'V=L' is shorthand in any set theory text for the Axiom of Constructibility.

[7] There are trivial examples if infinitely many constants are allowed.

Summing up, if a second order theory is complete and easily described (i.e. recursively axiomatized) or has an intended model which is 'small' (countable or Borel) then (at least[8] when $V = L$) it is categorical. More strongly, any countable structure that has an 'arithmetic' presentation (each primitive relation is definable by a formula of arithmetic) is second order categorical, provably in ZFC. Awodey and Reck [Awodey & Reck 2002a] point out that Carnap provided (as he realized) a false proof that every finitely axiomatized complete second order theory is categorical. The Solovay/Ajtai result above shows this question cannot have a positive answer in ZFC. The fact that the most fundamental structures were categorical[9] may partially explain why it took 25 years to clarify the distinction between syntactically complete and categorical. As reported in [Awodey & Reck 2002a], Fraenkel in the mid 1920s [Fraenkel 1928] distinguished these notions in the context of higher order logic without establishing that they are really distinct.[10]

One might argue that it is hard to actually prove category in ZFC. But for this argument to have much weight, one would have to get around two facts. (1) Consistently,[11] categoricity is easy to achieve. Even in ZFC, there are many examples of categorical structures: various ordinals, the least inaccessible cardinal,[12] the Hanf number of second order logic, etc. (2) We have just seen that second order categoricity tells us nothing about the internal 'algebraic' properties of the structure. So the fact that a second order theory is categorical provides little information.

Bourbaki wrote,[13]

Many of the latter [mathematicians] have been unwilling for a long time to see in axiomatics anything other else than a futile logical hairsplitting not capable of

---

[8] In fact (per Väänänen), for countable models $V = L$ could be replaced by the existence of a second order definable well-ordering of the reals; in ongoing work Väänänen and Shelah are weakening this requirement.

[9] We follow current model theoretic practice and label a structure with any property that is satisfied by its complete first order theory. We extend this practice by saying e.g. M is second order categorical when the second order theory of M is categorical. Indeed the recent results cited above show that under $V = L$ each of the fundamental structures had to be categorical.

[10] More precisely Awodey and Reck point out that in the 2nd edition of Fraenkel's book (1923) he had distinguished between categoricity and deductive completeness. In the 3rd (1928) edition he also clarifies the distinction between syntactic and semantic completeness. Corcoran [Corcoran 1980] points out that Veblen in 1904 distinguishes deductive and semantic completeness but confuses semantic completeness and categoricity. Corcoran relates the generally unsettled nature of these notions in the first decades of the twentieth century and raises a number of precise historical questions.

[11] I.e., in some extension of ZFC.

[12] See Proposition 1, ([Väänänen 2012], 106).

[13] Throughout the book, square brackets in quotations indicate my interpolation; round brackets are from the original.

fructifying any theory whatever. This critical attitude can probably be accounted for by a purely historical accident. The first axiomatic treatments and those which caused the greatest stir (those of arithmetic by Dedekind and Peano, those of Euclidean geometry by Hilbert) dealt with univalent theories, *i.e.* theories which are entirely determined by their complete systems of axioms; for this reason they could not be applied to any theory except the one from which they had been abstracted (quite contrary to what we have seen, for instance, for the theory of groups). If the same had been true of all other structures, the reproach of sterility brought against the axiomatic method, would have been fully justified. ([Bourbaki 1950], 230)

Bourbaki have missed (ignored?) the fact that there are two different motivations for axiomatizing an area of mathematics. Bourbaki focused on describing the properties of *classes of structures* that appear often in mathematics, specifically groups. Dedekind, Peano, Hilbert, and the American Postulate Theorists provided second order axiomatizations that were explicitly intended to describe certain *canonical structures*. From the Bourbaki standpoint of investigating the impact of certain unifying concepts on mathematics as a whole, one can reasonably ascribe sterility to such specific axiomatizations. However, the goal of these axiomatizers was to understand the given structure, not generalization. An insightful categorical axiomatization of a particular structure is expected only to explain the given structure, not to organize arguments about other structures. The other arguments mentioned above that provide many other categorical second order structures don't even have this benefit. Their real significance is to the understanding of second order logic, identifying some kinds of structures it can describe precisely (at least under $V = L$).[14] Even though (at least in L) there are many order categorical structures, this fact tells us little about such structures. They have fairly simple descriptions, but not in a way that the reflects the properties of the structure. There is no known mathematical consequence of the statement that the second order theory T is categorical beyond 'it has only one model'. We will see the situation is far different for first order logic and categoricity in power.

Internal Categoricity: The philosophical discussion of *internal categoricity* seems to begin with Parsons' observation [Parsons 1990b, Parsons 1990a] that an appropriate formalization of Dedekind's characterization of the

---

[14] The 'idea' of the arguments presented on FOM (see footnote 6) is that for well-ordered structures one can express in second order logic the assertion that a model is minimal (no initial segment is a model) provided that the axiomatization can be properly coded. The coding can be done in L. A similar approach (by Scott) proving that semantic completeness does imply categoricity for pure second order logic is Proposition 3 of [Awodey & Reck 2002b].

natural numbers is provable in second order logic. We explicate this notion from a model theoretic standpoint as in [Väänänen 2012, Väänänen & Wong 2015] (page 257).

As in Chapter 1.2, we work with a fixed vocabulary. Button and Walsh work with a notion of 'informal isomorphism' by, loosely speaking, closing under definitional equivalence. We discuss this further in Chapter 4.6, but it doesn't seem a central issue here: if A and B are isomorphic τ-structures and we can add a function F by explicit (second order) definition to one, the same definition expands the other and the expansions are isomorphic in the expanded language. However, like them and mathematicians in general, we now speak informally of *isomorphism types*.

Suppose the second order categoricity of a τ-structure A is formalized in an *applied logic* by a τ-sentence Ψ_A such that $A \models \Psi_A$ and such that any two models of Ψ_A are isomorphic. This provides, *up to isomorphism*, a specific isomorphism type (as in [Button & Walsh 2016, Parsons 1990a]) that Ψ_A refers to. To express the second clause formally, one works in *pure second order logic*. For a second order sentence $\Phi(\overline{B})$ with predicate symbols $\overline{B}$ write $\Phi(V, \overline{X})$ for the formula[15] asserting the $\overline{X}$ are relations on V satisfying Φ. For example in arithmetic, $\Phi(S, 0) := \forall X \forall x(X(0) \wedge (X(x) \to X(Sx))) \to \forall x X(x))$ becomes $\Phi(V, S, 0) : (\forall X \subset V)(\forall x \in V)[(X(0) \wedge (X(x) \to X(Sx))) \to \forall x X(x)]$.

Now the second order sentence Φ is *internally categorical* if $\vdash \Psi$ where Ψ is:[16]

$$((\exists V, \overline{X})\Phi(V, \overline{X}) \wedge$$
$$(\forall V, \overline{X}, V', \overline{X}')(\Phi(V, \overline{X}) \wedge \Phi(V', \overline{X}'))$$
$$\to (\exists F)\ ISO(F, (V, \overline{X}), (V', \overline{X}'))))$$

We have included the hypothesis $(\exists V, \overline{X})\Phi(V, \overline{X})$ because, as Parsons[17] points out, otherwise the implication could be vacuously true. Thus, in the formulation here, a full proof of internal categoricity requires a separate proof of consistency as given in [Väänänen & Wong 2015].

Does internal categoricity provide a specific reference for Φ as Φ_A did? Not really, in the first case we began with a specific structure A (specified in set theory) and proved every model of Φ_A is isomorphic to A. Here we have the absolute statement that any two models of Φ are provably isomorphic in L^{II}. The model theoretic view clarifies this distinction. Internal categoricity requires that for any set M and collection of subsets[18] $\mathcal{G}$, $(M, \mathcal{G}) \models \Psi$.

---

[15] Relativize the formula to V in the sense of page 275.

[16] $ISO(F, (V, \overline{X}), \Phi(V', \overline{X}'))$ means F is a τ-isomorphism between $\Phi(V, \overline{X})$ and $\Phi(V', \overline{X}')$.

[17] He offers several alternative justifications in Section 2 of [Parsons 1990a].

[18] That is, $\mathcal{G}$ is elements of the 'subset sort' in the Henkin semantics on page 114.

That is, any two models of Φ internal to the same $(M, \mathcal{G})$ are isomorphic. But different $(M, \mathcal{G})$ may have models that are not 'externally' isomorphic and even satisfy different sentences.

To clarify the point, consider determinacy of truth value as opposed to determinacy of reference (pages 68 and 257). As noted after Definition 1.4.4, for $L_{\omega,\omega}$ and $\text{Ł}_{\omega_1,\omega}$ *categoricity implies semantic and thus Post completeness*. But second order and internal categoricity do not imply Post completeness. To see this, just note that second order Peano arithmetic cannot be a complete theory with respect to the Henkin semantics, since it is recursively axiomatizable but not decidable. Thus internal categoricity does not determine truth values; it shows the sentence refers to a unique isomorphism type but does not identify that type or its theory. However, as in ([Button & Walsh 2016], 25) and [McGee 1997], one can argue that it makes truth values determinate without determining what they are.

3.2 $L_{\omega_1,\omega}$-categoricity

We argue that $L_{\omega_1,\omega}$-categoricity is more virtuous (Chapter 2.3) than second order categoricity and that it better meets the three goals for categoricity arguments proposed by Toby Meadows in 'What can a categoricity theorem tell us?'

(1) to demonstrate that there is a unique structure which corresponds to some mathematical intuition or practice;
(2) to demonstrate that a theory picks out a unique structure;
(3) to classify different types of theory. [Meadows 2013]

As with Detlefsen's questions, Meadows seeks to clarify the advantages of categoricity. Noting the triviality of first order categoricity, Meadows suggests that two paths to pursue are infinitary and higher order logic; he chooses higher, specifically second order. Here, we first review our discussion of second order logic from Chapter 3.1 in terms of Meadows' questions and then look at the infinitary option.

The distinction we made above between theory and axiom set is crucial. Clearly, for pure second order logic, there are only countably many finitely axiomatized theories (we'll say sentences) but continuum many theories. As we noted in Theorem 3.1.1, consistently all countable or Borel structures (e.g. arithmetic, the reals) are categorical. So the mere fact that a theory is second order categorical tells us little; exhibiting a specific axiom set that is categorical may demonstrate that those axioms catch the intuition – thus satisfying both goals (1) and (2). However, consider the discussion of insightful axiomatizations ending Chapter 3.1.

The third goal, to classify types of theory, is a priori similar to our project. In reply to Detlefsen's question IB (page 15), in Chapter 2.3 we decreed a type (property) of a theory to be *virtuous* if it has significant mathematical consequences for every theory having that property. Meadows offers two types of theory on page 527 of [Meadows 2013]: Shapiro's [Shapiro 1997] distinction between algebraic (expected non-categorical) and non-algebraic (expected categorical) theories.

There are however at least two reasons we might expect a theory to be non-algebraic. One is that it describes a canonical object like arithmetic. But many other examples on page 71 were simply the 'least ordinal with property X'. The categoricity is derived from well-ordering. Second order categoricity cannot make this distinction. But since well-ordering is not definable in $L_{\omega_1,\omega}$, the second type of example is not $L_{\omega_1,\omega}$-categorical.

Using $L_{\omega_1,\omega}$ provides some different opportunities for giving categorial axiomatizations of intuitions about fundamental structures. The arguments below are routine but we sketch them to highlight the differences among several cases.

By the theory of Euclidean geometry we mean the geometrical results in the *Elements*. As shown in Chapter 9, specifically Theorem 9.3.4, Hilbert gave a first order axiomatization (EG) of Euclidean geometry.[19]

Theorem 3.2.1 *Each of the following is categorical in $L_{\omega_1,\omega}$. The first three are each axiomatized by a single sentence; the other two by a family of continuum many sentences.*

(1) *Arithmetic $(\mathbb{N}, +, \times, 0, 1)$;*
(2) *Euclidean geometry;*
(3) *geometry over the real algebraic numbers*[20]*; $(\mathfrak{R}^{alg}, +, \times, 0, 1)$;*
(4) *the complete ordered field $(\mathfrak{R}, +, \times, 0, 1)$;*
(5) *geometry over the complete ordered field $(\mathfrak{R}, +, \times, 0, 1)$.*

Proof. Arithmetic is obviously axiomatizable by a single sentence.[21] The sentence ϕ_{EG} just says that the geometry is over the minimal Euclidean field

---

[19] We make these terms more precise in Chapter 9.
[20] $\mathfrak{R}^{alg}$ denotes the real algebraic numbers; this is the maximal field without imaginaries or transcendentals. See page 248 for some alternative possibilities for choosing an $L_{\omega_1,\omega}$-categorical countable model for geometry.
[21] E.g. specify the structure is a discretely order semi(no additive inverse)-ring such that every element a satisfies is an nth successor of the additive identity for some natural number n.

(Definition 9.4.3); each element of that field is first order definable over the empty set. Similarly, categoricity of (3), the prime model of Tarski's geometry, is straightforward by adding the Archimedean axiom to those for the coordinatizing real closed field.

The reals are slightly more complicated. In addition to the first order axioms of real closed fields, one must say that the set of rationals is dense in the model. We can define the set of rationals in $L_{\omega_1,\omega}$ since each rational number is first order definable and there are only countably many of them. Then, just say there is a rational number between any two members of the field. Finally one must say each cut in the rationals is realized. This requires $2^{\aleph_0}$-sentences, so it is only a *theory* not a single sentence that is categorical. Further, that theory has *no* countable models. The categoricity of real geometry follows from the bi-interpretability of (appropriate) fields and their geometries.[22]

Here are a number of simple observations distinguishing second order and $L_{\omega_1,\omega}$-categoricity that shed light on Meadows' questions.

(1) Theorem 3.2.1 provides two theories which are plausible formalization of our intuitions about geometry. That each is categorical makes it even more urgent to see how the axiomatizations differently explain the intuition. See Chapter 11.2, 'Against the Dedekind Postulate for Geometry'.
(2) Under Meadows, item (3) the logic $L_{\omega_1,\omega}$ gives us a more restricted family of 'non-algebraic' theories than second order; the examples derived from well-order are excluded.
(3) The $L_{\omega_1,\omega}$ axiomatization of Dedekind's real numbers alters our commitments in the underlying logic. The distinction between the Henkin and full semantics for second order logic in axiomatizing the reals is replaced by doubts over whether our model of set theory really 'knows' every cut in the rationals (each cut it does know gives an element of the field).

The contrast in Theorem 3.2.1 between Euclidean and real geometry emphasizes the importance of distinguishing axiom and theory in formulating Meadows' second question. This distinction is further emphasized by considering various $L_{\omega_1,\omega}$-axiomatizations of arithmetic.

As in our critique of axiomatizing the reals by laying out their construction from the natural numbers (Chapter 3.1), the axiomatizations in Theorem 3.2.1 are not very useful mathematically; they provide no tools for proving results about the structure. Button and Walsh

---

[22] See Chapter 4.5 for details on interpretations and Chapter 9.4 for the required bi-interpretation.

[Button & Walsh 2016] note a similar expansion of first order Peano (by saying 'everything is either zero, or the successor of zero, or the successor of that, etc.') gives an $L_{\omega_1,\omega}$-categorical axiomatization of arithmetic. Their axiomatization has the advantage over that just given of adding the axiom scheme of induction (although it is superfluous for establishing categoricity) with respect to first order formulas; it reflects an important proof technique.

Button and Walsh dismiss such an axiomatization by saying, 'But to grasp this proposal, we need to grasp that "etc."; and that looks exactly like the original[23] challenge of grasping the (or a) natural number sequence.' Somewhat more precisely, the issue is, 'how can we grasp the notion of a single infinite recursive disjunction in the metatheory?' While this step[24] may trouble some, surely this notion is easier to grasp than 'the collection of all subsets of the natural numbers'. We have shown that categoricity of $L_{\omega_1,\omega}$-sentences meets Meadows' goals better in two ways. Since it is weaker, it forces the axiomatizer to identify the 'real' cause of categoricity. It also captures a finer sense of 'non-algebraic theory': the categorical structure must be countable and minimal.

Definition 3.2.2 *A model M is $L_{\omega_1,\omega}$-minimal if and only if M has no proper $L_{\omega_1,\omega}$-elementary submodel.*

The model (ω, S) is a good example of an $L_{\omega_1,\omega}$-minimal structure. The next theorem shows the mathematical importance of the *push-through construction* (name from [Button & Walsh 2016]):

Fact 3.2.3 (Push-through Construction) *Let M be a structure and π a bijection between M and a set A. Then there is a structure M_1 isomorphic to M with domain A. Just transfer each relation on M to its image under π.*

Theorem 3.2.4 *An $L_{\omega_1,\omega}$ sentence ϕ is categorical if and only if its unique model is $L_{\omega_1,\omega}$-minimal.*

Proof. Suppose ϕ is categorical. Since $L_{\omega_1,\omega}$ satisfies the downward Löwenheim-Skolem theorem there is a countable model M of ϕ and we may as well assume that ϕ is the Scott sentence (footnote 10, page 173) of M (since the Scott sentence is equivalent to ϕ). Now suppose for contradiction

---

[23] This seems a good challenge (there is surely some circularity) to (1) of Theorem 3.2.1, but not to (2) through (5).

[24] Moore [Moore 1997] reports Henkin replied to objections to infinite Boolean operations by saying that in sufficiently rich set theory, 'we can develop a theory of symbols in a precise way so as to admit infinitely long formulas.' Henkin [Henkin 1955] suggests several possibilities for this development.

there is a proper $L_{\omega_1,\omega}$-elementary submodel M_0 of M. Now $M_0 \approx M$ so, by Fact 3.2.3, M_0 has a proper elementary extension M_1. (Let π_0 be the isomorphism between M_0 and M and extend it a bijection π with domain M and some range A. Then endow A with a structure M_1 isomorphic to M by push-through.) Construct a sequence of proper elementary extensions with M_{i+1} produced from M_i and M_{i-1} as we constructed M_1 from M and M_0. Let $M_\delta = \bigcup_{i<\delta} M_i$, when δ is a limit ordinal. As long as δ is countable, $M_\delta \approx M_0$ and so we can continue the construction. Letting $M^* = \bigcup_{i<\omega_1} M_i$, M^* is an uncountable $L_{\omega_1,\omega}$-elementary extension of M. So it satisfies ϕ which thus is not categorical since it has models in two cardinals.

We have shown that if the countable model M of ϕ is not minimal then it is not categorical. But if M is minimal then the Scott sentence of the countable model has no uncountable model N. Otherwise, applying the Löwenheim–Skolem theorem there would be a countable submodel of N isomorphic to and properly extending M, thus contradicting the minimality.

Thus, using our notion of virtue, axiomatizing by a sentence in $L_{\omega_1,\omega}$ has a slight advantage over second order; M is categorical for such a sentence if and only M is countable and is not isomorphic to a proper[25] $L_{\omega_1,\omega}$-elementary extension[26] of itself.[27] But, in contrast to first order categoricity in power, we see no further informative common trait of $L_{\omega_1,\omega}$-categorical structures.

By weakening the logic to $L_{\omega_1,\omega}$, we have obtained a bit more virtue and a slightly stronger answer to Meadows' questions. We will see in Chapter 7.3 that studying infinitary second order logic with the insights of first order classification theory we can find virtue for categoricity in $L^2_{\kappa^+,\omega}$. But first let us examine why categoricity in power is such a useful concept.

3.3 $L_{\omega,\omega}$: Categoricity in Power

Most people have an intuition for only a few infinite structures: the natural numbers, the rationals, and perhaps the linear order of the reals. As we

---

[25] We can recognize 'proper' only because we have separate grasps of the domain of the model and its isomorphism type. See Chapter 4.7.

[26] M is an elementary submodel of N if every formula with parameters from M is true in M if and only if it true in N.

[27] The examples might lead one to conjecture that every element is first order definable over the empty set; but the situation is vastly more complicated. See [Deissler 1977] and [Baldwin 2012].

noted on page 10 the strength of such an intuition depends heavily on the exact vocabulary intended: $(N, <)$ is clear; $(N, <, +, \cdot)$ is not.[28] Most mathematicians extend this to the complex numbers and then to a deeper understanding of various structures depending on their own specialization $(SL_2(\mathfrak{R})$, the complex projective line, certain small ordinals). Gödel analyzed the ability to visualize ordinals.[29] Giaquinto [Giaquinto 2008] explores this 'cognition of structures' in more detail with a stricter view of what people can grasp. He carefully examines, beginning with finite structures, the process of abstracting from a single diagram to equivalence classes called 'visual images' and then at the next level 'visual categorical specification' (where for example our normal picture of a binary tree of height 2 is equivalent to dividing a line segment in half and repeating on each subsegment). Then he argues that an infinite structure such as $(\omega, <)$ can be visualized by inductively repeating the picture. And he analyzes the visualization of ordinals. In contrast to finite and 'short' infinite well-orders, in his view a dense linear order does not admit such a 'visual grasp'. Our notion is more general; we assert that understanding how to repeat a construction can lead to a 'grasp' of an infinite object and in this sense we have similar grasps of $(\omega, <)$ and $(Q, <)$. That is, there is no distinction between a step adding points at the end to form $(\omega, <)$ and a step adding a point between each pair of consecutive elements to form $(Q, <)$. Our grasp of the reals is somewhat more tenuous since it assumes the filling of continuum many cuts, but most mathematicians don't agonize over their grasp of the linear ordering of the reals.[30]

However, all these structures have cardinality at most the continuum. There are few strong intuitions of structures with cardinality greater than the continuum. However, there is a crucial exception to this remark. It is rather easy to visualize a model that consists of copies of a single countable or finite object. Consider a vocabulary with a unary function f. Assert that $f(x)$ never equals x but $f^2(x) = x$. Then any model is a collection of 2-cycles. On the one hand we have the notion that there are models of arbitrarily large cardinality but we have no really different image distinguishing among the models of different large cardinalities. This situation generalizes

---

[28] All subsets of N^n definable in the first are simply constructed from intervals; there is no algorithm to describe those in the second.

[29] William Howard told me that Paul Cohen turned this into an intelligence test: for what ordinals can you visualize the termination of descending chains witnessing well-foundedness of the ordinal? Cohen claimed ϵ_0.

[30] As with Kossak's remarks about arithmetic, the picture of the solutions of a polynomial in six variables is less clear so replacing linear ordering by real field would be suspect.

when the number of disjoint copies of the same structure is replaced by the dimension of a vector space or field. Thus we might consider the class of structures A_κ, a direct sum of κ copies of Z_2. The isomorphism type of the model depends solely on the number κ of copies (and not at all on the internal structure of the cardinal κ). That is, the class is κ-categorical.

Categoricity is not a necessary condition for such a clear visualization: consider an equivalence relation with two infinite classes and fix a totally categorical theory and make each class a model of the given theory; the model is determined by two cardinals – the cardinality (or more precisely the dimension) of each class. Nor is it sufficient; our grasp of the complex field is certainly far less immediate. Nevertheless, we view this visualization of models that are categorical in power as one of a significant illuminating feature of such theories, although there are no explicit 'significant mathematical consequences of the visualization'.

Categoricity is trivial for first order logic;[31] all and only finite structures are categorical in $L_{\omega,\omega}$. The interesting notion is 'categoricity in power'. As we now explain, the cases of $\aleph_0$-categoricity or categoricity in some (equivalently all) uncountable cardinalities are increasingly virtuous properties. The upward and downward Löwenheim–Skolem theorems show that for first order theories, categoricity in power of an axiomatization Φ *implies* the theory generated by Φ is complete. Ryll-Nardzewski [Ryll-Nardzewski 1959] characterized first order theories that are $\aleph_0$-categorical. Unlike the second order case, the Ryll-Nardzewski theorem contains a lot of information. The technical terms in the following theorem can be found in Chapter 5.1 or in introductory texts in model theory (e.g. [Marker 2002]).

Theorem 3.3.1 (Ryll-Nardzewski[32]) *The following are equivalent.*

(1) *T is $\aleph_0$-categorical.*
(2) *T has only finitely many n-types for each finite n.*
(3) *T has only finitely many inequivalent n-ary formulas for each n.*
(4) *T has a countable model that is both prime and saturated.*

Wildly different kinds of theories are $\aleph_0$-categorical. The theory of an (infinite-dimensional) vector space over a finite field differs enormously from the theory of an atomless Boolean algebra, the random graph, or a dense linear order. But in the 1950s there was no systematic way to make

---

[31] Here by first order, we mean $L_{\omega,\omega}$. We discussed $L_{\omega_1,\omega}$-categoricity in Chapter 3.2.
    Categoricity in power is also a virtuous property for $L_{\omega_1,\omega}$ and we consider it in Chapter 7.1.
[32] Engeler and Svenonius each obtained the result independently at about the same time.

this difference precise. One distinction stands out: only the vector space is categorical in an uncountable power. Łos [Łos 1954] conjectured that a countable theory categorical in one uncountable cardinality would be categorical in all uncountable powers; this conjecture drove much later research in model theory.

In general, by a *structure theory* for T we mean the isolation of certain basic well-understood families of structures (e.g. geometries) usually identified by a single cardinal invariant, and systematically decomposing each model of T into these basic structures, which are called *strongly minimal*.

Definition 3.3.2 *Let T be a first order theory. An infinite definable subset X of a model is T is strongly minimal if every definable subset $\phi(x, \boldsymbol{a})$ of X is finite or cofinite (uniformly in $\boldsymbol{a}$). A theory T is strongly minimal if the set defined by $x = x$ is strongly minimal in T.*

A strongly minimal set can be viewed as a combinatorial geometry (Chapter 5.4) with $a \in \mathrm{cl}(B)$ (algebraic over X) if $\phi(a, \mathbf{b})$ for some formula with only finitely many solutions. The ability to capture such a basic mathematical structure by the syntactic notions in Definition 3.3.2 exemplifies the power of formalization.

Example 3.3.3 *The prototypical strongly minimal structures are:*

(1) *The integers under successor.*
(2) *Any infinite-dimensional vector space.*
(3) *The complex field.*

By Theorem 5.4.4 each model of a strongly minimal theory has a dimension[33] generalizing that in vector spaces. Recall that the main result in a linear algebra course is that a vector space is determined by its dimension.[34] $\mathfrak{R}^3$ is the standard example of a three-dimensional space. It is given by three basis vectors: $(1, 0, 0)$, $(1, 0, 0)$, $(0, 0, 1)$. While the pictorial aspect disappears in infinite dimension the cardinality of a basis remains as the determining factor; this shows the theory of F-vector spaces (for countable F) is categorical in every uncountable power. In particular, a

---

[33] Dimension is a natural generalization of the notion of two- and three-dimensional space. The dimension tells us how many coordinates are needed to specify a point. This dimension (for a countable language) and uncountable strongly minimal (more generally $\aleph_1$-categorical) structure is the same as the cardinality of the model; [Baldwin & Lachlan 1971] show that for countable models either every model has dimension $\aleph_0$ or there are models of infinitely many finite dimensions.

[34] We have just discussed how the dimension in the discrete case, 'number of components', determines each model.

vector space of dimension κ over a field F is the direct sum of κ copies of F. The definition of algebraically closed fields and defining a dimension theory for them is in [Steinitz 1910]. Categoricity in an uncountable power holds for every strongly minimal theory (Theorem 5.4.4) by the same argument. This understanding of categoricity is extended to arbitrary $\aleph_1$-categorical theories by the next theorem [Morley 1965a, Baldwin & Lachlan 1971, Zilber 1991]. Given two models M, N of such a theory of power κ with a strongly minimal set D, $D(N) \approx D(M)$ by the κ-categoricity of strongly minimal theories and this extends to an isomorphism between M and N by condition iv.

Theorem 3.3.4 *(Morley/Baldwin–Lachlan/Zilber) The following are equivalent:*

(1) *T is categorical in one uncountable cardinal.*
(2) *T is categorical in all uncountable cardinals.*
(3) *T is ω-stable and has no two cardinal models.*
(4) *Each model N of T is prime and minimal[35] over a strongly minimal set.*
(5) *Each model of T can be decomposed by finite 'ladders' of strongly minimal sets.[36]*

First order theories that are *totally categorical*[37] have a much stronger structure theory. Zilber's quest to prove that no totally categorical theory is finitely axiomatizable [Zilber 1984c, Cherlin et al. 1985] not only gave a detailed description of the models of such theories but sparked 'geometric stability theory'. Moreover, to eliminate the classification of the finite simple groups[38] from the proof, Zilber gave new proofs of the classification of two-transitive groups [Evans 1986, Zilber 1980, Zilber 1984a].

Because of Morley's theorem,[39] for any theory we can say '$\aleph_1$-categorical' for 'categorical in one (and therefore) all uncountable cardinals'. In addition [Baldwin & Lachlan 1971] shows that an $\aleph_1$-categorical theory has either 1 or $\aleph_0$ models in $\aleph_0$. By (4) and (5) of Theorem 3.3.4, strongly minimal sets are the building blocks of uncountably categorical theories.

The dimension of a model of an arbitrary $\aleph_1$-categorical theory is the dimension of the strongly minimal set over which it is prime by Theorem

---

[35] M is *prime over* $X \subset M$ if every embedding X into a model $N \supset X$ extends to an embedding of M into N fixing X. N is *minimal over* $D(N)$ if there is no model N' of T with $D(N) \subseteq N' \subsetneq N$.
[36] Zilber shows certain automorphism groups (the linking groups of the strongly minimal sets) are first order definable; this leads to the definability of the field in certain groups of finite Morley rank. See Chapter 6.1.
[37] T is categorical in every infinite cardinality.
[38] The [Cherlin et al. 1985] proof relied on the classification.
[39] Morley's theorem is the equivalence of (1) and (2) in Theorem 3.3.4.

3.3.4(4). Thus a theory is $\aleph_1$-categorical if and only if each model is determined by a single dimension.

Sections 5 and 6 of Pillay's survey, 'Model theory' [Pillay 2010], give an accessible and more detailed account of Zilber's [Zilber 1984c] refinement, 'geometric stability theory', of Shelah's general classification program for first order theories. Approaching the Bourbaki ideal (Chapter 2.4), Zilber made his *trichotomy conjecture*: all strongly minimal sets have a 'trivial' or 'vector-space like' (modular) or 'field-like' (nonmodular) geometry. Zilber describes his motivation:

The initial hope of this author in [Zilber 1984c] that any uncountably categorical structure comes from a classical context (the trichotomy conjecture), was based on the belief that logically perfect structures could not be overlooked in the natural progression of mathematics. Allowing some philosophical license here, this was also a belief in a strong logical predetermination of basic mathematical structures. As a matter of fact, it turned out to be true in many cases. Specifically for *Zariski geometries*, which are defined as the structures with a good dimension theory and nice topological properties, similar to the Zariski topology on algebraic varieties (See [Hrushovski & Zilber 1993]). Another situation where this principle works is the context of o-minimal structures [Peterzil & Starchenko 1998]. [Zilber 2005a]

Hrushovski found a counterexample [Hrushovski 1993] to Zilber's conjecture. His structure remains an outlier; but the *Hrushovski construction* led to many interesting developments.[40] [Hrushovski & Zilber 1993] launched a program to rescue the conjecture (and better attune model theory to algebraic geometry). They analyzed the counterexamples that were at first sight pathologies by showing exactly how close 'ample Zariski structures' are to being algebraically closed fields. Zilber [Zilber 2010] lays out a detailed account of the further development of Zariski structures. We provide more detail in Chapter 14.

Moreover, this dimension theory extends to more general classes (Chapters 5.4 and 5.5) than the $\aleph_1$-categorical ones. For ω-stable theories a dimension (Morley rank) can be defined on all definable subsets. It is similar to and specializes to, in the case of algebraically closed fields, the notion of dimension in algebraic geometry. Dimension is not a priori a syntactic notion, but those that appear in model theory and algebraic geometry are.

If a theory is viewed as axiomatizing the properties of a specific structure (e.g. the complex field) categoricity in power is the best approximation that first order logic can make to categoricity. But it turns out to have far more

---

[40] See [Baldwin 2004, Wagner 1994] for surveys and Chapter 7.1.

profound implications than categoricity for studying the original structure. If the axioms are universal and existential then the theory is model complete (page 60) and under slightly more technical conditions admits elimination of quantifiers [Lindström 1964]. Thus the global property of categoricity in $\aleph_1$ determines the complexity of definable sets in the models. What can be very technical proofs of quantifier elimination by induction on quantifiers are replaced by more direct arguments for categoricity.

3.4   The Significance of Categoricity (in Power)

We have posited a criterion for evaluating the virtue of a property of theories (in some logic): whether the property has significant mathematical consequences. That is, whether the theories or more importantly the models of the theories which have this property display other significant similarities.

We agree that from the foundational standpoint of clarifying our intuitions about a canonical object, categoricity of a second order *axiomatization* plays an important role. And this is a mathematical as well as philosophical role. For example, Dedekind provides a framework to prove results about real numbers.

We have argued that according to our criterion, categoricity of a theory is not very interesting for second order logic and trivial for first order logic. However, for first order logic, categoricity in power is very significant because each such theory is seen to possess a dimension function similar to prototypical examples such as vector spaces. The stability hierarchy provides both a classification of first order theories which calibrates their ability to support nice structure theories and the details of such a structure theory. The key to the structure is the definition of local dimensions extending the basic phenomena in theories which are categorical in power.

One slogan for model theory is the 'study of definable sets of a structure'. Our more general term formalization encompasses both the *definability of classes of models* and *definability within a particular model*. A key lesson of the last half century is how much the first study influences the second. Both exemplify the use of formal methods. Most current model theoretic work takes place in first order logic. Despite its limited expressive power, so far first order logic has been more useful than other logics in formalizing notions of traditional mathematics because the compactness theorem allows stronger applications of the formalization. Thus not only formalization but a particular property of the formalization plays a central role.

We employ the distinction between 'logical formalism' and 'axiomatization' made by Bourbaki.[41] But we think Bourbaki has missed the significance of 'logical formalism' for mathematics and reverses the relative importance of the two methods.

The examples in Chapter 6 all use a formal language as a tool to prove mathematical results. These applications have more real effect in mathematics than the use of formalization in seeking global foundations for mathematics. While mathematicians frequently study classes of models such as groups, model theory adds several layers: (i) studying classes of very similar structures (i.e. elementarily equivalent structures) can deepen the study of a particular structure; (ii) the stability theoretic classification of theories provides new links between theories in different content areas of mathematics and enables the transfer of results and methods.

Formal methods in model theory impact other areas of mathematics in several ways: (1) some areas study formally definable sets; the link is immediate once the area is formalized; (2) the definable sets are sufficiently rich that concepts that are ostensibly not (first order) formalizable can be studied by model theoretic means;[42] and (3) syntactic properties in the stability hierarchy provide connections across mathematics that are not evident without the logical perspective. Ehud Hrushovski's address (my emphasis) at the International Conference of Mathematicians in 1998 summarizes the issue:

Instead of defining the abstract context for [stability] theory, I will present a number of its results in a number of special and hopefully more familiar, guises: compact complex manifolds, ordinary differential equations, difference equations, highly homogeneous finite structures. Each of these has features of its own and the transcription of results is not routine; they are nonetheless *readily recognizable as instances of a single theory.* [Hrushovski 1998]

---

[41] See the quotation in Chapter 2.2.

[42] Two such situations are the study of chain conditions in rings discussed in Chapter 2.4 and arithmetic algebraic (Diophantine) geometry, Example 6.2.1, where several different formal theories interact.

The Paradigm Shift

In a 1967 letter to Hao Wang, Gödel explained why others had missed his proof of the completeness theorem,

This blindness (or prejudice, or whatever you may call it) of logicians is indeed surprising. But I think the explanation is not hard to find. It lies in a widespread lack, at that time, of the required epistemological attitude toward metamathematics and toward nonfinitary reasoning. ...

I may add that my objectivist conception of mathematics and metamathematics in general, and of transfinite reasoning in particular, was fundamental also to my other work in logic.

As we'll see this 'objectivist conception', at least in the sense of envisioning models, is central to model theory. Part II traces the historical roots of the paradigm shift and then its effect on doing and organizing mathematics. We begin in Chapter 4 by seeing how the influence of Tarski and Malcev in the 1930s along with Robinson and Henkin almost 20 years later distinguished model theoretic concerns from those that prompted Gödel's work. Then we see the development of the tools of quantifier elimination, model completeness, indiscernibility, and interpretation in the 1950s. In Chapter 5, with Morley and Vaught, properties of theories such as Stone spaces, saturated models, and categoricity in power come to the forefront. We then discuss the key ingredient: Shelah's syntactic hierarchy with its dividing lines culminating in the main gap theorem, specifying which theories are classifiable. Zilber's trichotomy for combinatorial geometry highlights the interaction of model theory with other areas of mathematics. Chapter 6 demonstrates the role of formalization in sharpening the notion of tame and the consequent deep interaction of model theory and algebra. First order analysis moves the impact beyond algebra. An interlude in Chapter 7 considers some reasons for generalizing infinitary logic and the interaction of first order and infinitary logic including Vaught's conjecture. Chapter 8 recounts the role of the paradigm shift in the separation of set theory from first order model theory. We discuss such issues as the 'identity of indiscernibles' and Voevodsky's univalent type theory. We examine the relationship of model theory with both axiomatic and combinatorial set theory, seeing a greater entanglement of infinitary logic with axiomatic set theory.

4 | What Was Model Theory About?

Pillay writes,

> The notion of truth in a structure is at the centre of model theory. This is often credited to Tarski under the name Tarski's theory of truth. But this relative, rather than absolute, notion of truth was, as I understand it, already something known, used, and discussed. In any case, faced with the expression truth in a structure there are two elements to be grasped. Truth of what? And what precisely is a structure? [Pillay 2010]

We defined the notions Pillay refers to in Chapter 1. In this chapter we consider his questions more closely and consider some key episodes in the development of the notion of a first order theory as a focal point.

4.1 The Downward Löwenheim–Skolem–Tarski Theorem

The meaning of 'contradictory' underwent a vast change in the early decades of the twentieth century. It is important to read the famous letter from Hilbert ([Frege & Hilbert 1980], 39) in the 1899–1900 Frege–Hilbert correspondence as written by the turn-of-the-century Hilbert not the later Hilbert:

> if the arbitrarily given axioms do not contradict one another with all their consequences, then they are true and the things defined by the axioms exist. This is for me the criterion of truth and existence. The proposition 'Every equation has a root' is true, and the existence of a root is proven, as soon as the axiom 'Every equation has a root' can be added to the other arithmetical axioms, without raising the possibility of contradiction, no matter what conclusions are drawn. ([Frege & Hilbert 1980], 39)

As we observed on page 33, the Hilbert writing this passage has not yet made the distinction between formal and informal language. Today we might replace 'do not contradict one another' by 'do not imply $0 = 1$ in an ambient formal system' and the word 'true' by 'satisfiable' and read the first sentence as an instance of Gödel's completeness theorem for first order logic. Such replacements are anachronistic in several respects; not only has a formal sentence entered the discussion but it assumes the first/second order distinction. Further, this interpretation presumes the modern distinction between truth and validity which was only made by Bernays [Bernays 1918] for propositional logic in 1918 and Behmann [Behmann 1922] for first order logic in 1922. But this is eventually the meaning that Hilbert [Hilbert & Ackermann 1938] gave to such a sentence.[1]

Philosophical concerns about the size of models forged the development of the (downward) Löwenheim–Skolem theorem in the work of Löwenheim and Skolem, who were both well-acquainted with the Schröder school of algebraic logic,[2] and later for Gödel–Malcev–Tarski.

Badesa [Badesa 2004]   analyzes the exact meaning of Löwenheim's theorem in *The Birth of Model Theory*. Here is the standard (strong) form of the theorem:[3] any countable collection of first order sentences that has an infinite model has a countable model (indeed a countable elementary submodel[4]) of the given model. In 1920 Skolem phrases the result (Theorem 2 of [Skolem 1967a]): every statement of a proposition[5] in normal form either is a contradiction or is already satisfiable in a finite or denumerably infinite domain. Although Hilbert has begun his proof theory program, making contradictory a precise *syntactic* concept, for Skolem (as van Heijenoort notes in his introduction), 'contradictory' is to be read as not-satisfiable.[6] While Skolem is concerned with the result

---

[1] A key confusion that arises here is that until the early twentieth century, valid was an adjective applied only to (truth-preserving) arguments. But as we have just said, we now speak of a valid sentence (true in all *models*) as well as a valid argument. In mathematical logic, a *deductive system is sound* if the rules of inference preserve truth. A number of online philosophy glossaries call an argument *sound* if it is valid and its hypotheses are true. For context, see [Mancosu & Zach 2015].

[2] Those looking for the origins of the first order/second order logic distinction should note both that Peirce defined first order logic and quantification over varying domains in the 1880s, and that he introduced the direct ancestor of modern notations for quantifiers (though Frege also had quantifiers). [Brady 2000, Moore 1988, Moore 1997, Putnam 1982].

[3] We will not discuss Löwenheim's account beyond the following. Conventional wisdom (dating from van Heijenoort's commentary on [Löwenheim 1967]) asserts that Löwenheim gave a flawed argument for the weak form that every satisfiable sentence has a countable model; there is a strong argument in [Badesa 2004] that Löwenheim gave a correct proof of the strong form. Badesa argues that the misinterpretation of Löwenheim's proof grew from the unclear distinction between syntax and semantics when Löwenheim wrote, while it was taken as given by van Heijenoort. There is some ambiguity, but Löwenheim probably meant the result to apply to what eventually became $L_{\omega_1,\omega}$ and certainly it is true in that generality.

[4] Recall that A is an *elementary submodel* of B ($A \prec B$) if every sentence with parameters from A that is satisfied by B is already satisfied in A.

[5] He means a proposition in first order logic, which, following Löwenheim, means in fact $L_{\omega_1,\omega}$. See statement of Theorem 5 in [Skolem 1967a].

[6] He expresses his skepticism about the Hilbert program on page 300 of [Skolem 1967b].

for arbitrary theories in the 1920 paper, in 1923 [Skolem 1967b] he is particularly focused on the theorem as it applies to set theory. He is careful in the second paper to make the proof in Zermelo–Fraenkel set theory without choice. His papers make several notable advances. He criticizes Zermelo for a deficient notion of a 'definite proposition' and remedies this by proposing that 'definite' should mean what we now call first order expressible (definable). He writes ([Skolem 1967b], 292), 'Furthermore, it seems to me clear that, when founded in such an axiomatic way, set theory cannot remain a privileged logical theory; it is then placed on the same level as other axiomatic theories.' He then proves that every satisfiable first order sentence has a countable model[7] and asserts: 'Thus, *axiomatizing set theory leads to the relativity of set-theoretic notions, and this relativity is inseparably bound up with every thoroughgoing axiomatization.*' By the relativity of set theoretic notions, he means that each set theoretic notion, e.g. cardinality, must be given an interpretation that depends on a particular model. That is, if M is a countable model of set theory then a set X such that M satisfies 'X is uncountable', nevertheless is countable from an external viewpoint. All of his mathematical statements are completely correct. He viewed this situation as paradoxical. But at this point one looks back with satisfaction at almost 100 years of progress in set theory and notes the extent to which this 'relativity of set-theoretic notions' has led to a vast clarification of our understanding of those notions. Treating set theory as 'just another theory' worked.[8] We discussed this approach for more traditional mathematical topics in Chapter 2.4.

In his 1920 proof [Skolem 1967a], Skolem introduces a powerful tool that appears repeatedly in model theory. He eliminates existential quantifiers for a theory T by expanding the vocabulary τ to τ^+, by adding for each formula $\phi(\mathbf{x})$ of the form $(\exists y)\psi(\mathbf{x}, y)$ a *Skolem function* $f_\phi(\mathbf{x})$ and by adding to T the axioms

$$(\forall \mathbf{x})[(\exists y)\psi(\mathbf{x}, y) \rightarrow \psi(\mathbf{x}, f_\phi(\mathbf{x}))]$$

to form a *Skolem theory* T^+, the *Skolemization* of T. Using the axiom of choice, each model M of T expands naturally to a model M^+ of T^+. For any $A \subset M \models T$, the *Skolem hull* of A in M^+ is the closure of A under the Skolem functions as interpreted in M^+. Note that it is only a matter of bookkeeping[9] to extend the argument to show (a) that any model M of

---

[7] See [Ebbinghaus 2007] for a modern exposition.

[8] Consider this introduction from Roitman's set theory text, 'These concepts – of first-order theory and of model – have had profound effects on set theory. ... One can make a case for the statement that modern set theory is largely the study of models of set theory.' [Roitman 1990].

[9] There are only countably many conditions to be met; we just have to order them properly.

a countable first order theory has a countable elementary submodel and (b) that any sentence of $L_{\omega_1,\omega}$ is true in a countable submodel. But in the last case an *elementary* submodel may not be possible. (See Chapter 7 and [Ebbinghaus 2007].) The countable model satisfies the given sentence but not every $L_{\omega_1,\omega}$-sentence true in it transfers down; consistent *theories* in $L_{\omega_1,\omega}$ may not have countable models.

4.2 Completeness, Compactness, and the Upward Löwenheim–Skolem–Tarski Theorem

In this section we discuss the historical difficulty in making the separation (Chapter 1.3) between the notions of theory and logic. With the Hilbertian goal of grounding all of mathematics and establishing 'the solvability of all mathematical problems' there was little motivation in this direction in the 1920s. With Gödel's results the distinction between logic and theory becomes more central. He establishes a form of completeness for *first order logic*: every valid sentence is provable. But this is weaker than the result for propositional logic: every sentence is provable or refutable. We distinguished between a consistent set of sentences and a complete theory in Definition 1.4.4. The contrast between Presburger's 1930 proof that $\mathrm{Th}(\mathbb{N}, +)$ is a deductively complete theory and the incompleteness theorem for $\mathrm{Th}(\mathbb{N}, +, \times)$ highlights this distinction.

It is a model theoretic commonplace that the compactness theorem for first order logic is a trivial consequence of Gödel's completeness theorem. For example, Chapter 2 of [Marker 2002] quotes the completeness theorem and gives a one line proof of compactness as corollary.[10] However, several direct model theoretic proofs of compactness, e.g. using ultraproducts, are well-known. Such a fully semantic proof emphasizes the acceptance of Gödel's separation of syntax and semantics. Franks[11] [Franks 2014] takes Gödel's 'blindness' comment (page 87) as a failure to appreciate alternate approaches:

Gödel never considered that others' logical vision might be, rather than defective, simply different—that their inability to see their way to the completeness theorem

---

[10] While the compactness theorem doesn't appear explicitly in Gödel's thesis [Gödel 1929], in the paper [Gödel 1930], he states and proves first order compactness (modulo soundness of first order deduction) for *countable* sets of sentences. However, as we'll see below, it is unlikely that he was fully aware of its significance. See [Dreben & van Heinenoort 1986] for the distinction between the thesis and the published paper and for the connections with Skolem's two proofs of the downward Löenheim–Skolem theorem.

[11] Franks [Franks 2014, Franks 2010b] emphasizes the contrast between Gödel's notion of completeness and those of Skolem, Herbrand, and especially Bolzano and Gentzen.

derived from their focus being held elsewhere. But thinking about logic in the terms that defined Gödel's contribution was not universal, perhaps not even common, in the early twentieth century. His early writing plays a major part in an implicit argument that the correspondence of proof and truth, of logical form and content, is a proper way of thinking about logical completeness. [Franks 2014]

Thus, Skolem is missing the duality between syntax and semantics; both hypotheses and conclusion of the Löwenheim–Skolem theorem are semantic. Many nuances appeared in discovering the consequences of the completeness theorem. Gödel writes:

But one might perhaps think that the existence of the notions introduced through an axiom system is to be defined outright by the consistency of the axioms and that, therefore, a proof [of completeness] is to be rejected out of hand. This definition ... however, manifestly presupposes the axiom that every mathematical problem is solvable. Or, more precisely, it presupposes that we cannot prove the unsolvability of any problem. For, if the unsolvability of some problem (in the domain of real numbers, say) were proved, then, from the definition above, there would follow the existence of two non-isomorphic realizations of the axiom system for the real numbers, while on the other hand we can prove the isomorphism of any two realizations. We cannot at all exclude out of hand, however, a proof of the unsolvability of a problem if we observe that what is at issue here is only unsolvability by certain precisely stated formal means of inference. For all the notions that are considered here (provable, consistent, and so on) have an exact meaning only when we have precisely delimited the means of inference that are admitted. ([Gödel 1929], 61–63)

In his first sentence, Gödel notes that if consistency is taken as in the Hilbert quote opening Chapter 4.1, then there is no need to prove completeness. But in his paper he restricts both the collection of sentences and the means of inference to obtain his illuminating result. In exploring the philosophical background of Gödel's thesis, Kennedy [Kennedy 2011] and Goldfarb [Goldfarb 1999] discuss the influence of Carnap and the work Gödel does in disentangling categoricity from (syntactic or semantic) completeness. In this excerpt Gödel clearly understands that he is referring to *first* order syntactic completeness and *second* order categoricity of the reals.

Gödel clarifies the distinction further in his Königsberg lecture:[12]

I would furthermore like to call attention to a possible application of what has been proved here to the general theory of axiom systems. It concerns the concepts of 'decidable' and 'monomorphic' ... One would suspect that there is a close connection between these two concepts, yet up to now such a connection has

---

[12] See [Goldfarb 1999].

eluded general formulation. Indeed, quite a few monomorphic[13] axiom systems are known – Euclidean geometry for example – for which we have no idea whether they are decidable. In view of the developments presented here it can now be shown that, for a special class of axiom systems, namely those which can be expressed in the restricted functional calculus, decidability always follows from monomorphicity[14] ... If the completeness theorem could also be proved for higher parts of logic (the extended functional calculus), then it would be shown in complete generality that decidability follows from monomorphicity. And since we know, for instance, that the Peano axiom system is monomorphic, from that the solvability of every problem of arithmetic and analysis expressible in Principia Mathematica would follow. Such an extension of the completeness theorem is, however, impossible, as I have recently proved ... [Gödel 1930]

Here Gödel specifically calls on (second order) Peano to make the point that his incompleteness theorem shows that categoricity of a second order theory does not imply decidability of the first order theory. And he asserts that decidability is open for Euclidean geometry. But Euclidean geometry can be thought of in several ways. If we mean the full second order theory, then it includes the definition in pure second order logic of arithmetic so it is rather straightforwardly incomplete by the incompleteness theorem. If we mean Hilbert's second order axiomatization of the geometry over the real numbers then, in the same sense, its second order consequences are undecidable. On the other hand the first order properties of this geometry are exactly the 'Cartesian geometry' that Tarski showed decidable only a year later. However, the first order theory of the Euclidean plane over a fixed Pythagorean field is undecidable (Remark 10.2.3, [Ziegler 1982]). Moving from logics to theories provides much more detailed information (Chapter 9).

Vaught ([Vaught 1986], 873) partially agrees with Pillay's reservation on who first defined truth, as he notes that Gödel's proof uses an intuitive basis for truth. While Gödel does intuitively define ([Gödel 1929], 69) the notion of truth of an arbitrary sentence in an arbitrary structure, in his proof he defines truth for atomic formulas in a structure and a procedure for satisfying π_2-sentences. Satisfaction for an arbitrary sentence is transferred through syntactically provable equivalence to a π_2-sentence.[15] But Vaught argues that as 'all of mathematics' could be done in axiomatic set theory,

---

[13] Monomorphic is a synonym for categorical (Definition 1.4.3); the theory has exactly one model.

[14] Goldfarb points out ([Goldfarb 1999], 191) that Gödel was unaware in these 1930 remarks that a first order sentence can only be categorical if its only model is finite.

[15] For more detail, see [Baldwin 2017a].

it was natural to ask, and answer, whether this 'all' includes the notion of truth. There is a deeper import to Tarski's result. Gödel's proof is by contraposition. He shows that a sentence that is not refutable has a model (indeed on the natural numbers). In contrast, Tarski's truth definition empowers his 1936 concept of logical consequence in [Tarski 1956]: Σ logically implies ϕ if every model of Σ is a model of ϕ. Thus Tarski makes a contribution beyond his truth definition to the modern understanding of completeness; it is only with his notions of truth and semantic consequence that the 'semantic completeness equals deductive completeness' formulation becomes possible.[16]

The Tarski in the section title refers to what is sometimes called the upwards Löwenheim–Skolem theorem: any (countable collection of) first order sentence(s) Σ that has a model, has a model in every uncountable power. Vaught [Vaught 1986] writes[17] that Tarski proved this theorem in his seminar of 1927–28. The first published proof[18] is in 1936 [Malcev 1971b]. A striking point is that the upwards theorem[19] was not immediate to Gödel after he had proved the completeness theorem. The proof requires a genuinely model theoretic idea. To show there is a model of a set of τ-sentences Σ with cardinality κ, *expand the vocabulary*[20] τ *by adding κ constants*; let the theory Σ' be the result of adjoining to Σ the sentences $c_i \neq c_j$ for each $i < j < \kappa$. Now by compactness there is a model of Σ' and a fortiori of Σ with cardinality at least κ. Notably, in 1941 Malcev formulated [Malcev 1971b] a modern, more informative version of the result: every infinite domain for any system of First Order Predicate Logic

---

[16] [Robinson 1951] contains the first explicit publication of the generalized completeness theorem, a (possibly uncountable) theory is deductively complete if and only if it is semantically complete. The result is implicit in Henkin's thesis (1947), extracted in [Henkin 1949]. Compare [Dawson 1993], [Henkin 1996], [Baldwin 2017a], and [Franks 2014].

[17] More precisely, Vaught points out that the *Fundamenta* editors made this assertion for (an) uncountable model in 1934. But Tarski never published his proof and Vaught gives reference to several places in which he (Vaught) has speculated about the details.

[18] Malcev proves first the compactness theorem for propositional calculus (with arbitrarily many propositional variables) and then reduces first order logic (*explicitly with equality*) to that case. Malcev also gives the first explicit statement of the modern compactness theorem but not until 1941 [Malcev 1971a].

[19] I thank Rami Grossberg for pointing out that while Gödel's proof of completeness is in ZF (indeed, needing only König's lemma as a vocabulary is specified in advance), there is a subtlety about upward Löwenheim–Skolem: that a sentence with an infinite model has a model in every infinite cardinal requires choice; but that it has arbitrarily large models requires only the Boolean prime ideal theorem. I have no information on whether this actually affected Gödel's reasoning. See [Felgner 1971] for the role of the axiom of choice.

[20] We stressed in Chapter 1.2 the importance of specifying a particular vocabulary for the areas being studied. This is a basic example of the power of augmenting that vocabulary in appropriate ways. Adding constants is as harmless as it gets; yet it gives here a strong result.

(FOPL) formulas can be properly extended to another satisfying domain. Getting a proper elementary extension in the same cardinality apparently is first done by 'Skolem ultrapowers' in [Skolem 1934].

Henkin's proof of the completeness theorem was an important step towards the modern conception. Henkin asserts an advantage of his new proof:

> In the first place an important property of formal systems which is associated with completeness can now be generalized to systems containing a non-denumerable infinity of primitive symbols. While this is not of especial interest when formal systems are considered as logics – i.e., as means for analyzing the structure of languages – it leads to interesting applications in the field of abstract algebra. [Henkin 1949]

The paper introduces the vocabulary as including 'for each number $n = 1, 2, \ldots$ a set of functional symbols[21] of degree n.'

Many mathematicians (e.g., Pasch, Hilbert, Noether, van der Warden) had adopted the fixed vocabulary practice as geometry became more carefully formalized and notions of groups, rings, and fields developed. But the incorporation of this requirement into logic (as opposed to the vocabulary with predicates and functions of all arities and, indeed, all orders) was much later. Unlike Gödel's, Henkin's proof takes place in the original vocabulary and this empowers the shift. The abstract notion of 'structure' (for a given vocabulary) was first formalized in 1935 by Garrett Birkhoff [Birkhoff 1935]; both Tarski [Tarski 1946] and Robinson [Robinson 1952] refer to that paper.[22] Robinson specifies a vocabulary for the particular topic. Certainly, Tarski makes the step by [Tarski 1954, Tarski & Vaught 1956]. The evolutionary background of these developments in the 1930s is stressed by [Schiemer & Reck 2013]. Both Henkin and Gödel ([Gödel 1929], 101) are aware that the completeness proof routinely transfers between pure and applied first order systems. The

---

[21] To modern eyes, the term 'functional symbol' is misleading; the semantics make clear these are what we would now call relation symbols. Moreover, the modern specification of a particular vocabulary is presaged on page 161 of [Henkin 1949] with 'Let S_0 be a particular system determined by some definite choice of primitive symbols.' The article is part of his continuing work on both second order logic and the theory of types; the function variables occur to accommodate these extensions. In his thesis, Henkin explicitly formulated the compactness theorem (for uncountable vocabularies), introduced the method of diagrams (independently discovered and baptized by A. Robinson), and used these methods in an 'applied logic', fixing the vocabulary for rings, to give a new proof of the Stone representation theorem and some further algebraic applications [Henkin 1953, Henkin 1996]. He published the last section of his thesis [Henkin 1953] with the modern notation including functions with no distinction of pure and applied logic.

[22] We also address this issue in Chapter 1.2 and in footnote 13 of that chapter.

issue is, 'which should be the focus?' Henkin published the last section of his thesis [Henkin 1953] with the modern notation of formalization by specifying a vocabulary using primitive symbols directly connected to the specific subject with no distinction of pure and applied logic.

We briefly detour from first order logic to consider certain extended logics that have completeness theorems. An abstract formulation of the completeness theorem was popular in the late 1970s: the set of first order validities is recursively enumerable. This formulation was discussed by the authors but did not make it into [Barwise & Feferman 1985]. A reason is that it does not persist so nicely to infinitary logics. There are (infinitary) proof systems for the logics $L_{K,\omega}$. But these systems do not demonstrate that the validities are recursively enumerable or even Borel.[23] This weakness in the proof system also prevents these infinitary logics from satisfying the full compactness theorem and thus they fail the upward Löwenheim–Skolem theorem.

Continuous logic of metric structures [Ben Yaacov et al. 2008] is currently the most important approach among model theorists[24] to the study of functional analysis. Using ultraproducts one (e.g. [Ben Yaacov et al. 2008]) shows the logic satisfies the compactness theorem. Ben Yaacov and Pederson introduce an interesting distinction between two 'precursors' of continuous logic in their paper proving a completeness theorem for this logic.

Historically, two groups of logics precede continuous first order logic. On the one hand, continuous first-order logic has *structural precursors*. The structural precursors are those logics which make use of machinery similar to that of continuous first-order logic yet were never developed to study complete metric structures. Such structural precursors include Chang and Keisler's continuous logic [Chang & Keisler 1966], Lukasiewicz's many-valued logic [Hájek 1998], and Pavelka's many-valued logic [Pavelka 1979]. ... continuous first-order logic is an improved variant of Chang and Keisler's logic. On the other hand, continuous first-order logic has *purposive precursors*. The purposive precursors are those logics which were developed to study complete metric structures yet do not make use of machinery similar to that of continuous first-order logic. The purposive precursors of continuous first-order logic include Henson's logic for Banach structures and compact abstract theories (cats). Continuous first-order logic does not suffer from several shortcomings of these logics. Importantly, continuous first-order logic is

---

[23] In fact, the validities of $L_{\omega_1,\omega}$ are Σ_1-definable on the hereditarily countable sets $(\langle H(\omega_1), \in \restriction H(\omega_1)\rangle)$ and Σ_2 on $\langle H(\kappa), \in \restriction H(\kappa)\rangle$ for $L_{\omega_1,\omega}$ when κ is uncountable ([Dickmann 1985], 328).

[24] The nonstandard analysis school works in parallel with most of model theory.

less technically involved than the previous logics and in many respects much closer to classical first-order logic. [Ben Yaacov & Pederson 2010]

These authors see the main outlines of continuous logic being developed in the applied contexts before the explicit statement/proof of the completeness theorem by experts in the field. This viewpoint supports my remark in Chapter 1.3 that for our purposes a system of deduction is the least important aspect of a formalization. From the standpoint of the authors just quoted, one was found to clean up loose ends after some 30 years of preliminary work. However, José Iovino has provided a fuller story.[25] It begins in 1929 with Lukasiewicz logic, propositional logic with values in the closed interval $[0, 1]$. Chang gave two proofs of the completeness of this logic in 1958–59. Predicate Pavelka logic is first order logic along with the proviso that for each rational number r there is a sentence ϕ^r whose truth value in every model is r and this logic was proved complete in the 1990s. Caicedo and Iovino show (Proposition 1.18 of [Caicedo & Iovino 2014]) that positing this family of 'constants' is equivalent to the requirement that the valuation of each symbol in continuous logic is uniformly continuous. Thus the main distinction between Ben Yaacov–Usvyatsov continuous logic and Pavelka logic is that the first focuses on complete metric spaces and the latter on so-called 1-Lipschitz structures. And they are both included in a common logic (formulated as $\mathcal{L}^0$ in [Caicedo & Iovino 2014]) satisfying completeness, that is the special case of Chang–Keisler continuous model theory whose connectives are sup and inf and whose truth value space is $[0, 1]$.

There are a number of other strong logics that satisfy varying notions of compactness ($\mathcal{L}$ is κ-compact if a set of sentences Φ such that every subset of Φ of cardinality less than κ is satisfiable then Φ is satisfiable). The most prominent fully compact ($\kappa = \omega$) logic is cofinality logic.

In summary, Gödel's focus in the completeness theorem is on logics, not theories. But Tarski's proof of the decidability of the real field signaled a new direction almost immediately. It became clear over the next few decades with the clear formulation of 'computable' and the study of decidability and undecidability of distinct theories that a divide (into separate theories) and understand (if not conquer) strategy gives much finer information than the general completeness result.

---

[25] The next few paragraphs are largely a paraphrase of an email from Iovino. More detailed references are in [Caicedo & Iovino 2014].

4.3   Complete Theories

As we noted on pages 15–16, Detlefsen's Question IV asked:

Question II (philosophical question): Given that categoricity can rarely be achieved, are there alternative conditions that are more widely achievable and that give at least a substantial part of the benefit that categoricity would? Can completeness be shown to be such a condition? If so, can we give a relatively precise statement and demonstration of the part of the value of categoricity that it preserves?

One virtue is clear: showing a first order axiomatization is complete determines the truth value of each derived sentence, solving the reference issue in Chapter 3.

The very concept of *complete theory* contributes to the beginning of the paradigm shift. Tarski uses the concept in [Tarski 1950] but calls it an *arithmetic type*, a class of structures elementarily equivalent to a given structure. The terminology is fully established only by Robinson's 1956 monograph *Complete Theories* [Robinson 1956].

We argue that the notion of a complete theory provides a formidable mathematical tool: one that is not sterile but allows for the comparison in a systematic way of structures that are closely related[26] but not isomorphic. So we argue that *completeness* of a non-categorical theory gives substantial benefits that approximate (especially in combination with further properties) the mathematical benefits of categoricity in power, which we expounded in Chapter 3.3, but are completely impossible for categorical theories.

We now give several examples that illustrate the mathematical power of the notion of complete theory, demonstrating that completeness is a *virtuous property*. Axiomatic theories arise from two distinct motivations. We emphasized in Chapter 3.1 the importance of a (usually second order) axiomatization of single significant structure such as $(\mathbb{N}, +, \cdot)$ or $(\mathfrak{R}, +, \cdot)$. The other is to find the common characteristics of a number of structures; theories of the second sort include groups, rings, fields, etc. In the second case, little is gained simply from knowing a class is axiomatized by first order sentences.[27] Although in Example 1.4.2 we exhibited a mathematical

---

[26]   One natural approach to specifying 'closely related' is to say 'elementarily equivalent' with respect to some logic. We focus here on first order but there is important work for various infinitary logics.

[27]   One does know such properties as closure of the class under ultraproduct, the ability to use compactness arguments, and, as illustrated by some 1950s results in Chapter 14, semantic equivalents to the syntactic form of the axioms.

impact of formal axiomatizability, in general, the various completions of the theory simply provide too many alternatives. But for complete theories, the models are sufficiently similar that information can be transferred from one to another. Kazhdan illuminates the key reason to study complete theories:

> On the other hand, the Model theory is concentrated on [the] gap between an abstract definition and a concrete construction. Let T be a complete theory. On the first glance one should not distinguish between different models of T, since all the results which are true in one model of T are true in any other model. One of [the] main observations of the Model theory says that our decision to ignore the existence of differences between models is too hasty. Different models of complete theories are of different flavors and support different intuitions.[28] So an attack on a problem often starts [with] a choice of an appropriate model. Such an approach lead to many non-trivial techniques for constructions of models which all are based on the compactness theorem which is almost the same as the fundamental existence theorem.
>
> On the other hand the novelty creates difficulties for an outsider who is trying to reformulate the concepts in familiar terms and to ignore the differences between models. [Kazhdan 2006]

The last sentence points to a central aspect of model theory; it studies distinctions between models that are elementarily equivalent. Necessarily, these distinctions must be expressed in set theory or some extension of first order logic. Notions such as the realization or omission of types, the saturation of models, and the categories in the stability hierarchy are tools in first order model theory; they are about, not in, first order logic.

The theory of algebraically closed fields is *model complete* and the close connection between it and its completions is mathematically significant.

Example 4.3.1 (Algebraic Geometry) A model theoretic maxim[29] asserts, 'algebraic geometry is the study of definable subsets of algebraically closed fields (page 63)'. More precisely the fundamental goal of algebraic geometry is to solve systems of polynomial equations. A set defined by a conjunction of equations is a *variety*. A relation on a field is *constructible* if it is defined by a Boolean combination of equations. A *projection* of a subset X of M^n to $M^m \subset M^n$ for $m \leq n$ maps an element of X to its coordinates in M^m. Here are two equivalent statements of the same result showing the connections of formal definability to algebraic geometry.

---

[28] Examples are atomic, homogeneous, and saturated models.

[29] For some limits on this maxim see [Macintyre 2003b] and [Manders 1984] and for a more supportive view the appendix to [Hrushovski 2002].

Theorem 4.3.2 (Chevalley–Tarski Theorem[30]) *Chevalley: Over an algebraically closed field, the projection of a constructible set is constructible.*

Tarski:[31] The theory of algebraically closed fields admits elimination of quantifiers.

Thus quantifier elimination (Definition 4.4.1) is a different phrasing of a fundamental tool of algebraic geometry. But there are too many first order definable sets; geometers study unions of varieties and this class is not closed under negation. This problem is addressed model theoretically by the notion of Zariski geometry.[32]

The next two examples use the *syntactic* concept of a complete theory to transfer results between structures in a way that was impossible or ad hoc without the formalism. The theory of algebraically closed fields is not complete; all that is missing is to specify the characteristic.[33]

Theorem 4.3.3 (Minor Principle of Lefschetz) *Let ϕ be a sentence in the language $\mathcal{L}_r = \{0, 1, +, -, \cdot\}$ for rings, where $0, 1$ are constants and $+, -, \cdot$ are binary functions. The following are equivalent:*

(1) ϕ *is true in every algebraically closed field of characteristic 0.*
(2) ϕ *is true in some algebraically closed field of characteristic 0.*
(3) ϕ *is true in algebraically closed fields of characteristic p for arbitrarily large primes p.*
(4) ϕ *is true in algebraically closed fields of characteristic p for sufficiently large primes p.*

Proof. This follows from the completeness of algebraically closed fields of characteristic 0 and Gödel's completeness theorem.

In Chapter 7 we explore logicians' attempts to better reflect the Lefschetz principle using infinitary logic.

It frequently turns out that important information about a structure is only implicit in the structure but can be manifested by taking a saturated elementary extension of the theory. In particular, within a saturated model

---

[30] The version of this theorem for the reals is also known as the Tarski–Seidenberg theorem [Seidenberg 1954]. Tarski announced the quantifier eliminability of the real closed fields in [Tarski 1931]. He apparently became aware that his argument extended to the complex numbers when Robinson proved the quantifier eliminability of algebraically closed fields in [Robinson 1954]. There are rumors that the *Bourbakiste* Chevalley (1945) was aware of Tarski's proof for the reals.

[31] This result was established independently by Abraham Robinson [Robinson 1952].

[32] The *Zariski topology* on a field declares as closed the solution set of equations. Hrushovski and Zilber [Hrushovski & Zilber 1993] define a Zariski geometry in a general model theoretic context by defining Zariski topology in which the solutions of equations form the closed sets and then axiomatizing the family of closed sets in M^n for all n. See 307.

[33] A field has characteristic p if for every x, $px = 0$ and characteristic 0 if there is no such p.

the syntactic types[34] over a subset A can be realized as orbits of automorphisms of the ambient saturated model that fix A.

The germs of this idea are seen in the role of types over sets in giving a general formulation of the notions of *generic point*. The notion of a generic point on a variety X defined over a field k is a rather amorphous and complex concept ([Weil 1962], 27) for much of the twentieth century. In the model theoretic approach a generic point a of X is a point in an extension field of k. More precisely, if $\bar{k}$ is the algebraic closure of k, a is a realization in an elementary extension of $\bar{k}$ of a non-forking extension of the type of minimal Morley rank that contains the formula X. Weil's notion of a universal domain is, in model theoretic terms, an $\aleph_1$-saturated model of the theory of algebraically closed fields.

Example 4.3.4 The **Ax–Grothendieck** theorem asserts that an injective polynomial map on an affine algebraic variety over $\mathbb{C}$ is surjective. The model theoretic proof[35] in [Ax 1968] observes that the condition is axiomatized by a family of 'for all – there exist' ($\forall\exists$) first order sentences ϕ_i (one for each pair of a map and a variety). Such sentences are preserved under direct limits and the ϕ_i are trivially true on all finite fields. So they hold for the algebraic closure of F_p for each p (as it is a direct limit of finite fields). By Theorem 4.3.3, $T = \text{Th}(\mathbb{C})$, the theory of algebraically closed fields of characteristic 0 is axiomatized by a schema Σ asserting each polynomial has a root and, for each p, stating that the characteristic is not p. Since each ϕ_i is consistent with every finite subset of Σ, it is consistent with Σ and so in T, as the consequences T of Σ form a complete theory.

---

[34] Let A be a subset of a model M and $b \in M$. The syntactic type of b over A in the sense of M is $\{\phi(x, a) : a \in A, M \models \phi(b, a)\}$. A key reason for Shelah introducing the universal domain is to remove the dependence on the ambient M. Saturated models are defined in Definition 5.2.2.

[35] There is a non-model-theoretic proof in the spirit of Ax which replaces model completeness by multiple references to the Nullstellensatz [Kang 1993]. Ax [Ax 1968] was apparently unaware of Grothendieck's proof [Grothendieck 1966]. He cites other work by Grothendieck and not this, but says 'This fact seems to have been noticed only in special cases (e.g. affine space over the reals by Białynicki-Birula and Rosenlicht [Białynicki-Birula & Rosenlicht 1962]).' The argument in the real case is more complicated, but it easily yields the result for the complexes. In fact, there is no model theoretic proof of the real case. We can express the compactness argument in Ax's proof by noting $\mathbb{C}$ is an ultraproduct of locally finite fields. However, $\mathfrak{R}$ is not such an ultraproduct because $(\exists z)(x + z^2 = y)$ defines a linear order of $\mathfrak{R}$. So this must be true for a family of fields indexed by a member of the ultrafilter if such an ultraproduct representation existed. But no locally finite field can be linearly ordered. Moreover, in this situation there are 'too many' definable sets. The natural variant for o-minimal theories that Nash functions (ie. real analytic functions that are semi-algebraic) satisfy the theorem fails. $f(x) = \dfrac{x}{(x^2+1)^{\frac{1}{2}}}$ is a one-to-one map from $\mathfrak{R}$ into $(-1, 1)$.

Note that surjectivity implies bijectivity is false. A model theorist might sense the failure since injectivity is universally axiomatized and so passes to substructure, while surjectivity is ∀∃ axiomatized and so does not pass to substructures; an algebraist would immediately note that the map $x \mapsto x^2$ is a counterexample in, for example, the complex numbers.

Tao ([Tao 2009]) gives an algebraic proof. He makes extensive use of the Nullstellensatz and notably misses[36] the simpler direct limit argument to go from the finite fields to the algebraic closure of F_p. These distinct approaches reflect the different perspective of logic on such a problem.

Example 4.3.5 (Division Algebras) *The study of finite-dimensional division algebras provides a similar example.*

Theorem A: (Real division algebras): Any finite-dimensional real division algebra[37] must be of dimension 1, 2, 4, or 8. This is proved (by Hopf, Kervaire, and Milnor in 1958) using methods of algebraic topology (Bott periodicity).[38]

Theorem B: (Division algebras over real closed fields): Any finite-dimensional division algebra over a real closed field must have dimension 1, 2, 4, or 8.

Jacobson ([Jacobson 1964], 314) showed Theorem B follows from Theorem A by the completeness of the theory of real closed fields. No other proof is known.

This kind of direct transfer between models of complete (so necessarily formalized) theories is one benefit of formalization. Here is another example of the virtue of the right choice of a complete first order theory. In a rough sense undecidability seems to disappear when a structure is 'completed' to answer natural questions (e.g. adding inverses and then roots to the natural numbers). Here is a specific measure of that idea which is more fully developed in the work of Manders (Chapter 5.2).

Example 4.3.6 (Ruler and Compass Geometry) [Beeson 2008] notes that the theory of 'constructible geometry' (i.e. the geometry of ruler and compass) is undecidable. This result is an application of Ziegler's proof [Ziegler 1982] that any *finitely axiomatizable theory* in the vocabulary

---

[36] He gracefully acknowledges this simplification in reply to a comment.

[37] An algebra over a field is a vector space over the field with a multiplication. 'Division algebra' means the multiplication has inverses. 'Real division algebra' appears in the literature with several meanings. For Frobenius the multiplication is associative and for Hurwitz it means multiplication by unit vectors is distance-preserving. But Hopf, Kervaire, and Milnor mean only that the multiplication is continuous.

[38] Frobenius had shown the bound of 4 for associative division algebras in 1878.

$(+, \cdot, 0, 1)$ of which the real field is a model is undecidable. The complete theory is tractable while none of its finitely axiomatized subtheories are.

While I have given only a few simple examples, note that all the works in 'algebraic model theory' of A. Robinson, Ax–Kochen, Macintyre, van den Dries, and the entire school are using formal methods in the sense here described. Investigating the definable sets in models of a complete theory promotes the understanding of the area axiomatized.

4.4   Quantifier Complexity

As quantifiers are alternated, formulas become more difficult to understand. In this section we address the epistemological issue of accessibility by discussing the relation between two notions of quantifier complexity[39] as a measure of the complexity of a theory T: (i) the complexity of the axioms for T and (ii) an assertion that each formula is equivalent on the models of T to a formula of some bounded complexity. Because of the commonplace that no one can understand a formula with more than, let's say, three alternations of quantifiers, if either of these complexities is low, we have a better understanding of the theory. The simplest theories in sense (ii) satisfy the following definition.

Definition 4.4.1 *A theory T admits* elimination of quantifiers *if every formula $\phi(\mathbf{x})$ is equivalent in T to a formula $\psi(\mathbf{x})$ with no quantifiers.*[40]

There are two immediate advantages to proving a theory T admits elimination of quantifiers in an easily understood[41] vocabulary: if the axioms are recursive, decidability is immediate; further all first order definable subsets of models of T are analyzable by the basic mathematical concepts of T. The work in the 1920s bears out the comments of Manders, quoted in the introduction, highlighting emphasis on 'reliability' of general patterns of reasoning as the central concern of traditional epistemology.[42] Mancosu and Zach [Mancosu & Zach 2015] trace the notion of quantifier eliminability back to

---

[39]   The complexity of a formula is the number of quantifier alternations of its prenex normal form.

[40]   Sometimes, the specification that ψ have the same free variables as ϕ is omitted. But this seems wrong-headed. The actual equivalence is then $\phi(\mathbf{x}) \leftrightarrow \forall \mathbf{y} \psi(\mathbf{x}, \mathbf{y})$. And the negation of ϕ is equivalent to an existential formula. So this is an equivalence to a Δ_0-formula and holds only on models of T. Quantifier elimination in our sense means that each formula has unique interpretation on each substructure of any model of T.

[41]   The consequences explicate what I mean by 'easily understood'.

[42]   While the focus in this work is on the epistemology of model theory, such advances have taken place in other areas of logic; see for example [Sieg 2013].

Behmann[43] in 1921 and note that Behmann is drawing on what is called the 'solution problem' in the algebraic tradition of Peirce and Schroeder. In particular, he obtains, independently from but later than Löwenheim and Skolem, a decision procedure for pure monadic second order logic. Behmann himself seems to regard his main problem as formulating the problem of deciding all sentences of a logic. The elimination of quantifiers reduces the decidability problem to that of more intelligible formulas. Both Langford's quantifier elimination [Langford 1926–27] for dense linear orders and Presburger's proof of the completeness of a natural axiom system for the additive structure of the natural numbers yield a decision method for a theory.[44]

Frequently, although quantifiers cannot be eliminated in the original vocabulary, they can be after a mild extension of the vocabulary. Thus in 1928 Presburger [Presburger 1930] achieved quantifier elimination for the natural numbers with addition and successor by adding predicates for divisibility by n for each natural number n. The goal of the paper[45] was to prove the completeness of Presburger's axioms. Since it is relatively easy to compute the truth values of the quantifier-free formulas this yielded the decidability of Presburger arithmetic as well. It is only with Tarski's [Tarski 1931] study of the real field that quantifier elimination becomes a tool for investigating problems of traditional mathematics. Tarski shows that every definable (with parameters) subset of the reals is defined as a finite union of intervals.[46] This extends to quantifier elimination of formulas in any finite number of variables.

Robinson discovered in the early 1950s that a near approximation to quantifier elimination was also very useful. We'll introduce some more general notation to explain this approximation. A formula is called $\forall_n$ ($\exists_n$) if its prefix normal form has n blocks and the first is universal (existential). A set of sentences is $\forall_n$ if each axiom is. In particular then, in algebraic situations an $\exists_1$-formula, $\exists \mathbf{y} \phi(\mathbf{y}, \mathbf{x})$, asserts the solvability of a family of equations (and inequations) with parameters $\mathbf{x}$.

Work in the early 1950s showed two striking connections of the quantifier complexity of an axiomatization with (a) semantic properties of its models and (b) the quantifier complexity of its definable sets.

---

[43] Their paper contains both the translation and original of a Behmann lecture.

[44] Presburger adds to the basic vocabulary $(+, 1)$ predicates $P_n(x)$ for x is divisible by n. He notes the decision problem is open if multiplication is considered.

[45] See the background as well as the translation in [Stansifer 1984]. Tarski felt the result too obvious an application of the method of quantifier elimination and Presburger never received a doctorate.

[46] This is later generalized to o-minimality; see Definition 6.3.1.

Theorem 4.4.2

(1) *(Chang–Łos–Suszko) A theory is $\forall_2$ or $\forall\exists$-axiomatizable   if and only if the class of its models is closed under unions of increasing chains of submodels.*

(2) *(Robinson) A theory T is* model complete *if it satisfies either of the following equivalent conditions.*

   (a) *If $A \subset B$ are models of T then $A \prec B$.*
   (b) *Every formula is equivalent in T to a $\exists_1$-formula.*

Here are some important examples of these notions in mathematical practice. The Chevalley–Tarski Theorem 4.3.2 asserts the theory of algebraically closed fields is model complete[47] (page 60). This identifies the definable sets with the *constructible* sets in algebraic geometry [Hartshorne 1977]. The theory of real closed fields in the language of rings is not quantifier eliminable. But with the added symbol for 'less than' the theory of real closed ordered fields admits elimination of quantifiers. The connection with these classical structures motivated Robinson's interest.

Theorem 4.4.2 shows that every model complete theory is $\forall_2$-axiomatizable. Chang later illustrated the deep relation between quantifier complexity and the number of models by showing that if a theory is $\forall_2$-axiomatizable and κ-categorical (for any infinite κ) then it is model complete. For two reasons we introduce what might seem a technical definition. First, it is a specific example of an amalgamation property, a notion which will continue to play a major role. Second, this test itself has many applications.

Definition 4.4.3 *A first order theory T is* submodel complete *(substructure complete) if for any submodel (substructure) A of models M, N there is an amalgamation, a pair of embeddings f, g of M and N respectively into a third model $\hat{M}$ such that f and g agree on A and that f is an elementary map.*

Shoenfield[48] [Shoenfield 1967] gave a criteria for quantifier elimination that Sacks [Sacks 1972] generalized to two parallel results:

Theorem 4.4.4 (Shoenfield's Lemma) *A first order theory T admits elimination to $\exists_1$ formulas (quantifier elimination) if and only if it is submodel (substructure) complete.*

---

[47] It is not complete until we specify the characteristic.
[48] The proof idea using saturated models is in Theorem 4.3.28 of [Marker 2002].

We stress two important points concerning low-quantifier rank. First, it provides a tool for proving fundamental algebraic results[49] that date from around 1930 and, in the 1950s, provides a uniform statement of these results as examples of quantifier elimination. Second, such a reduction in quantifier complexity makes the study of definable sets accessible to algebraic techniques. Robinson gave a uniform account of this phenomenon.

Definition 4.4.5

(1) T_1 *is a* model completion *of T if $T \subset T_1$, T_1 is model complete, and for every $M \models T$, $T^* \cup \{\phi(\mathbf{m}) : \mathbf{m} \in M\}$ is complete.*

(2) *The theory Dcf_0 of characteristic 0 differentially closed fields is the model completion of the theory of characteristic 0 differential fields.*[50]

The notion of model completion[51] was investigated by the Robinson school specifically for allowing insight into mathematical concepts. Manders summarizes:

Model completion is another model theoretically analysable strategy with a recognizable historical role. The notion arose in Robinson's work on representing elements of ordered fields as sums of squares; and in his application the key elements of a reconceptualization strategy are clear: a problem is posed for arbitrary ordered fields; he moves to a real-closed field, where the issue is more tractable, and then develops a translation theory allowing inferences for the original fields. Moves to model completions are now recognized in a variety of contexts: algebraic closure, real or p-adic closure, and projective closure of affine space, shown to be a model completion (when accompanied by model completion of the underlying ring structure) in [Manders 1984]. In each case, questions are recognized as more accessible in the model complete setting, more so where we even have elimination of quantifiers. Model completions tend to unify concepts, and eliminate case distinctions. [Manders 1987]

We have indicated only a small part of the vast literature using the notions of quantifier elimination and model completeness to study algebra from a model theoretic viewpoint; we place the notion in a broader context in Chapter 5.2. The big idea is that formulas of low quantifier complexity are easier to understand (more accessible) and that one can systematically study the reduction in complexity by considering theories rather than logics. Crucially, the goal of this work is epistemic: to make complicated formulations more accessible.

---

[49] Examples include the Artin–Schreier theorem, the Hilbert Nullstellensatz (e.g. [Marker 2002]), and the differential (Ritt) Nullstellensatz.

[50] For characteristic p, the somewhat weaker condition 'model companion' applies.

[51] See Chapter 6.3.

4.5   Interpretability

Interpretation is a fundamental and wide ranging model theoretic notion with nineteenth century roots in, for example, the Poincaré model for hyperbolic geometry. Borovik and Nesin express the deep ties between the formal notion of 'interpretation' and algebra:

> The notion of interpretation in model theory corresponds to a number of familiar phenomena in algebra which are often considered distinct: coordinatization, structure theory, and constructions like direct product and homomorphic image. For example a Desarguesian projective plane[52] is coordinatized by a division ring; Artinian semisimple rings are finite direct products of matrix rings over division rings; many theorems of finite group theory have as their conclusion that a certain abstract group belongs to a standard family of matrix groups over ... All of these examples have a common feature: certain structures of one kind are somehow encoded in terms of structures of another kind. All of these examples have a further feature which plays no role in algebra but which is crucial for us: in each case the encoded structures can be recovered from the encoding structures definably. ([Borovik & Nesin 1994], 29)

In Chapter 6.1 we expand on the mathematical uses of interpretation. To ground our discussion of some of the philosophical uses of this notion, we begin with a short account based on Sections 5.3 and 5.4 of [Hodges 1993], illustrated by examples from Hilbert's *Grundlagen der Geometrie*. Crucially, we restrict to interpretations that are definable entirely by *first order* formulas and for simplicity we assume the vocabularies are relational.

Definition 4.5.1 *An (n-dimensional) interpretation Γ of a vocabulary τ into a vocabulary σ consists of two items:*

(1) *A σ-formula $\partial_\Gamma(x_0, \ldots x_{n-1})$*
(2) *for each relation symbol $R(y_0, \ldots y_{m-1})$ a formula $R_\Gamma(\mathbf{x}_0, \ldots \mathbf{x}_{m-1})$ where the $\mathbf{x}_i$ are disjoint n-tuples of distinct variables.*

Item (1) defines the (possibly n-dimensional) set which is the domain of the interpreted structure; item (2) defines the relations. An interpretation of two vocabularies is just an assignment of atomic formulas; but it is easily proven that formulas of $L_{\infty,\omega}$ (page 45) are preserved (5.3.2 of [Hodges 1993]).

Hilbert argues as follows for the consistency of Axiom groups I–IV (the first order axioms of the *Grundlagen*; see Notation 9.3.3). He starts with the

---

[52] Theorem 12.2.4.

field he calls Ω, generated by 1 and closed[53] under the field operations and $\sqrt{1 + \omega^2}$, where ω is any element of the field. Then he defines a geometry with elements of Ω^2 as 'points' and elements $\langle u, v, w \rangle \in \Omega^3$ as 'lines' where $\langle x, y \rangle$ is on $\langle u, v, w \rangle$ if $xu + yv + w = 0$. That is, he describes the usual Cartesian plane over Ω.

In modern terms, Hilbert is interpreting (as in Definition 4.5.1) the vocabulary for geometry into that for fields. The domain formula is the disjunction of the formulas $\partial_\Gamma^P(x, y)\colon x = x \wedge y = y$ and $\partial_\Gamma^L(u, v, w)\colon uvw \neq 0$ for points and (nontrivial) lines respectively. Easily, incidence is given by $xu + yv + w = 0$. In a similar way one can interpret relations for congruence,[54] betweenness etc. (See Notation 9.3.1.) Now Hilbert observes this structure satisfies the geometric axioms and concludes the consistency of the 'first order axioms' from his ability to construct Ω. But Hilbert is still taking sentences as being in natural language.

We can reformulate this argument as an interpretation of theories with the following definition; we give a more concrete version in Definition 12.5.3.

Definition 4.5.2 *An (n-dimensional) interpretation Γ of a vocabulary τ into a vocabulary σ*

(1) *yields an interpretation (definition)[55] of the τ-structure B in the σ-structure A if B is isomorphic to the τ structure defined on $\Gamma(A^n)$ by Γ. We call this structure $f_\Gamma(A)$.*

(2) *Let T be a τ-theory and U a σ-theory. We say T is interpreted (defined) in U if*

   (a) *For[56] every $A \models U$, $f_\Gamma(A) \models T$;*
   (b) *for every $B \models T$ there is an $A \models U$ with $f_\Gamma(A)$ isomorphic to B.*

Often (1) and (2) are distinguished as *semantic* and *syntactic* interpretations. This is a difference in method not effect.[57] We follow Hodges in omitting computability requirements; they can be added when decidable rather than merely model theoretic notions are to be preserved. T_1 and T_2 are *mutually interpretable* if each is interpreted in the other; they are *bi-interpretable* if those interpretations are inverse to each other.

---

[53] Thus he constructs a minimal Pythagorean field; see [Hartshorne 2000].

[54] For example, two segments are congruent if they have the same length (computed as Pythagorean distance in the field between the end points).

[55] In current usage, if the universe of the image of B is a subset of A^n the precise term is 'definable in'. The modern notion of 'interpretability' extends the notion to allow quotients; this has important mathematical consequences (Section 4.f of [Poizat 1987; Poizat 2001].)

[56] Hodges doesn't have this clause, but we follow [Tarski et al. 1968].

[57] Clearly (2) yields (1); but (1) yields (2) for Th(A) and Th(B). Schiemer [Schiemer 2016] discusses ways in which Hilbert's constructions can be viewed.

In the sense of Definition 4.5.2, Hilbert has given an interpretation of (first order) Hilbert geometries satisfying the parallel postulate[58] into the theory of Pythagorean fields (Definition 9.4.3). But this is an anachronistic statement; Hilbert does not speak of a theory of fields. He writes ([Hilbert 1962], 26), 'From these considerations, it follows that every contradiction resulting from our system of axioms must also appear in the arithmetic related to the domain Ω.' He is simply showing that geometry is consistent in the most direct way, exhibiting a model.[59]

Similarly, to show the independence of the parallel postulate, he refers to a 'well-known manner' and explains:

As the individual elements of a geometry of space, select the points, straight lines, and planes of the ordinary geometry as constructed in §9, and regard these elements as restricted in extent to the interior of a fixed sphere. Then, define the congruences of this geometry by aid of such linear transformations of the ordinary geometry as transform the fixed sphere into itself. By suitable conventions, we can make this non-euclidean geometry obey all of the axioms of our system except the axiom of Euclid (group III). Since the possibility of the ordinary geometry has already been established, that of the non-euclidean geometry is now an immediate consequence of the above considerations. ([Hilbert 1962], 27)

Again, in 1899 Hilbert doesn't explicitly interpret into another theory even for independence proofs. But, he is asserting that the parallel postulate is not deducible from the other axioms. Thus there is an implicit assumption that the (unformulated) rules of inference are sound (another anachronism).

The second interpretation in the *Grundlagen* is of at least equal importance for model theory: the interpretability of the theory of fields into geometry. We explore at length, in Chapter 9.4, the vast epistemological difference between the meanings Hilbert and Euclid give to essentially the same construction for 'multiplying' line segments. Euclid is constructing another segment – the fourth proportional[60]; Hilbert is defining an operation on congruence classes of segments that is associative and commutative. Hilbert's motivation for this use of interpretation is not to show independence or consistency but rather to enable the use of algebraic methods in proving geometric theorems via coordinatization. The basic idea reappears in Hrushovski's Theorem 5.6.4 that interprets classical groups into *any* structure that satisfies certain purely model theoretic

---

[58] See Notation 9.3.3.

[59] Hilbert was not concerned in 1899 with anything like the proof strength of finding this model. But it is provable in RCA_0 [Simpson 2009].

[60] Euclid's multiplication of two segments gives a rectangle.

conditions. In the original application Hilbert shows that any plane that interprets a field can be embedded in *n*-space. In Chapter 12.3 we expound this result and use it to discuss the purity of certain proofs.

4.6   What Is a Structure, Really?

Button and Walsh [Button & Walsh 2017] suggest that isomorphism between structures is too fine an equivalence relation for certain philosophical concerns. They propose to define two structures to be *informally isomorphic* if they are equivalent up to definitional expansion. That is, $A \approx_{inf} B$ if there are definitional expansions of A and B to a common vocabulary that are isomorphic. The initial vocabulary of A and B is irrelevant. In this section we explore some variants of this notion that have appeared in model theory. We begin with a basic notion.

Definition 4.6.1 (Explicit Definition) *Let $\mathcal{L}$ be a logic.*

(1) *If $\phi(\mathbf{x})$ is a $\mathcal{L}(\tau)$-formula, R_ϕ is a new relation symbol[61] not in τ, and T is a theory such that T satisfies*

$$(*) \qquad (\forall \mathbf{x})[\phi(\mathbf{x}) \leftrightarrow R_\phi(\mathbf{x})];$$

we say R_ϕ is $\mathcal{L}$-explicitly defined by ϕ in T. Letting τ' denote $\tau \cup \{R_\phi\}$, a τ'-structure $A' = (A, R_\phi)$ that satisfies $(*)$, is an expansion by definitions of A.

(2) *If no $\mathcal{L}$ occurs we mean $\mathcal{L}$ is first order logic.*

Button and Walsh point out that this notion is *sensitive to the logic in which definitions are allowed*. We stress this point with reference to the Spivak example of Chapter 1.2. The structure (N, s) for the vocabulary with one unary function is extremely well-behaved: its first order theory is categorical in every uncountable power. If we add the addition function, we lose categoricity but still have a superstable theory; in both cases the first order theory of the structure is decidable. But if we again expand by recursion to add multiplication the first order theory becomes unstable and undecidable. The first two structures are tame, the last is wild. As Hodges warns ([Hodges 1993], 61), 'In general, recursive definitions define symbols on a particular structure, not on all models of a theory. There is no guarantee

---

[61] Explicit definition of functions and constants is slightly more complicated; see
     ([Hodges 1993], 59).

that they can be turned into explicit definitions in a first order theory.' For this reason we restrict ourselves to *first order explicit* definition.

Two misleading examples illustrate why this warning is necessary. First, if the recursive definitions are to a (second order) categorical structure, then from a *second order standpoint*, Shapiro observes that the warning is unnecessary ([Shapiro 1991], 119), since the recursive definitions are explicit in second order logic. Secondly, expansions of $(N, +, *)$ by recursive definitions are in fact first order definable by a positive aspect of the Gödel phenomena. Nevertheless, beware of Pierce's paradox (page 38) and over-generalizing; these are very special cases.

Tarski[62] and others frequently sought to to find a minimal set of 'primitive concepts' to formalize a topic. And arguments for the naturality of various notions are part of the justification for a particular choice of primitives. Nevertheless, we will explore two types of definitional expansion that are crucial for model theory: (a) adding explicit definitions to obtain quantifier elimination and (b) adding 'imaginary elements' to form M^{eq}.

Explicit definition can reduce complexity in a way that can be measured formally – by reducing quantifier complexity as discussed in Chapter 4.4. In developing a first order theory to describe a given mathematical structure or class of structures, a model theorist would like to have the resulting theory T admit elimination of quantifiers. Further, as we stressed in Chapter 1.2 in discussing the process of formalization, T should be formulated with concepts which are basic and central to the topic under study, enabling Manders' goal of clarity. Sometimes, with judicious choice, a small family of well-understood predicates can be added that suffice for the quantifier elimination. Two examples are Presburger arithmetic (page 105) and Macintyre's [Macintyre 1976] celebrated proof of quantifier elimination for the p-adic numbers; he adds predicates $P_n(x)$ meaning x is an nth power.

Elimination of quantifiers can be shown in two ways. The first is an induction on quantifier rank with possible simplifying lemmas such as Shoenfield's lemma (Theorem 4.4.4). This technique is essential for the epistemic purposes discussed in Chapter 4.4. In contrast, quantifier elimination can be easily established by fiat. Morley's trick [Morley 1965a] makes a definitional extension of a first order theory by adding a predicate symbol for each definable relation: add explicit definitions (page 111) for each formula $\phi(\mathbf{x})$, a predicate $R_\phi(\mathbf{x})$, and an axiom $R_\phi(\mathbf{x}) \leftrightarrow \phi(\mathbf{x})$.

---

[62] See [Givant & Tarski 1999] for a summary of Tarski's work on this area for geometry.

But now the vocabulary contains many essentially incomprehensible relations.[63] At the time this appeared only a technical convenience. In retrospect, it allows one to focus attention on the kind of property whose significance we explore in Chapter 5.3; properties of first order formulas that do not depend on quantifier complexity. Thus, most studies in pure model theory adopt the convention that this expansion has taken place. The definition of infinitely many formulas by recursive introduction of quantifiers of indefinite length has been done in one step by fiat. There is no attempt to ground the meaning of each formula from the primitive concepts. This defeats the goal of quantifier elimination in studying specific theories: every definable relation is a Boolean combination of *well-understood* relations. Yet quantifier elimination by fiat is a powerful tool at the foundation of stability theory. It provides positive results for all theories that apply to 'algebraic examples'. But it loses the analysis of definable sets in terms of the primitive notions.

Thus there are two approaches to the choice of vocabulary in modern logic. For applications, the vocabulary is carefully chosen to reflect the primitive notions of the subject at hand – with perhaps judicious choice of additional symbols to reduce quantifier complexity. For general first order model theoretic study the vocabulary is chosen to have a predicate for each first order definable relation. Shelah reduced the study of complete sentences in $L_{\omega_1,\omega}$ to the study of atomic models of a complete first order theory by adding predicates for $L_{\infty,\omega}$-definable relations. See [Baldwin 2009, Chapter 6].

We turn now to systematic augmentation of a structure M to a structure M^{eq} by a family of explicit definitions. Before giving the formal definition of M^{eq}, we briefly digress to describe many-sorted logics. The notion of a many-sorted structure is familiar in philosophy from the proof of the completeness of second order logic using the Henkin (general) semantics: an ordinary structure is augmented by a second sort and a binary relation. Assert that each element of the second sort S_2 represents a subset of the first sort.[64] Thus, first order quantification over the second sort represents quantification over *a subset* of the subsets of the first. The *full second order*

---

[63] Of course, it may be possible to unpack individual formulas. But instead of working with the original language or with a list of clearly understood additional concepts, Morley's procedure replaces each formula of quantifier rank at least one by an arbitrary abbreviation.

[64] That is, add an *extensional relation* R between the two sorts. I.e.
$\forall y_1, y_2 S_2(y_1) \wedge S_2(y_2)[\forall x(S_1(x) \rightarrow (R(y_1,x) \leftrightarrow R(y_2,x))) \rightarrow y_1 = y_2]$. The comprehension axiom scheme $\exists X(\forall x)(Xx \leftrightarrow \phi(x))$ is consistent as we require that X not free in ϕ. In a *general*, i.e. *normal*, model quantifiers may range over any $\mathcal{G} \subseteq \mathcal{P}(M)$ containing the definable sets. That is $(M, \mathcal{G})$ satisfies comprehension.

semantics requires that *every* subset of the first sort is represented by a point in the second. Enderton [Enderton 2007] wrote, 'The main feature of the general semantics is a result of the "nothing but" type: second-order logic with the general semantics is nothing but first-order logic (many-sorted) together with the comprehension axioms. Thus a sentence is valid in the general semantics iff it is logically implied (in first-order logic) by the set of comprehension axioms.'

Definition 4.6.2 *A many-sorted structure contains a number of unary predicates or sorts. The logic is altered so that each relation or function symbol specifies for each argument or value the sort from which it must come. Similarly each variable is assigned a sort. There is a distinct existential (universal) quantifier for each sort.*

A many-sorted vocabulary or structure[65] behaves much like an ordinary one, with a significant exception. If there are infinitely many sorts, non-standard models would (by compactness) have elements that don't realize any of the given sorts. Those elements are just ignored. Technically we omit the type of an element realizing a nonstandard type. This is easily expressible in $L_{\omega_1,\omega}$ if there are countably many sorts and it is straightforward to interpret the many-sorted logic into a single sort. Many-sorted logic now plays an important role in the model theory of compact complex manifolds [Moosa 2005]. The first order theory of modules (Example 1.2.4) illustrates the difficulty of deciding whether a many-sorted approach is warranted.

Shelah introduced *imaginary elements* in [Shelah 1978] for an ostensibly technical reason,[66] but expansion of M to M^{eq} to represent them has become a routine step in model theory.[67] An important consequence is that for any M, M^{eq} is closed under definable quotients. (Suppose H is a normal subgroup of G and both are definable in a structure M. Then by the next definition, the structure G/H where $E(g_1, g_2)$ holds if they are elements of G that are in the same H-coset is definable in M^{eq}.)

Definition 4.6.3 *Let E be a definable (over the empty set) equivalence relation on M^n for some n. M^{eq} is the many-sorted structure obtained by adding a sort U_E for each such equivalence relation E and a map from M^n to U_E taking a to the equivalence class a/E. The elements a/E are dubbed 'imaginary'.*

---

[65] A current survey by Väänänen is found at http://tsinghualogic.net/events/2014/easllc/wp-content/uploads/2014/03/China_MSL_20142.pdf.

[66] Obviously, imaginary is an allusion to adding i to the real numbers. There are points missing from the structure so he adds them. But this is only metaphorical; Shelah's imaginaries are less intrusive; adding them does not affect the stability classification of the theory.

[67] The 'eq' is for 'equivalence relation'. Each equivalence class is an 'imaginary' element.

The passage to M^{eq} plays a central role in geometric stability theory. Poizat [Poizat 1983] emphasized that the ability of a theory to *eliminate imaginaries* made that theory more accessible. Crucially for applications, such important theories as algebraically and differentially closed fields do not need the expansion in Definition 4.6.3. So these important algebraic theories automatically have the desirable properties that M^{eq} was designed to guarantee.

We have seen unfettered explicit definition prevents clear understanding of the definable relations on specific structures. Pillay regains the spirit of the Button–Walsh notion of 'informal isomorphism' when he writes:

In any case, central objects of (first-order) model theory are structures and their first-order theories, as well as possibly incomplete theories. What is the notion of sameness for structures? As these structures (and theories) may be in different vocabularies, isomorphism does not make sense. The right notion of sameness is bi-interpretability. [Pillay 2000]

He explains that he takes interpretability as interpretability in the expansion in T^{eq}, given by Definition 4.6.3, because closure under definable quotient is desired.[68]

At first sight it seems extension by first order explicit definition is totally harmless. No new concept can be introduced;[69] it is just a kind of abbreviation. And if no new properties that are not provable in the base theory are introduced, this step should surely not infringe on the purity of an argument. It is a 'logical' not a 'mathematical' step. However, we will see in Chapter 12 that if arbitrary first order explicit definitions are allowed in a proof, the notion of such a proof being pure is almost meaningless.

We saw in Chapter 2.1 that both Lakatos and Manders were concerned with how defined concepts are introduced. Manders argues that the proper domain of epistemology includes the analysis of 'Grasping, understanding, and "seeing"'. In order to make this analysis, the process of, e.g. changing

---

[68] This requirement has a rigorous formulation. In studying categorical logic, Makkai and Reyes [Makkai & Reyes 1977] establish a precise correspondence between first order theories and their notion of a Boolean logical category (page 303). Further, such a category is a pretopos if and only if it is closed under quotient by equivalence relations. They show the existence of a *pretopos* completion of any T (or rather its associated category). In the context of model theory, this completion corresponds to T^{eq} and the Makkai–Reyes approach provides a natural universality condition characterizing the extension. See [Harnik 2011] for an extremely accessible account.

[69] We think of the concepts formalizable in a theory as exactly those expressed by the formulas. Making an explicit definition is focusing attention on an existing concept, not adding a new one. See the discussion in Chapter 12.1.

vocabularies must be explicit. He describes two objections to that analysis from the standpoint of 'commonplace realism':

> commonplace realism counts all structures set theoretically definable from a given one as existing; the first objection counts *the ideal cognitive agent as acting in all conceptual settings set theoretically definable from the one in which she acts.* Commonplace realism counts any relation definable from given primitives as part of any structure interpreting those primitives; the second objection *counts the ideal cognitive agent as in possession of all concepts expressible in a formal language whose primitives she possesses.* I will call this, in either version, attributing conceptual omnipotence to the cognitive agent. ([Manders 1987], 200–201)

We agree that an effective epistemological analysis requires the rejection of both italicized objections. Thus we must reject 'commonplace realism' and take into account, as Manders posits, a cognitive agent that can reconceptualize (another Manders term) a mathematical situation. Our study of various formalizations of the same intuitive situation provides a tool for identifying and comparing such differing conceptualizations. This stance underlies our strictures on recursive definition on page 112.

Pillay's position on bi-interpretabiity is consistent with this rejection. We may want to consider two structures as equivalent even if we don't fully understand the definable sets. Indeed, the success of artificially simplifying the class of definable sets underpins modern model theory since Morley.

4.7   When Are Structures 'Equal'?

The natural equivalence relation for a model theorist is isomorphism. Two τ-structures are isomorphic if there is a bijection between their domains preserving the function and relation symbols of τ. But the notion that 'isomorphism implies identity'[70] is anathema. At first glance this seems to directly conflict with the ideology of 'univalent foundations'. We explain why the distinction between identity and isomorphism is fundamental to model theory and then see that the conflict with category theoretic foundations is only apparent.

We noted in Theorem 3.2.4 the importance of considering isomorphic structures that are not identical in characterizing those $L_{\omega_1,\omega}$-sentences that are categorical. Another important example comes from a famous problem

---

[70] The actual axiom is better phrased as 'isomorphism is isomorphic to identity' (e.g. [Burgess & Tsementzis 2016, Awodey et al. 2013]). This is a slogan of the univalent foundations program [The Univalent Foundations Program 2015].

in abelian group theory (motivated by a problem in complex analysis) and raised by the topologist J. H. C. Whitehead[71] around 1950. Call A a Whitehead group if every short exact sequence of abelian groups

$$0 \to Z \to B \to A \to 0$$

splits. That is, if $f : B \to A$ is a surjective (i.e. onto) group homomorphism whose kernel is isomorphic to the group of integers Z then B is isomorphic to the direct sum of Z and A. It is easy to see that any free abelian group is a Whitehead group. The conjecture proposes that every Whitehead group is free.

Elias Stein proved the conjecture for countable A in 1951. Shelah proved in 1974 that the full conjecture is independent of ZFC. He showed that if every set is constructible ($V = L$), then every Whitehead group is free; if Martin's axiom and the negation of the continuum hypothesis both hold, then there is a non-free Whitehead group. Since these conditions are known to be jointly consistent with ZFC, the problem is undecidable in ZFC.

We want to give a very broad picture[72] emphasizing one key point: the result is about abelian groups; the proof *depends on* the set theoretic properties of the construction of the group. This stands in contrast to the standard mantra: a mathematician doesn't care whether the reals are Dedekind cuts or Cauchy sequences. Here we detect a failure of isomorphism in the limit model (even though the stages along the way are *isomorphic but not identical*) by studying its *exact set theoretic construction*.

More precisely, an uncountable group A is constructed as a union of countable groups A_α and properties of ω_1 are used to investigate when the isomorphism type of the limit depends on the ambient set theory. Construct a chain $\langle A_\alpha : \alpha < \omega_1 \rangle$ such that each A_α is a countable free Abelian group and for each limit ordinal δ, $A_\delta = \bigcup_{\alpha < \delta} A_\alpha$. Let $A = \bigcup_{\alpha < \omega_1} A_\alpha$.

For our purposes here, read the phrase 'a set $S \subseteq \omega_1$ is *stationary*' (Definition 8.3.1) as 'S is *not small*'. Now let $E = E^A$ be the set of limit ordinals δ such that some countable subgroup of A/A_δ is not free. The crucial ZFC lemma is: A is free if and only if E is not a stationary subset of ω_1. Shelah shows under $V = L$ that in a Whitehead group of cardinality $\aleph_1$, E must not be stationary; this gives the result.[73] In contrast, under $2^{\aleph_0} > \aleph_1$ and Martin's axiom he shows there is a Whitehead group which is not free.

---

[71] See http://en.wikipedia.org/wiki/Whitehead_problem.

[72] See [Eklof 1976] for a more detailed but accessible account.

[73] In particular, if A and B are constructed as above, E^A is stationary, and E^B is not, then each $A_\alpha \approx B_\alpha$ but A and B can fail to be isomorphic.

We have given two examples showing that the distinction between isomorphism and identity is critical in model theory and in mathematics.[74]
This concrete realization also grounds our counting of the number of models of T with cardinality κ. Each model of cardinality κ is isomorphic to one
whose universe is the cardinal κ. And then it is an easy calculation to show
that there are only $\kappa^{|\tau|}$ τ-structures with universe κ. So when we say T has
the maximal number of models in κ this means T has 2^κ models.

These examples do not conflict with the 'isomorphism is identity' slogan
in homotopy type theory (HoTT) as the homotopy type theorist would
analyze the Whitehead group example by considering not only the category
of abelian groups but also the 'carrier category' (corresponding to our ZFC
background) and the groups are not isomorphic in the carrier category
(page 36) and so not required to be identical. Here is an explanation[75]
for the confusion that often arises when model theorists discuss this issue
with experts in type theory. The inductive definition of truth in a structure
demands that the equality symbol be interpreted as identity:

$$M \models a = b \text{ iff } a^M = b^M.$$

That is, model theory uses an *extensional*[76] definition of equality while
constructive type theory (Martin-Löf or HoTT) makes an *intensional*
definition.

These examples illustrate one advantage of local formalization: the ability
to contrast the behavior of structures in different vocabularies. The second
stage of the paradigm shift has not yet emerged. Thus we move to the next
chapter.

---

[74] A still more naive example is to consider the Klein 4 group with elements $\{a, b, c, e\}$, the first
three having order 2 and the last the identity. Each of the first three generate isomorphic but
not identical subgroups. Tsementzis [Tsementzis 2016] explains how to formalize this situation
in HoTT while arguing that neither ZFC nor Lawvere's 'elementary theory of the category of
sets' can meet a precise criterion he proposes for a structural foundation while univalent
foundations can.

[75] William Howard pointed this out to me. See [Angere 2017] for an independent account.

[76] Two predicates are extensionally equal if they hold of the same objects. The notion of
intensional equality is more disputed in the literature. Two terms are defined or proved to be
equal; they will then hold of the same objects; but the converse fails. The evening star and the
morning star are extensionally equal (both being Venus) but not intensionally so.

5 | What Is Contemporary Model Theory About?

On page 12, we transcribed Maddy's Second Philosopher's questions about the role of set theory to questions about model theory – What sort of activity is *model* theory? How does *model* theoretic language function? What are *models* and how do we come to know about them? We see in this chapter that as the paradigm changes there is increasing emphasis on the study of theories and the role of types[1] as tools in assigning invariants to models and eventually these developments lead to an entanglement with classical mathematics.

As background for the key concepts of the stability hierarchy in Chapter 5.3, we start with another founding paper of modern model theory, Vaught's 'Denumerable models of complete theories' [Vaught 1961], delivered in Warsaw in 1959.

5.1 Analogy to Theorem to Method

MacLane ([MacLane 1986], 37) describes analogy as 'finding a common structure ... underlying different but similar phenomena.' In his example the concept of vector space is the common notion underlying geometry, linear equations, and linear differential equations. A main line of this book is the discovery of 'well-defined notion of dimension' as the common structure underlying a flock of mathematical areas. The result is not merely analogy as inspiration but analogy manifested as actual mathematical theorems. A fundamental tool for this work is itself a product of an analogy turned into a theorem and then into a model theoretic method. As Schlimm [Schlimm 1985] explains, successive analogies of propositional logic with algebra by Boole, of Boolean algebras with rings by Stone, and of deductive systems with Boolean algebra by Tarski led to a remarkable unification of topology and logic.[2] Even more remarkable is that Tarski's topological

---

[1] I don't mean, as in the theory of types, but as infinite descriptions of elements in Chapter 5.1.
[2] For further accounts see Chapter 4.2 of Corfield [Corfield 2003], Grosholz [Grosholz 1985], and Johnstone [Johnstone 1982]. As Corfield reminds us, Stone not only was one of the founders of lattice theory, his background was in functional analysis. The Stone representation theorem in

description of syntactic objects was transformed in the 1950s into a powerful method for studying semantics. First, we explore the Boole–Stone–Tarski axis; then we pass to the more modern incarnation in model theory.

Theorem 5.1.1 *There is a functorial 1–1 correspondence between Boolean algebras and totally disconnected Hausdorff spaces.*

The exact nature of this correspondence is crucial. To describe it we introduce the notion of the Stone space of a Boolean algebra and relate it to logic. Consider propositional logic with a set V of k propositional variables. These variables generate a *Boolean algebra*,[3] B_k, that is generated by the variables with constant symbols 0, 1 denoting T, F and operations $\wedge$ (conjunction), $\wedge$ (disjunction), and $\neg$ (negation).

The 2^k rows in a truth table for a propositional formula each represent a possible assignment of true or false to each variable. They represent the 2^k possible worlds for this vocabulary and logic. Each world (row), W, determines an ultrafilter[4] on the set of formulas. Namely W maps to the formulas true in W. This space of ultrafilters is called the *Stone space* of the Boolean algebra. Each ultrafilter is determined by a function from V to 2 so there are 2^k ultrafilters. But since B_k is finite each ultrafilter p is principal (p is generated by $\bigwedge p$). And each principal generator is an atom of B_k so B_k has 2^k atoms.

The proof of Theorem 5.1.1 yields Stone's representation theorem that every Boolean algebra B is isomorphic to a Boolean algebra of sets. Use the map that sends an element $b \in B$ to the set of all ultrafilters f on B with $b \in f$. Thus $|B_k| = 2^{2^k}$. The definition of the natural *Stone topology* on $S(B)$, which has basis $U_b = \{p \in S(B) : b \in p\}$ for $b \in B$, implies the image of each b is a clopen[5] set.

1936 comes after both von Neumann's work on the lattice of subspaces of operator algebras and Garrett Birkhoff's 1935 characterization of the subspaces of a finite-dimensional projective geometry as irreducible, finite-dimensional, complemented modular lattices.

[3] Simply put, a Boolean algebra is any structure in the specified vocabulary which satisfies the basic laws of propositional logic such as De Morgan's law. Boolean algebras and ultrafilters are defined in any graduate logic text. Excellent references are http://plato.stanford.edu/entries/boolalg-math/ and http://plato.stanford.edu/entries/modeltheory-fo/.

[4] Again, in this context, an ultrafilter is a maximal consistent set of formulas. Here we blur two concepts, the essence of analogy. From the standpoint of logic, the set is consistent. From the standpoint of algebra an ultrafilter is a filter, a set closed under $\wedge$ and $\rightarrow$ (implication is a derived operation) that is maximal but not containing 0.

[5] That is, U_b is both closed and open. A topological space is a totally disconnected Hausdorff one if every pair of points are separated by disjoint clopen sets. Stone's analogy was to a topology on the space of ideals of a ring.

Lindenbaum and Tarski realized the significance of this notion for first order theories. For a first order τ-theory T consider the Boolean algebra $F_0(T)$ of τ-sentences identified up to equivalence in T. The Boolean operations are given by conjunction, disjunction, and negation. The Stone space $S_0(T)$ is just the set of complete extensions of T. Now ϕ maps to the clopen set U_ϕ, the set of complete extensions of T that contain ϕ. Generalize further. Instead of considering the sentences of T, consider the Boolean algebra $F_n(T)$ of all formulas[6] with n-free variables (for $n < \omega$), again up to equivalence in T. The Stone space of that Boolean algebra is denoted by $S_n(T)$ and we call the elements of $S_n(T)$ n-types. Now the key step[7] is the following *semantic interpretation* of 'type'.

Example 5.1.2 Consider a structure, say $\mathbb{Q} = (Q, <, a_1, a_2, a_3, \ldots)$ where a_n names the rational integer n. For $b \in \mathbb{Q}$ define the *complete type* of b over the empty set to be $\{\phi(x) : \mathbb{Q} \models \phi(b)\}$.

We say *complete* because for each formula, either it or its negation is true of a. Indeed, the type is an element of the Stone space of the Boolean algebra of 1-ary formulas. We say b *realizes* $p \in S(A)$ if $\phi(b)$ is true for each formula $\phi(x)$ in p. To understand this example let's explore a bit further. What are the possible types[8] over the empty set ($\emptyset$) in Example 5.1.2? Since the a_n are in the vocabulary, types over $\emptyset$ may mention them.

Suppose I now ask what are all the complete types over Q? The types over the empty set described in the footnote are not complete any longer – because a realization might be in Q. Now there is one type for each element of Q and one for each cut in the ordering. There are $2^{\aleph_0}$ types over the countable structure Q. Analogously to the propositional logic case, each type over Q is an ultrafilter on the Lindenbaum algebra (the Boolean algebra of formulas with one[9] free variable x and parameters from Q).

We formalize these notions.

Definition 5.1.3 *Let T be a complete first order theory. For any set A contained in a model M of a fixed theory T*

---

[6] Lindenbaum–Tarski algebra.

[7] The omitting types theorem, usually attributed to Henkin and Orey [Henkin 1954, Orey 1956], is described in terms of ω-consistency in both papers. Neither mentions the word type in this sense. The semantic notion is in full force in [Ryll-Nardzewski 1959] and [Vaught 1961].

[8] Answer: The countably many types are determined by the following formulas: $x = a_i$, $a_i < x < a_{i-1}, x < a_1$, and one that needs infinitely many formulas: $x > a_i$ for all i. Note that the type, $x < a_1$ is realized by 0, but also by $-10,000$.

[9] For illustration, we used singletons. The definition applies *mutatis mutandis* to any natural number.

(1) *let* $\mathrm{Th}(M, A)$ *denote the collection of all true sentences in the structure obtained by expanding M by using a constant symbol to name each element of A;*

(2) *a maximal set of formulas $\{\phi(\mathbf{x}, \boldsymbol{a}) : \boldsymbol{a} \in A\}$ with $\mathbf{x}$ of length n that is consistent with $\mathrm{Th}(M, A)$ is called a* complete n-type *over A;*

(3) *the* Stone space *of A, $S_n(A)$ is the collection of complete n-types over A; $S(A)$ usually denotes $S_1(A)$, sometimes, $\cup_{n<\omega} S_n(A)$;*

(4) *$S_n(A)$ has a natural topology with subbasis $U_\phi = \{p : \phi \in p\}$.*

Somewhat awkwardly, the type of $\mathbf{b}$ over A apparently depends on the choice of a particular M containing both A and $\mathbf{b}$. One virtue of the universal domain discussed in Chapter 5.2 is to remove this dependence.

Ryll-Nardzewski's theorem (T is $\aleph_0$-categorical if and only if each $S_n(T)$ is finite: Theorem 3.3.1) employs this semantic use of types. In his classic 1961 paper [Vaught 1961], Vaught makes the decisive move to the study of complete first order theories. He proves the following theorem.

Theorem 5.1.4 (Vaught) *The following conditions are equivalent:*

(1) *Each $S_n(T)$ is countable;*

(2) *T has a countable, saturated model.*

The proof (Theorem 2.3.7 of [Chang & Keisler 1973]) of (1) implies (2) just requires the compactness theorem and a prescient bookkeeper to organize the construction. The converse is immediate. Condition (2) is distinctly semantical; a countable model is (weakly) *saturated* (Definition 5.2.2) if it realizes every type over (the ϕ) every finite set. Vaught then comments on the meaning of the word 'syntax':

One is tempted to say, by analogy with the discussion in the last paragraph of 3 [where he asserts that the condition: finitely many n-types for each n is syntactic], that [the] condition [numbered 5.1.4(1) here] is purely syntactical. Indeed, in [5.1.4(1)], no reference to any semantical concept, such as 'model', is made. However, a little thought convinces one that a notion of 'purely syntactical condition' wide enough to include [Theorem 5.1.4(1)] would be so broad as to be pointless. [Vaught 1961]

We vigorously dispute 'so broad as to be pointless.' The objection is perhaps a bit easier to take if rephrased as, 5.1.4(1) is a *condition on syntax*. Now, a single formula is clearly a syntactic object; so we consider the set of such, the Lindenbaum algebra (more generally $F_n(T)$), as a syntactic object and thus the assertion that each $S_n(T)$ is countable is a condition on a syntactic object. Without pettifoggery on the exact meaning of *syntactic*, the crucial point is that conditions on the cardinality and topology of Stone spaces play crucial roles in analyzing the class of models and the definable relations on models. Chapter 5.3 demonstrates that the conditions involved are very concrete, specifically that they are set theoretically absolute (page 46).

The crucial advances are Vaught [Vaught 1961], where types over finite sets come to the fore, and then Morley [Morley 1965a] shows the importance of types over arbitrary sets. While the $S_n(\emptyset)$ are all determined by T, the properties of the $S_n(A)$ for infinite A, depend also on the cardinality of A.

Mostowski [Mostowski 1937] used the natural topology[10] on $S_0(T)$ to deduce that it was always countable or had cardinality $2^{\aleph_0}$. So, as with my argument that the cardinality of a Stone space is a syntactic notion, there is an entanglement with descriptive set theory.[11] As descriptive set theory is the second of Grosholz's examples in 'Two episodes in the unification of logic and topology', it is fitting to support Thesis 2 by quoting her remark illustrating the origins of the systematic comparison of theories.

Thus, even a philosopher who is interested in the formal and universal function of logic within mathematics must also attend to logic as a subject matter in its own right, one mathematical field among others. Moreover, many of the developments within logic and in twentieth century mathematics as a result of the influence of logic can be explained best by regarding logic as a field with its own characteristic objects, problems and procedures. For growth in mathematical knowledge often occurs as the result of identifying, testing and elaborating partial structural analogies between mathematical fields, and this is also true for logic. ([Grosholz 1985], 148)

Stone built on an analogy between propositional logic and rings[12] to introduce the notions of homomorphism and ideal into the study of Boolean algebras and proved a theorem not only giving a uniform representation for Boolean algebras but placing a topology on the space of ideals to define the Stone space. Lindenbaum and Tarski extended this result from propositional to first order logic and identified the Stone space as a space of completions of a theory or more generally as a space of

---

[10] More precisely, he showed $S_0(T)$ is a *perfect set*, closed and contains all its limit points. Every uncountable *perfect set* of reals has power $2^{\aleph_0}$. Perfect set is an absolute notion while cardinality is not.

[11] Descriptive set theory is the study of (first and second order) definable subsets of the real line and, more generally, Polish spaces. The Wikipedia article is a good survey with excellent references, although it omits the major connections between descriptive set theory and large cardinals.

[12] Schlimm [Schlimm 1985] emphasizes that crucial to this advance is to realize ring addition corresponds to symmetric difference, not join.

n-types. Ryll-Nardzewski and Vaught turned this insight into a method by interpreting *n*-types semantically as description of points in models. Thus, Vaught characterized in Theorem 5.1.4 those theories that have countable saturated models; he deduced the fascinating fact that a complete theory cannot have exactly two countable models. We next explore the, so far mysterious, notion of a saturated model.

5.2   Universal Domains

Shelah[13] introduces the modern form of a universal domain or 'monster model' – every theory has a sufficiently large saturated model – into model theory in [Shelah 1978]. At first sight the notion appears a mere technical convenience. But in fact it enables a shift in view that is fundamental to the paradigm shift and echoes two similar shifts in view in algebraic geometry. Manders [Manders 1989] distinguishes such shifts as either *context changing* or *context internal*.

Weil's notion of universal domain in [Weil 1962] (first published in 1946) represents both shifts in view. For a fixed characteristic *p* he moves from an arbitrary field to an *algebraically closed* field (context changing). Then he requires the algebraically closed field to have *infinite dimension* (context internal) so it will contain any possible solution of a system of polynomial equations over (a finite extension of) the prime field.[14] In the 1950s Abraham Robinson provided, via model completeness, a unified account including ordered fields and differential fields [Robinson 1959a] of Weil's passage to an algebraically closed field. In Chapter 4.4 we discussed Manders' [Manders 1987] account of the epistemological advantages of reducing quantifier complexity and seeking model completions of theories. In [Manders 1989] he pursues his philosophical argument for the methodological advantages of this tool by focusing on theory change.

I wish here to outline just such a theoretical justification for a variety of domain extensions from the history of mathematics. I shall show how domain extensions unify concepts, in a technical sense which covers the widely cited advantages of simplification and clarity for the use of complex numbers, projective geometry, and other examples from the mathematical and physical literature. [Manders 1989]

---

[13] ([Shelah 1978], 7).

[14] A *prime field* of characteristic *p* (possibly 0) is the unique field which can be embedded in every field of that characteristic. Following Vaught, a *prime model* of a complete theory *T* is one that can be elementarily embedded in every model of *T*.

We discussed Manders' notion of an accessibility property on page 60. 'Accessibility' is a measure of clarity, the ability to understand the definable sets. Manders approaches the problem of finding a domain where all conceivable solutions to an interesting family of syntactically given problems appear. He is thinking about single formulas; we extend the problem to simultaneously solving infinite sets of formulas (types). He makes a crucial observation about the epistemological significance of 'domain extension'.

Model-completion-like domain extensions play a significant role in mathematical intellectual advance. We have some theoretical understanding of how and why. What are the epistemological consequences of this theory?

Domain extensions are moves from one theory context to another; not context-internal inferences. To accept negative quantities, one had to give up traditionally fundamental properties of quantity, such as the monotonicity of proportionalities. In moving from real to complex algebra, one must give up 'truths', statements concerning the ordering, solvability criteria for equations, such as the discriminant formula for the quadratic equation. One must embrace different 'truths', such as the solvability of every equation in one unknown. These domain extensions are truth-destroying, not deductive, truth-preserving inferences.

Nonetheless, domain extension does not start a completely new intellectual enterprise, abandoning one subject in favor of a new one, say, moving from checkers to chess, even moving from Euclidean to hyperbolic geometry. Rather, a domain-extension move brings an existing intellectual enterprise forward, realizing general formal conditions, which allow a more systematic understanding of the previous theoretical setting. [Manders 1989]

In our context the type of domain extension[15] discussed by Manders is a transformation of a portion of algebra from one formalization (e.g. fields) to another (algebraically closed fields). The syntactic relation between these two formalizations is given by a general commandment: consider as solved all equations that are solvable (in some extension). Thus, Manders fleshes out the original motivations of Robinson in introducing model completeness.

Finding the model completion is a *context-changing* operation; the ambient theory changes. In contrast, finding a saturated extension is 'context-internal'; the model changes but not the theory. Having fixed a theory which admits quantifier elimination, perhaps by fiat (page 113), can we extend the domain of an arbitrary model M to one that solves problems as fundamental but more complicated than those specified by single formulas? For example, suppose in algebraic geometry we want not a solution to a

---

[15] We discuss the vast impact of such extensions on number theory in Example 6.2.1.

polynomial equation but a 'generic' solution. Then, as noted on page 102, we must realize a specific type over the parameters of the equation. We now discuss how to find a model of the same theory (hence context-internal) which solves all such problems.

In a line dating from Fraïssé [Fraïssé 1954] and Jónsson [Jónsson 1956, Jónsson 1960] universal domains are constructed from a purely semantic (algebraic) standpoint. A class K of *countably many* finite models closed under the natural relations of amalgamation,[16] joint embedding,[17] and substructure yield a countable structure M which is both ultrahomogeneous[18] (any two isomorphic finite structures are automorphic[19] in M) and universal (any finite member of K can be embedded in M). Crucially, Jónsson advances beyond the countable and considers κ-universal structures for arbitrary κ. We formulate his notion of *universal-homogeneous* in the modern form for abstract elementary classes[20] as discussed in Chapter 14.

Definition 5.2.1 *M is μ-model homogeneous for K if for every $N \prec_K M$ and every $N' \in K$ with $|N'| < \mu$ and $N \prec_K N'$ there is a K-embedding of N' into M over N.*

Extending the syntactic side from realizing formulas to realizing types, the semantic and syntactic lines meet in the fundamental paper of Morley and Vaught [Morley & Vaught 1962].

Definition 5.2.2 *A structure M is κ-saturated for $K = \text{Th}(M)$ if for every $A \subset M$ and $p \in S(A)$ with $|A| < \kappa$, p is realized in M. M is* saturated *if it is $|M|$-saturated.*

Now the crucial equivalence is:

Theorem 5.2.3 *A structure M is saturated if and only if it is homogeneous-universal for the class of models of $\text{Th}(M)$ with cardinality less than $|M|$ under the relation of elementary substructure. Moreover, any two elementarily equivalent saturated models M and N of the same cardinality are isomorphic.*

On the one hand, the notion of homogeneous-universal is an algebraic property of a class of models; it talks about the embedding of one model in

---

[16] For any embeddings of a structure $A \in K$ into B and C there are embeddings of B and C into a D such that the composition of the maps agree on A.

[17] Any two structures in K have a common extensions in K.

[18] The standard definition of countably homogeneous is in Vaught's terms: two finite sequences that realize the same type are automorphic.

[19] That is, there is an automorphism of M mapping one onto the other.

[20] We wrote $\prec_K$ to indicate flexibility in the interpretation of the substructure relation; here we will interpret it either as ordinary substructure or elementary submodel.

another. On the other, saturation is a property of a single structure M (and its complete theory); it describes the relation of M to types over subsets of M. The basic idea for the proof goes back to Hausdorff's [Plotkin 1990] proof that any two countable dense linear orders are isomorphic. We sketch the argument for the countable case of Theorem 5.2.3.

A *back-and-forth* system, $\mathcal{S}$, is a set of sets S_n where each $S_n \subseteq M^n \times N^n$ such that each $(\mathbf{a}, \mathbf{b}) \in S_n$ and each $c \in M$, there is a $d \in N$ such that $(\mathbf{a}c, \mathbf{b}d) \in S_{n+1}$ (forth) and conversely each d gives a c (back).

Define a 'back-and-forth' between M and N as follows: if $\mathbf{a} \in M$ and $\mathbf{b} \in N$, $\langle \mathbf{a}, \mathbf{b} \rangle$ is in the back-and-forth system $\mathcal{S}$ if they (each computed in their own model) have the same type over the empty set (Definition 5.1.3). Now there are two ideas. First (Hausdorff), the existence of $\mathcal{S}$ implies M and N are isomorphic. Second (Vaught), if M and N are countably saturated, $\mathcal{S}$ can be constructed. The proof of Theorem 5.2.3 is the nontrivial generalization to uncountable cardinalities.

Since amalgamation (for elementary submodel) is true for any complete theory, the existence of saturated models in κ depends on the number of types over sets of size less than κ. The existence was technically complicated in [Morley & Vaught 1962] except under the GCH. But GCH is not needed in the stable case (Theorem 8.1.7): one gets saturated models in all regular cardinals (and 'special models' are a good approximation in limit cardinals).

Here are several ways (Section 4.3 of [Marker 2002]) that saturation contributes to finding accessibility properties that clarify definitions.

Theorem 5.2.4 *Let M be saturated and $A \subset M$ with $|A| < |M|$.*

(1) *If a subset $X \subseteq M^n$ is definable with parameters from M and is fixed set-wise by automorphisms fixing A then it is definable with parameters from M.*

(2) *If $b \in M$, b is algebraic over A if and only if $\mathrm{tp}(b/A)$ is realized only finitely many times in M.*

These results might appear technical but they yield a crucial link between semantic and syntactic properties. These techniques provide a criterion for quantifier elimination (Theorem 4.4.4), which is the crucial lemma to show elimination of quantifiers for differentially closed fields (page 107).

There were many uses of saturated models in the late 1960s on an ad hoc basis [Ax & Kochen 1965] and as tools for simplifying the preservation results of the previous generation.[21] I began this section by saying that 'the monster model' represents a change in view echoing two similar changes

---

[21] The first real textbook in model theory, [Chang & Keisler 1973], is fully based in the first stage of the paradigm shift. The chapter on saturation has two sections on preservation and

in view in algebraic geometry. The two changes in algebraic geometry are the passage to algebraically closed fields of various characteristics (to gain quantifier elimination) and Weil's notion of universal domain. In the general case, Morley's 'quantifier elimination by fiat' accomplishes the first; what is the model theoretic analog of the second? Most papers in general first order model theory start with a statement[22] such as, 'We work in a sufficiently large monster model $\mathbb{M}$.' By 'sufficiently large monster', they mean, given a particular problem, $\mathbb{M}$ is chosen to be κ-saturated for a κ bigger than all the parameters of the proffered problem. The choice of κ is studied in Theorem 8.1.7; it is greatly simplified by classification theory. There is an explicit 'context-internal' component of the universality here; the model will be closed under all the normal model theoretic operations on submodels of smaller cardinality by Theorem 5.2.3. In specific cases there is a hidden context-changing aspect that reflects Manders' 'moves from one theory context to another'; this is accomplished by choosing an appropriate, usually model complete, theory in which to work so the definable sets become accessible. That is, the model theoretic development now imposes new guiding principles for the process of formalization of a particular topic as in Chapter 2.1; choose the primitives so the quantifier complexity of the definable sets will be low.

A weaker but very useful notion arose later.[23] M is κ-*homogeneous* (homogeneous if $|M| = \kappa$) if $f: A \mapsto B$ is elementary with $|A| = |B| < \kappa$ and $p \in S_1(A)$ is realized in M then so is $f(p)$. Now, every model has a homogeneous elementary extension, and if $M \equiv N$ realize the same types over $\emptyset$ and both are homogeneous they are isomorphic.

In this section we argued for Thesis 1 both (in mathematics) by showing formalization attains mathematical results and (in philosophy) by showing the epistemological goal of clarity is obtained by context-changing and context-internal domain extension.

5.3   The Stability Hierarchy

In this section we discuss some key notions which provide sufficient conditions to separate, as Davis (see page 148) put it, 'the wild infinite from the tame mathematical world.' Crucially, we work in a first order

---

interpolation theorems, one on fields (Ax–Kochen) and one classifying the complete theories of Boolean algebras. There are 24 pages on Morley–Shelah and 45 on relations with set theory.

[22] This originated as a 'mere' convention in [Shelah 1978].

[23] The '1-element at a time' idea is attributed to Keisler in [Morley & Vaught 1962].

theory with a fixed vocabulary. Then, we impose syntactic conditions on that theory. We consider *stability* here and later *o-minimality* (Definition 6.3.1) and NIP (page 130) as approximations to tameness.

As described in Chapter 3.3, these general conditions arose from the study of *categoricity in power*. Łos conjectured that a countable first order theory that is categorical in one uncountable power is categorical in all powers. In this section, we investigate a significant impact of formalization that arose in Morley's proof of the Łos conjecture: the ability to define ranks on formulas. Morley began the careful analysis of the relations between such a rank on n-ary formulas and the cardinality of the $S_n(A)$ for varying[24] A. He defined a certain rank on types in $S(A)$ which had to be evaluated by looking at models extending A. By using the universal domain one can avoid this extension step. We define the Morley rank of formulas directly rather than passing through types as in Morley's development.

Definition 5.3.1 (Morley rank) *All formulas in this definition have parameters somewhere in the universal domain.*

(1) $R_M(\psi(\mathbf{x})) \geq 0$ *if* $(\exists \mathbf{x})\psi(\mathbf{x})$.
(2) *For* $\alpha = \beta + 1$, $R_M(\psi(\mathbf{x})) \geq \alpha$ *if and only if there exist an infinite family of formulas* $\psi_i(\mathbf{x})$ *which are pairwise inconsistent and* $\forall \mathbf{x}[\psi_i(\mathbf{x}) \to \psi(\mathbf{x})]$.
(3) $R_M(\psi(\mathbf{x})) \geq \delta$ *for a limit ordinal* δ *if and only for each* $\alpha < \delta$, $R_M(\psi(\mathbf{x})) \geq \alpha$.

$R_M(\psi(\mathbf{x})) = \beta$ *for the least* β *such that* $R_M(\psi(\mathbf{x})) \not\geq \beta + 1$. *If there is no such* β, ψ *is unranked.*

Morley rank is defined here in terms of combinatorial properties of formulas. This description was reverse engineered from Morley's topological formulation, a modification of the Cantor–Bendixson rank on the Stone space. Morley [Morley 1965a] called a theory T *totally transcendental* if every formula has an ordinal rank; he proved that if every formula is ranked then for every A, $|S(A)| = |A|$. We'll see in Theorem 5.3.7 that for countable theories an equivalent but more descriptive term than totally transcendental is ω-stable.[25] He also showed every $\aleph_1$-categorical theory is totally transcendental. A surprising agreement with algebraic geometry was discovered much later [Berline 1982]. The triple identity in algebraically closed fields

---

[24] For infinite A, either $|S_n(A)| = |A|$ for all n or (under GCH) it is $2^{|A|}$ for all sufficiently large n. For the role of GCH, see page 197.
[25] We say *stable* as the cardinality of the Stone space $S_n(A)$ is the same as that of the model A.

of Morley's topological rank on the Stone space with the Weil rank on the space of formulas with the Zariski topology and the algebraic rank of Krull was one force motivating the growth of geometric stability theory. These model theoretic ranks continue to interact with ranks defined in other areas of mathematics (e.g. differential algebra [Freitag 2015]).

One might wonder why I chose Shelah's rather than Morley's work as the marker of the paradigm shift. That placement does not deny the incredible originality and impact of Morley. His categoricity proof demonstrated the critical role of the topology of Stone spaces, introduced the stability spectrum, exploited indiscernibles for the proof that $\aleph_1$-categorical theories are ω-stable, used rank to construct indiscernible sequences, and made the first sophisticated use of saturation with the assistance of the new notion of a prime model over a possibly infinite set.

The distinctive feature is the shift from a magnificent study of one class ($\aleph_1$-categorical) of theories to the discovery of a hierarchy of theories. Shelah generalized Morley's notion by placing each complete first order theory into one of four virtuous classes. Crucially the stability hierarchy is defined by syntactic conditions.[26] For example,

Definition 5.3.2 *A formula $\phi(\mathbf{x}, \mathbf{y})$ has the* order property[27] *in a model M if there are $\mathbf{a}_i, \mathbf{b}_i \in M$ for $i, j < \omega$ such that*

$$M \models \phi(\mathbf{a}_i, \mathbf{b}_j) \text{ iff } i < j.$$

A complete first order T is unstable *if and only if there is a formula $\phi(\mathbf{x}, \mathbf{y})$ that has the* order property *in some model M of T.*

But existentially quantifying out the $\mathbf{a}_i, \mathbf{b}_i$, ϕ is unstable in T just if for every n the sentence $\exists x_1, \dots x_n \exists y_1, \dots y_n \bigwedge_{i<j} \phi(x_i, y_j) \wedge \bigwedge_{j \geq i} \neg\phi(x_i, y_j)$ is in T. This last is clearly a syntactic condition. Equivalently, Shelah defines a variant of Morley's rank, called *formula rank*, defining $R_\phi(\psi)$ by requiring

---

[26] See Chapter 5.1.

[27] There are two refinements of this notion. $\phi(\mathbf{x}, \mathbf{y})$ has the *strict order property* in a model M if there are $\mathbf{b}_i \in M$, for $i < \omega$, such that

$$M \models (\forall\mathbf{x})\phi(\mathbf{x}, \mathbf{b}_i) \rightarrow \phi(\mathbf{x}, \mathbf{b}_j) \text{ iff } i < j.$$

$\phi(\mathbf{x}, \mathbf{y})$ has the *independence* property in a model M if for every $n < \omega$ there are $\mathbf{b}_i \in M$, for $i < n$ and $\mathbf{a}_s$ for $s \in 2^n$, such that

$$M \models \phi(\mathbf{a}_s, \mathbf{b}_i) \text{ iff } i \in s.$$

If T is not stable there is either a formula (precisely, a $\phi(x, \mathbf{y})$) with the strict order property or one with the independence property. If no formula has the independence property the theory is said to be NIP (not the independence property) or dependent. (There is a finite version of the strict order property analogous to the one for the independence property.)

the ψ_i in condition (2) of Definition 5.3.1 to all be substitution instances of a fixed formula ϕ. This apparently minor variant is momentous, because it allows the application of the compactness theorem to, for example, show this rank, if bounded, must be finite, while Morley's rank is bounded by ω_1.

A hidden existential quantifier in this definition disguises some of the significance. A theory T is unstable if there is a formula with the order property. This formula may change from theory to theory. In a dense linear order one such is $x < y$; in a real closed field one is $(\exists z)(x + z^2 = y)$, in the theory of $(\mathbb{Z}, +, 0, \times)$ one is $(\exists z_1, z_2, z_3, z_4)(x + (z_1^2 + z_2^2 + z_3^2 + z_4^2) = y)$. In the theory[28] of $(\mathbb{C}, +, \times, \exp)$, one first notices that $\exp(u) = 0$ defines a substructure which is isomorphic to $(\mathbb{Z}, +, 0, \times)$ and uses the formula from arithmetic. It is this flexibility, grounded in the formal language, which underlies the wide applicability of stability theory. In infinite Boolean algebras an unstable formula is $x \neq y$ & $(x \wedge y) = x$; here the domain of the linear order is *not* definable.

There are a number of variants on the notion of rank which serve different purposes. In particular, by choosing correct notions, each level in the stability hierarchy can be defined by saying every type is given an ordinal value by an appropriately chosen rank. Morley established a crucial link between these rank notions and the number of types over infinite sets.

Definition 5.3.3

(1) *The* stability spectrum function *of T is (if omitted, $m = 1$)*

$$g_T^m(\lambda) = \sup\{|S_m(M)| : M \models T, |M| \leq \lambda\}.$$

(2) *The complete theory T is λ-stable (stable in λ) if for every $M \models T$, and every $A \subset M$, $|A| \leq \lambda$ implies $S_1(M, A) \leq \lambda$.*

Theorem 5.3.4 *[Morley 1965a] If T is totally transcendental (equivalently ω-stable) then T is stable in every infinite cardinality.*

The key step is to notice that if M is the Ehrenfeucht–Mostowski model (page 184) on a sequence of indiscernibles with order type a cardinal λ, then for each $A \subseteq M$ with $|A| \leq \lambda$, $S_1(M, A) \leq \lambda$. Shelah generalizes this theorem by relativizing Morley's idea to ϕ-types[29] and shows there are only a few possibilities for the spectrum function.

---

[28] Here exp denotes complex exponentiation, e^z, where z is a complex number.
[29] $S_\phi(A)$ is the collection of incomplete types that contain only instances of ϕ but contain either $\phi(\mathbf{x}, \mathbf{a})$ or $\neg\phi(\mathbf{x}, \mathbf{a})$ for each $\mathbf{a} \in A$ and so are 'complete for ϕ'.

Theorem 5.3.5 (The Stability Hierarchy: Semantic Version) *Every countable complete first order theory lies in exactly one of the following classes.*

(1) *(unstable) T is stable in no λ.*
(2) *(strictly stable) T is stable in exactly those λ such that $\lambda^\omega = \lambda$*
(3) *(strictly superstable) T is stable in exactly those $\lambda \geq 2^{\aleph_0}$.*
(4) *(ω-stable) T is stable in all infinite λ.*

The stability hierarchy can also be described in terms of rank.

Theorem 5.3.6 (The Stability Hierarchy: Syntactic Version) *Every countable complete first order theory lies in one of the following classes.*

(1) *(unstable) T has the order property; some formula $\phi(\mathbf{x}, \mathbf{y})$ defines a linear order on M^n.*
(2) *(stable) For every formula ϕ, there is a rank R_ϕ so that for every formula ψ, $R_\phi(\psi) < \omega$.*
(3) *(superstable) There is a global rank R_C (for n-inconsistency) such that $R_C(\psi) < \infty$ for all ψ.*
(4) *(ω-stable) There is a global rank R_M (with respect to inconsistency) such that $R_M(\psi) < \infty$ for all ψ (total transcendence).*

Now the crucial point for applications is:

Theorem 5.3.7 *On countable theories the (strict[30]) hierarchies in the last two theorems are the same.*

Crucially from a methodological standpoint, the theorems are proved simultaneously by trading off between the syntactic and semantic viewpoints.

The stability hierarchy is essentially orthogonal to decidability. There are continuum many strongly minimal theories[31] and so most are not decidable. Two prototypic examples of unstable theories are both decidable as they are recursively axiomatized and $\aleph_0$-categorical: the random graph has the independence property and the theory of atomless Boolean algebras has both the strict order property and the independence property.

---

[30] The second hierarchy is cumulative; make it strict by subtracting off the previous class.
[31] This is easy; let T_X assert f is a bijection and $(\exists x)f^n(x) = 0$ if and only if $n \in X$. The choice of X changes only the algebraic closure of the empty set; the rest of any model is a collection of copies of Z with successor.

5.4 Combinatorial Geometry

We now develop Thesis 4 from the Introduction. Three different sorts of geometry impact model theory in three rather distinct but connected ways. We discussed Hilbert's two uses of interpretation in the study of geometry in Chapter 4.5. A second impact is in real and complex algebraic geometry. In that case, the central point is that these subjects can be seen as investigating definable sets and, in that sense, as a special case of model theory.

The third and most central impact comes from the notion of a *combinatorial geometry* (equivalently *matroid* or *abstract independence relation*) that yields a notion of dimension generalizing that concept in such examples as Euclidean geometry, vector spaces, and transcendence degree in fields;[32] it provides a way to assign a dimension to a mathematical structure. An important contribution of model theory is to find unifying principles explaining why such geometries arise in many different contexts. *Geometric stability theory* explores the connections between different geometries in a model and is crucial to the development of model theoretic classification theory and to applications. This perspective installs geometry as a 'mother-structure' that Bourbaki missed in their attempt to organize mathematics via a hierarchy of structures (Chapter 2.4).

As pointed out in Section III.17 of [Gowers 2008], the man-in-the-street notion of dimension says that an object 'living' in a plane is two-dimensional and one in 3-space is three-dimensional. But a sphere (i.e. the surface of a ball) presents a difficulty to this intuition; clearly the ball is in 3-space but walking on the surface of the earth makes it seem two-dimensional. A mathematical way to clarify this issue is to say the surface of the ball is two-dimensional because it takes only two 'coordinates' (longitude and latitude) to specify a point while the ball is three-dimensional.

Clearly this notion is useful for organizing certain types of phenomena. Less noticed is that certain subjects, notably number theory and set theory, can *not* support a notion of dimension. The existence of a (definable) pairing function means one cannot distinguish 2-space and 3-space in arithmetic. What then are the fundamental properties of dimension and what

---

[32] Success has many fathers. Logicians tend to point to van der Waerden [Van der Waerden 1949] in 1930, who was working from lectures by E. Artin and E. Noether. But http://jointmathematicsmeetings.org/meetings/national/jmm/MAA-ShortCourse.pdf points to other basic papers on the properties of independence in the 1930s with such authors as Garrett Birkhoff and Whitney. MacLane [MacLane 1936] makes the connection between matroids and fields. Hugo Hadwiger is sometimes credited with the name 'combinatorial geometry'.

does 'having a dimension theory' buy us? For this, consider the basic mathematical notion of a combinatorial geometry, described as a *closure system*.

The most basic example of a closed set[33] as in the following Definition 5.4.1 is the notion of a subspace in a vector space. Note that each vector a on a line is given by a single coordinate, the field element α that multiplies the unit vector $\mathbf{u}$ to give the point (vector) $a = \alpha\mathbf{u}$.

Thinking less geometrically we can consider a group and let the closure of a finite set of elements X be the smallest subgroup containing X. In a typical group, this notion of closure satisfies the first two axioms below but not the third (exchange) (a may be not in the subgroup generated by a^3).

Definition 5.4.1 *A pre-geometry is a set G together with a dependence relation*

$$cl : \mathcal{P}(G) \rightarrow \mathcal{P}(G)$$

satisfying the following axioms.

   A1. $cl(X) = \bigcup\{cl(X') : X' \subseteq_{fin} X\}$;
   A2. $X \subseteq cl(X)$;
   A3. If $a \in cl(Xb)$ and $a \notin cl(X)$, then $b \in cl(Xa)$;
   A4. $cl(cl(X)) = cl(X)$.
   We call the structure a *geometry* if:
   A5. Points are closed, $cl(a) = \{a\}$.
   If X is infinite $a \in cl(X)$ if and only if there is a finite subset X_0 of X with $a \in cl(X_0)$.

Pre-geometries will occur naturally[34] and to obtain a geometry we take the quotient of G by $a \sim b$ if and only if $a \in cl(b)$. In the quotient geometry the points are lines of the pre-geometry and the lines the planes. A clear basic axiomatic exposition is the initial chapter of [Beutelspacher & Rosenbaum 1998].

We say a subset X of G is *independent* if for each element $x \in X$, $x \notin cl(X - \{x\})$. The axioms for a combinatorial geometry are designed so the following theorem is proved exactly as in the special case for vector spaces in any linear algebra text.

Theorem 5.4.2 *Each geometry has a unique dimension, the cardinality of a maximal independent set.*

---

[33] The use of the scalar multiplication makes the line 1-dimensional in accord with intuition. In Euclidean geometry we say a line is generated by two points, a plane by three. This anomaly in geometry can be resolved by naming a point.

[34] E.g. algebraic closure in field or a vector space as described in footnote 33, where the closure of a point is the line, the closure structure is a pre-geometry, rather than a geometry.

There are two complementary notions that can define a combinatorial geometry. We formulated the axioms in terms of dependence (closure). Equally well, the axioms can be formulated as properties of independent sets.[35] We will see below important trade-offs between the two formulations.

Note that Definition 5.4.1 is a mathematical (formalism-free) concept (Chapter 14). The next definition and theorem provide formal (syntactic) conditions on a theory for its models to be combinatorial geometries under an appropriate notion of closure. Thus *formalization* strides onto the stage.

Definition 5.4.3 *Let* $B \subset M \models T$ *where* T *is a complete theory.* $a \in$ acl(B) *(algebraic closure) if for some* $\phi(a, \mathbf{y})$ *and some* $\mathbf{b} \in B$, $\phi(a, \mathbf{b})$ *holds and* $\phi(x, \mathbf{b})$ *has only finitely many solutions in M. An acl-basis for a set X is a maximal independent subset of X for this notion of closure.*

Theorem 5.4.4 *A complete theory T is strongly minimal if and only if it has infinite models, algebraic closure induces a combinatorial pre-geometry on models of T, and any bijection between acl-bases for models of T extends to an isomorphism of the models.*

The second condition is often rendered as 'the pre-geometry is homogeneous'; it is equivalent to say all independent sets of the same cardinality κ realize the same type (in κ-variables). By Theorem 5.4.4, the syntactic condition about the number of solutions of formulas leads to the existence of a geometry and a dimension for each model of the theory. We want now to consider the notion of an 'invariant' in this context. The key idea is to generalize the notion of geometry on a strongly minimal set in two important ways.[36]

The first innovation is to drop the requirement (A4 of Definition 5.4.1) that $cl(cl(A)) = cl(A)$. This is done by, apparently innocently, providing axioms on an independence relation rather than a dependence relation. With a closure relation, to say A is independent from B over C is to say the basis[37] for A over C remains a basis over B. In the absence of an obvious

---

[35] Precisely, pre-geometries correspond to matroids. [White & Nicoletti 1985] gives nine equivalent formulations. Some are in terms of independence; others dependence. The initial chapter of [Beutelspacher & Rosenbaum 1998] has a basic axiomatic exposition.

[36] We choose to describe these notions axiomatically; Shelah's original definition was syntactic/combinatorial (Definition 8.5.5). Lascar and Poizat gave an equivalent and apparently more friendly definition in [Lascar & Poizat 1979] but the equivalence held only for stable theories. Twenty years later when the field moved on to simple theories, the original definition was seen to be 'the right one'; it generalized.

[37] A set X is a *basis* for A if X is independent and $cl(X) = A$. That is, X generates A; but in our general setting we don't always have a nice notion of 'generating'.

notion of closure we must move from a binary relation 'A is independent from B' (A is not in the closure of B) to a ternary relation 'A is independent from B over C' (where for simplicity we take $C \subseteq B$), meaning A is no more 'constrained' by B than by C. The key point is

Theorem 5.4.5 *Every stable theory admits such an independence relation, non-forking.*[38] *It lacks only transitivity of closure to be a geometry.*

Shelah remedies this weakness by transforming the technical definition[39] of regular type to the equivalent (Theorem V.1.14 of [Shelah 1978]):

Theorem 5.4.6 *Let $p \in S(A)$ be a regular type and X the set of realizations of p. Then for $a \in X$ and $B \subset X$, the relation $\mathrm{tp}(a/B)$ forks over A defines a combinatorial geometry on X.*

Shelah found a family of local geometries in a stable theory and tools for relating them to enable his proof of the main gap, Theorem 5.5.1. Zilber classifies the geometries in Example 3.3.3. The first is discrete (or trivial) because $cl(ab) = cl(a) \cup cl(b)$. The second is modular because the lattice of closed subsets of the geometry is a modular lattice.[40] And the third is field-like (roughly, bi-interpretable with a field). The *Zilber trichotomy conjecture* (page 83) asserts that every strongly minimal set[41] is of one of these three types. Even though the trichotomy conjecture fails in general, the categories are independently important. For example, Zilber proved that every strongly minimal set in an $\aleph_0$-categorical theory is locally modular and this is crucial for his non-finite axiomatizability proof. Buechler [Buechler 1991] proved *Buechler's dichotomy*: in a superstable theory every *minimal stationary type*[42] is either locally modular or strongly minimal. We explore the 'philosophical problem' (finite axiomatizability) that led to this conjecture in Chapter 13.2, the surprising mathematical consequences of its failure in general in Chapters 3.3 and 14, and of its successes in Chapter 6.3.

---

[38] See page 286. The rather technical definition appears in any stability theory text. Chapter II of [Baldwin 1988a] describes the abstract properties.

[39] A stationary type p is regular if it is orthogonal to every forking extension.

[40] The conjecture is a bit more general; the second class is *locally* modular (a parameter may be needed to obtain modularity). With respect to intersection and 'subspace generated', the closed subsets behave like the subgroups of a group: $x \leq b$ implies $x \vee (a \wedge b) = (x \vee a) \wedge b$.

[41] The conjecture makes sense for regular types in stable theories and in o-minimal theory for formulas on which algebraic closure is a geometry. The trichotomy holds in o-minimal theories [Peterzil & Starchenko 1998] and in various area of (differential) field theory [Chatzidakis 2015] and this has enormous consequences for number theory and differential equations.

[42] A type is *minimal* if it has a unique non-algebraic completion.

Model theoretic problems led to the development of algebraic and geo-metric methods. Kueker conjectured that if every model of a first order theory is ω-saturated then the theory T is either $\aleph_0$ or $\aleph_1$-categorical. The proof was divided into cases according to the stability class of T. Buechler proved the superstable case; for the stable case, Hrushovski relied on his own discovery that purely model theoretic hypotheses on a structure could force groups to be definably embedded in that structure. Further model theoretic conditions give more control over those groups (Example 6.2.1). A *stable* theory T is *1-based*[43] if for any A, B in the monster model, A is independent from B over acl(A) $\cap$ acl(B). Hrushovski and Pillay proved such groups were 'close' to abelian:

Theorem 5.4.7 *Suppose T is 1-based and G is a definable group. Then G is interpretable in a module and is weakly normal (Definition 6.2.5).*

An important model theoretic innovation is to systematically investigate the relations among geometries on pieces of the same structure. Then the structure is determined up to isomorphism by a family of dimensions rather than just one. And, as we'll see in Chapter 5.5, the family is not just a set of dimensions but rather a tree of dimensions.

5.5 Classification: The Main Gap

In this section, after some general background on the notion of classifica-tion in mathematics, we sketch out in broad strokes the classification pro-gram in model theory. We apply the notion of algebraic independence from Chapter 5.4 here to count models and later to provide structure theories in other areas of mathematics. Gowers begins Section 1.6, 'Classifying', of *The Princeton Companion to Mathematics* as follows:

If one is trying to understand a new mathematical structure such as a group or man-ifold, one of the first tasks is to come up with a good supply of examples. Sometimes examples are very easy to find, in which case there may be a bewildering array of them that cannot be put into any sort of order. Often, however, the conditions that an example must satisfy are quite stringent, and then it may be possible to come up with something like an infinite list that includes every single one. For example, it can be shown that any vector space over a field $\mathbb{F}$ of dimension n over a field $\mathbb{F}$ is isomorphic to $\mathbb{F}^n$. This means that just one positive integer n is enough to determine the space completely. In this case our 'list' will be $\{0\}, \mathbb{F}^1, \mathbb{F}^2, \mathbb{F}^3, \mathbb{F}^4, \ldots$

---

[43] This simple formulation assumes that we are working in T^{eq} (Definition 4.6.3).

In such a situation we say that we have a classification of the mathematical structure in question. [Gowers 2008]

He concludes his next section, 'Identifying Building Blocks and Families', by writing 'Sometimes, instead of trying to classify all mathematical structures of a given kind, one identifies a certain class of "basic" structures out of which all others can be built in a simple way.' Strikingly, Gowers chose to illustrate classification in all mathematics by one of the fundamental examples in model theoretic exposition (Theorem 5.4.4); the strongly minimal theory of vector spaces over a fixed field. The classification of each model of any strongly minimal theory by their dimension makes them the prototypic building block in model theory (Chapter 3.3).

Gowers has laid out a general program for any area of mathematics; several words must be made more precise to focus on a particular area. What are 'mathematical structures of a given kind'? What are the 'building blocks' and what does it mean to be 'built in a simple way'?

One of the most basic problems in mathematics is to count objects with a certain property. How many groups are there with 60 elements? How many times will two curves given by polynomials of degree d cross? How many points are in a finite projective plane with n points on a line? A spectrum function is a common name for such a counting function. We defined the stability spectrum function in Definition 5.3.3. By *the spectrum function* of a theory T, we mean the function $I(T, \kappa)$ which gives the number of isomorphism types of models of T with cardinality κ.

We consider in detail one example of classification. The given kind is all models of a countable[44] complete first order[45] theory. More precisely, either provide such a classification of models for a theory T or show none is possible for T. Thus the classification program has two parts. Decide whether a theory admits a classification theory. If it does, identify the building blocks and the method of building. If it doesn't, explain why not.

Morley's *spectrum conjecture* (Chapter 3.3) asserts that a countable first order theory should have at least as many models in κ^+ as in κ (for uncountable κ). The natural approach to this theorem is uniform in κ. One should just somehow extend each model M of cardinality κ to an M' of cardinality κ^+ so that $M \not\cong M$ implies $M' \not\cong M'$. But that is not the way it was proved.

---

[44] Much of the theory works for uncountable languages; we restrict to countable vocabularies for simplicity.

[45] Again, there is work in more general logics; but it is far from being as successful.

Shelah's strategy[46] to prove Morley's spectrum conjecture proceeds by splitting theories into a finite number of classes[47] and providing for each class a (schema) of explicitly defined (increasing) functions [Hart et al. 2000] such that for each T in the class one of those functions serves as $I(T, \kappa)$.

Hart [Hart 1989] gave the first full proof that the function is increasing by combining the two strategies. For the non-classifiable side, Hart invokes Shelah's non-structure results elaborated in Chapter 8.3 to show any theory satisfying any of unstable, unsuperstable, DOP, OTOP, or deep has 2^κ models in any cardinal κ.

Hart then relied on the earlier calculation of the spectra for ω-stable theories to assume non-ω-stability. In the remaining case he proves that enough models of cardinality κ are *extendible* in the following sense: a model M of size κ is extendible if it has κ components with a specific isomorphic type A at some particular location in its decomposition. If one increases two non-isomorphic extendible models to cardinality κ^+ by increasing the number of copies of A, they remain non-isomorphic; this procedure gives a non-decreasing spectrum function. The original proof scheme, listing *all* the spectrum functions,[48] was realized only in [Hart et al. 2000]; the proof requires studying three more dividing lines that we won't explore here. But we sketch the main ideas which are summarized in the Hart, Hrushovski, and Laskowski paper[49] that completed the classification project:

Shelah found a number of dividing lines among complete theories. The definitions of these dividing lines do not mention uncountable objects, but collectively they form an important distinction between the associated classes of uncountable models. On one hand, he showed that if a theory is on the non-structure side of at least one of these lines then models of the theory embed a certain amount of set theory; as a consequence their spectrum is maximal (i.e., $I(T, \kappa) = 2^\kappa$ for all

---

[46] The 22 page statement of results introducing [Shelah 1970] outlined the main results in model theory between 1970 and 1975 and in particular a strategy for the solution of the conjecture. We sketch the eventual fulfillment of this sketch below. To stress the ground-breaking nature of this work, I report that Keisler gave me a mimeographed preprint of this paper at the AMS meeting in New Orleans in 1969.

[47] The technical definitions of *dimensional order property (DOP)* and the *omitting types order property (OTOP)* are on page 176. These are set theoretically absolute mathematical properties of configurations of models and types of the theory. Their negations are denoted NDOP and NOTOP.

[48] While proving the main gap, Shelah dealt with 'sufficiently nice' models and the spectra were not completely described on small cardinalities. He had proved the 'eventual' result completely.

[49] An excellent background with examples of the various spectrum functions appears in the companion paper [Hart & Laskowski 1997]. Åsa Hirvonen's slides [Hirvonen 2015] give a more detailed survey of the argument.

uncountable κ). This is viewed as a negative feature, ruling out the possibility of a reasonable structure theorem for the class of models of the theory.

On the other hand, for models of theories that are on the structure side of each of these lines, one can associate a system of combinatorial geometries. The isomorphism type of a model of such a theory is determined by local information (i.e., behavior of countable substructures) together with a system of numerical invariants (i.e., dimensions for the corresponding geometries). It follows that the uncountable spectrum of such a theory cannot be maximal. Thus, the uncountable spectrum of a complete theory in a countable language is not maximal if and only if every model of the theory is determined up to isomorphism by a well-founded, independent tree of countable substructures. [Hart et al. 2000]

If a theory is superstable, with NOTOP and NDOP there are enough regular types with a good dimension theory to guarantee that there is a good decomposition theorem for the models. The systematic representation of a model as prime over a tree of (independent) submodels is a fundamentally new mathematical notion. The invariants of a model will be given by the dimension in the sense of Theorem 5.4.6 for certain systems of regular types and by the shape of the decomposition tree.

'Sufficiently nice' models (page 142) that satisfy the structural side of these dividing lines (NDOP and NOTOP) can be decomposed as a tree of countable submodels indexed by some tree I. The root of this tree, M_0, is a prime[50] model. For $s \in I$, each M_s has a set of up to κ extension extensions $M_{s^\frown i}$ which are independent over M_s; M is prime over $\bigcup_{s \in I} M_s$. The theory is *shallow* if there is a uniform bound over all models on the rank of the tree;[51] essentially, *deep* means this tree is not well-founded. If a superstable theory satisfies NDOP and NOTOP and is shallow the theory is called *classifiable*. As syntactic conditions on theories the properties are all absolute.

Think of each successor oval M_η (and its predecessor) in the tree[52] (Figure 5.1) as a countable model that is 'prime'[53] over $a_\eta \cup M_{\eta^-}$ where a_η realizes a regular (page 136) type over the predecessor M_{η^-} and independent from its siblings. The tree is short (countable height) but quite wide (width κ). And the original model M is prime over the union of the tree.[54]

---

[50] See page 124.

[51] This bound is the *depth* of the theory.

[52] Taken from [Baldwin 1988a].

[53] Depending on the exact case there are several relevant notions of prime – choices for the precise meaning of 'generate'.

[54] The diagram is taken from [Baldwin 1988a] and reprinted with permission of the Association of Symbolic Logic.

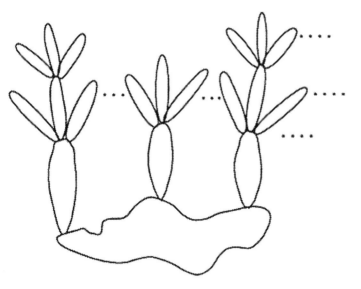

Figure 5.1 Decomposing a model

If there is a classification of the sort described and the tree is well-founded, $I(T, \aleph_\alpha) < \beth_\delta(\alpha)$ for some $\delta < \omega_1$ which depends[55] on T. Thus if one shows that $I(T, \kappa) = 2^\kappa$ for sufficiently large κ then there *can be no classification of the prescribed sort*. There are several alternative definitions of *classifiable*. For example, Hart et al. [Hart et al. 2000] consider deep theories classifiable. That is, they extend the definition of classifiable to allow trees of invariants which are *not* well-founded. For our purposes here, the more restrictive notion of classification allows an easier statement of the 'main gap'. But the dispute over the 'proper' notion has a certain philosophical tone. Is a 'well-founded' tree of invariants as natural as an arbitrary tree?

The following theorem [Shelah 1990] established the 'main gap' for the spectrum functions of first order theories. In this section, we have expounded the structure direction by discussing how to count models under stability theoretic constraints. In Chapter 8.4, we see how the other side of these dividing lines implies the theory has the maximal number of models.

Theorem 5.5.1 (Main Gap) *Let T be a countable complete first order theory. If T is superstable without the* omitting types order property *or the* dimensional order property *and is shallow then each model of cardinality*

---

[55] δ is the supremum of the Kleene–Brouwer ([Rogers 1967], 396) ordinals of possible representing trees.

λ is decomposed into countable models indexed by a tree of countable height and width λ. Thus, for any ordinal[56] $\alpha > 0$, $I(T, \aleph_\alpha) < \beth_\delta(|\alpha|)$ (for a countable ordinal δ depending on T); otherwise $I(T, \aleph_\alpha) = 2^{\aleph_\alpha}$.

There are in fact twelve schemas of spectrum functions [Hart et al. 2000]. At one extreme there are the constant functions[57] 1 and $\beth_2$, and at the other $2^{\aleph_\alpha}$. In an intermediate spectrum such as $\min(2^{\aleph_\alpha}, \beth_{\beta-1}(|\alpha + \omega|))$, the β denotes the depth of the decomposing tree, $|\alpha + \omega|$ is the number of cardinals below $\aleph_\alpha$ and thus the number of possible dimensions of a component. There are variants depending on whether the type at the final node is trivial or not.

This summary skims over the delicate parts of the argument. Consider two regular types p, q over a model of M_s in the decomposition. The relation between the dimension of p and q in M (the number of realizations of each) is very intricate and the various possibilities are what lead to the different cases of the spectra. We discuss this connection (various sorts of orthogonality of types) in the next section.

Lakatos [Lakatos 1976] expounds the long evolution of the notion of *regular polyhedron* and *Euler characteristic* so as to make the formula $V - F + E = 2$ true in full generality. Similarly to the Euler case, the question of exactly what precise type of model is sufficiently nice that one can assign invariants to it was vexed. It resulted in a twenty year delay in proving the theorem in all its detail (Chapters 5.6 and 8.4). The key point is that the fine structure of the spectrum function relies on the set M on which the regular types are based being 'sufficiently' closed. Shelah's original work ([Harrington & Makkai 1985] and [Shelah 1990]) used the notion that M is $\aleph_\epsilon$-saturated.[58] To replace this saturation hypothesis by just 'model' is the work of [Hart et al. 2000].

The significance of the classification program is emphasized by two mathematical applications of the techniques: the work of Hrushovski, Itai, and Solkovic counting the number of countable models of the theory of differentially closed fields, which led to the discovery of new types of differential varieties, and a similar discovery in compact complex manifolds by Scanlon, Pillay, and Moosa.

---

[56] $|\alpha|$ is the cardinality of α. Just as $\aleph_\alpha$ denotes the αth iteration of cardinal successor, $\beth_\alpha$ denotes the αth iteration of cardinal exponentiation. In both cases take sups at limit ordinals. Thus $\beth_0(\kappa) = \kappa$, $\beth_{\alpha+1}(\kappa) = 2^{\beth_\alpha(\kappa)}$, $\beth_\delta(\kappa) = \sup_{\alpha<\delta} \beth_\alpha(\kappa)$ for limit δ.

[57] The first is given by uncountably categorical theories; the second by putting such a theory in each non-principal type of the theory of infinitely many independent unary predicates.

[58] The model must realize all strong types over finite sets; this is a slightly weaker hypothesis than $\aleph_1$-saturated and (as it turned out) significantly stronger than an arbitrary model.

Väänänen[59] summarized the main gap:

Shelah's Main Gap represents a Liberal Version of Hilbert's Vision that axioms can describe a structured infinite universe completely, 'categorically'. For the Main Gap shows that apart from a clearly distinguished class of theories, representing the non-structure case, we do get categoricity in uncountable cardinalities relative to invariants expressible in the infinitary logic $L_{\infty,\kappa}$. So Hilbert missed the Main Gap dichotomy for the obvious reason that model theory was totally undeveloped in Hilbert's time. But we can say that Hilbert had the right idea on the structure side of the Main Gap.

5.6 Why Is Model Theory So Entwined with Classical Mathematics?

Famously, during the panel[60] on 'The Prospects for Mathematical Logic in the Twenty-first Century', Pillay proclaimed, 'There is only one model theory.' Pillay was discussing two camps of model theorists (variously characterized as applied/pure, algebraic/set theoretic, external/internal, East Coast/West Coast[61]). He was not writing off finite model theory, models of arithmetic, infinitary logic, model theory in computer science, etc. but commenting on a particular situation. As he relates, the paradigm shift described in this book was gradually absorbed by the model theoretic community during the 1970s and 1980s. Mathematical results of Zilber and Hrushovski made this unification inevitable. Specific groups and fields arose from the relationship between pairs of strongly minimal sets.

Zilber showed certain subgroups of the automorphism group of an $\aleph_1$-categorical structure [Zilber 1984c] were definable in the structure. Using this he concluded that if each (any) strongly minimal set in an arbitrary $\aleph_1$-categorical theory has a trivial geometry, then the theory is almost strongly minimal (i.e. the universe is the algebraic closure of a strongly minimal set[62] ([Poizat 2001], Theorem 2.25)). Thus, we obtain the strongest kind of structure theorem from a property of the geometry.

---

[59] Slides from his 2003 talk, 'Model theory in the latter half of the 20th century' in San Sebastian. See Chapter 7.3.

[60] The quote is from the oral presentation; his more nuanced opinion appears in [Buss et al. 2001].

[61] At the height of the vogue for this meme, 'East Coast' was anchored by Robinson at Yale and 'West Coast' by Tarski and Vaught at Berkeley; unfortunately for the metaphor, Morley was at Cornell in Ithaca, New York.

[62] This theorem explains why in attempting to construct an $\aleph_1$-categorical that was not almost strongly minimal with (Z, S), I (and Herb Gaskill) found that the result 'always' failed to be

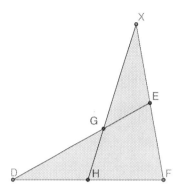

Figure 5.2 The group configuration

Hrushovski defined groups in stable theories. He introduces the notion of a *group configuration*, a particular figure in a combinatorial geometry, that yields a definable group, via a construction whose idea goes back to Euclid's geometric definition of multiplication.[63] Compare the diagram in Figure 5.2 with the one for the construction of the fourth proportional to segments A, B, C on page 216. We have moved from affine to projective geometry. There is one new point: X is the point at infinity, where lines that are parallel in affine space meet. In the diagram, segment A is represented as DF, B as DG and C as DH. Now if DH is taken as the unit, DE has length $DG \cdot DF$.

Hrushovski [Hrushovski 1986, Hrushovski 1989] generalizes from the strongly minimal set setting of Zilber to studying realizations of regular types (Theorem 5.4.6 and [Hyttinen et al. 2005]) to homogeneous quasi-minimal structures. Hrushovski shows that if the geometry of a regular type is not discrete (page 136), then there exists a group configuration: a diagram as above, where the dimension of all six points is 3; any triple of non-collinear points are independent; each point has dimension 1, and each line has dimension 2. Via quite complicated arguments,[64] using many tools of geometric stability theory, he now defines a group.

If one takes the same diagram as the group configuration but requires that each of D, H, F have dimension 2, each of x, G, E has dimension 1,

---

ω-stable. Lachlan then created a fundamental example in [Baldwin & Lachlan 1971] from whole cloth. A natural example, $(Z_4^\omega, +)$, was identified only later.

[63] Just after page 221, we give Hartshorne's version of Hilbert's definition of both multiplication and addition in an affine Euclidean plane.

[64] In particular, he generalizes an idea of Artin [Artin 1957] to regard each element A of the line to code the translation by distance OA. Then the argument concerns the behavior of these translation functions. Hrushovki's argument is more difficult as he must deal not with translation *functions* but with *germs of functions*.

{A, B, C} has dimension 4 but each of the other lines has dimension 3, the result is called a field configuration. As the name suggests, such a configuration implies the existence of a definable field.

So far, we have found the group and the field. To characterize the group, Hrushovski requires a definable action: a pair of a group and a set X such that each element of the group is represented as a permutation of X. The group action arises to answer a central model theoretic concern.

Suppose we have two types $p, q \in S(M)$; a fundamental difficulty is to determine how realizing one of these types in an extension of M affects the realization of the other. We use the word dominate to explain this relation.

Definition 5.6.1 *Consider types $p, q \in S(M)$.*

(1) *If any extension of M that realizes p also realizes q, we say p* dominates *q.*
(2) *Two types $p, q \in S(M)$ are* bidominant *if each dominates the other.*

It would simplify matters if we could add to Definition 5.6.1: (3) if neither type dominates the other they can be realized completely independently. That is not true; the groups and fields arise precisely from this subtlety. In the next examples P and Q are strongly minimal formulas; the unique non-algebraic type containing the formula $P(x)$ is said to be *generic*[65] for P.

Example 5.6.2

(1) If T is the theory of two disjoint strongly minimal sets P and Q, then the (infinite) cardinalities of P and Q can vary arbitrarily.
(2) Let A be a strongly minimal abelian group, say $(\mathbb{Q}, +)$. Let P be pairs $(a, 0)$ with $a \in A$ and Q be pairs $(a, 1)$ with $a \in A$. The group operation on P is in the vocabulary, but instead of the group operation on Q, we have the ternary relation R on $P \times Q^2$ which is the graph of a family of maps, indexed by P, from Q to Q: $f_{(a_1,0)} : Q \to Q$ where $f_{(a_1,0)}\{(a_2, 1)\} = \{(a_1 + a_2, 1)\}$.

We have used the *regular action* of an abelian group on itself to define an action of P on Q.

In the first example the sets (or more precisely their generic types) don't influence each other; in the second they are closely connected. The formal definitions reflecting these phenomena use the notion of independence as described just before Theorem 5.4.5.

---

[65] More generally, for each type $p \in S(M)$ and each $N \succ M$, p has a unique extension p' to $S(N)$ such that every realization of p' is independent from N over M. This distinguished extension is called *non-forking* (or *generic*).

Definition 5.6.3

(1) *Two types p and q (over the same domain), are said to be* weakly orthogo-nal[66] *(almost orthogonal) written $p \perp^w q$ ($p \perp^a q$) if $p \cup q$ is a complete type (any realizations of $\mathbf{a}$ of p and $\mathbf{b}$ of q are independent).*

(2) *Two types p and q (possibly over different domain) are said to be* orthog-onal *if any two non-forking extensions p', q' of p, q to a common domain are almost orthogonal.*

(3) *For any ordinal α, r^α denotes the type of α independent realizations of r.*

Equivalently, $p \perp q$ if for any independent sequences $\mathbf{a}$ of realization of p and $\mathbf{b}$ of realization of q, $\mathbf{a}$ and $\mathbf{b}$ are independent. This is exactly what failed in Example 5.6.2(2). The types p and q over M are almost orthogonal: any a realizing p and b realizing q are independent; but $p \not\perp q^2$.

Now the surprising fact is that there are very few ways[67] in which this situation can occur and they all involve definable groups. We state the the-orem in the generality of stable theories to indicate both the austerity and broad coverage of the hypothesis

Theorem 5.6.4 (Hrushovski) *Let T be a stable theory. Let $\tilde{p}$ and $\tilde{q}$ be nonorthogonal stationary, regular types and let n be maximal such that $\tilde{p}^n$ is almost orthogonal to $\tilde{q}^\omega$.*

Then there exist p almost bidominant to $\tilde{p}$ and q dominated by $\tilde{q}$ such that:

$n = 1$ *q is the generic type of a (type) definable group that has the regular action on the realizations for p.*

$n = 2$ *q is the generic type of a (type) definable algebraically closed field that acts on the realizations for p as an affine line.*

$n = 3$ *q is the generic type of a (type) definable algebraically closed field K that acts on the realizations for p as a projective line, i.e. by the action of $PGL_2(K)$.*

$n \geq 4$ *is impossible.*

The hypotheses are purely model theoretic. There is no assumption that a group or ring is even interpretable in the theory. But the conclusion gives precise kinds of group and field actions that are *definable* in the given structures.

---

[66] Weak and almost orthogonality agree if p and q satisfy the technical condition of stationarity.

[67] Laskowski found such groups and investigated the connection with weak orthogonality in his thesis which appears as [Laskowski 1988]. The main result was to extend Baldwin–Lachlan to uncountable theories: Let $|T| = \aleph_\alpha$ denote the number of formulas of L modulo T-equivalence. Then if T is categorical in some $\kappa > |T|$, $I(T, \aleph_\alpha) = \aleph_0 + |\alpha|$.

There are really two mysteries here. Where did these fields and matrices come from? Why is there any bound on n, particulary such a small one?

The orthogonality and nonorthogonality conditions give rise to a group configuration if $n = 1$ and to a field if $n = 2$ or 3. The case $n > 3$ is ruled out by a short analysis of the group representation (page 146 of [Hrushovski 1989]). Theorem 5.6.4 classifies the possible groups actions in stable theories.

Theorem 5.6.4 was proved as a tool for purely model theoretic problems and has been applied to the spectrum problem[68] [Hart et al. 2000] and the Vaught conjecture for superstable theories [Buechler 2008]. Perhaps most striking is the following connection between the number of models and algebra ([Hart et al. 2000], 209): any model of a complete theory whose uncountable spectrum is $I(\aleph_\alpha, T) = \min(2^{\aleph_\alpha}, \beth_{d-1}(|\alpha + \omega| + \beth_2))$ for some finite $d > 1$ interprets an infinite group.

One formulation of the pure/applied model theory distinction[69] was between internal (inward looking) and external (outward looking) perspectives. Much of this book emphasizes the external aspect, the role of model theory in identifying links across various areas of mathematics.

Poizat summed up the classification program in a review of a proceedings volume, entitled *Classification Theory* [Baldwin 1988b]:

Possibly, the editor found it too delicate to use a more ambitious term: this could give an impression of arrogance, or of a will to exclude people working in more traditional matters ... Those of us who are free from that kind of scruple will restore to this book the only name it deserves: Model theory. [Poizat 1990]

---

[68] Laskowski tells me the existence of the group is used but buried quite deeply in the proof.
[69] Kechris' contribution to [Buss et al. 2001] makes a similar distinction for set theory.

6 | Isolating Tame Mathematics

Martin Davis[1] wrote,

Gödel showed us that the wild infinite could not really be separated from the tame mathematical world where most mathematicians may prefer to pitch their tents.

We will now see how modern model theory avoids the Gödel phenomena; the key is to formalize topics locally by axioms which catch the relevant data but avoid accidentally encoding arithmetic and, more generally, pairing functions. We do not attempt a general definition of tame but provide a number of examples of sufficient model theoretic conditions. For more details, see page 160 and [Teissier 1997].

The most basic examples are when mathematicians are already studying definable relations on a class of structures and the natural axiomatization of the area yields a tame theory. In studying real or complex algebraic geometry, the formalization is automatic; Steinitz (ACF) and Artin-Schreier (RCF)[2] defined concepts that happen to be first order; these theories provide the framework for much of the development of the geometries. The theories are $\aleph_1$-categorical (Chapter 3.3; indeed, interpretable in a strongly minimal structure, Example 4.3.1) and o-minimal (Chapter 6.3), respectively.

One of the earliest algebraic discoveries linking algebraic structure with stability properties echoes the Bourbaki assertion of the importance of groups. An ω-stable group cannot have a descending chain of 'definable subgroups'.[3] This condition extends to stable groups (for uniformly definable chains) and the distinction between the stability classes is signaled by the size of the allowed quotient groups.[4] This principle is now seen to apply to different algebraic structures and gives a unified explanation for finding various kinds of radicals.[5] (See [Baldwin 1979] for a very early account

---

[1] Foundations of Mathematics Listserve Aug 27, 2015 www.cs.nyu.edu/pipermail/fom/2015-August/018934.html.

[2] A field is real closed if is formally real (-1 is not a sum of squares) and every odd degree polynomial has a solution.

[3] In rings this descending chain condition is on ideals.

[4] E.g. in a superstable group there may be descending chains if each quotient is finite.

[5] 'Radical' is a technical term used in the structure theory of rings.

of this phenomenon and [Altinel & Baginski 2014], [Aldama 2013], and [Freitag 2015] for recent updates.)

6.1 Groups of Finite Morley Rank

In this section we give a short case study of one research area that serves both as a tool for unifying studies in several areas of mathematics and for isolating the role of basic concepts. In particular, it is seen that *finite* plays a dual role in the study of finite groups: both as the size of structure and as a dimension on which one can do inductions. In the case at hand, the study is extended to infinite groups by introducing a broader definition of 'dimension', Morley rank (Chapter 5.3), and requiring it to be finite.

We sketch how the resources of formalization, in particular, interpretation and the stability apparatus, contribute to the study of a class of groups that were in fact approached both from the group theory standpoint by Borovik and from the model theoretic standpoint by Cherlin and Zilber; Poizat saw the concepts were the same.[6] This generalization of the study of finite groups illustrates the ways that formalization interacts with traditional mathematics and exhibits the power of formalization to introduce a generalizing principle. Considering groups of finite Morley rank provides a framework which includes both finite groups and algebraic groups over algebraically closed fields, thereby illuminating the role of finiteness conditions in each case.

Define a group of finite Morley rank (FMR) to be an infinite group (possibly with additional structure) that is ω-stable with finite rank.[7] Macintyre proved in the early 1970s that an $\aleph_1$-categorical, indeed any ω-stable field, is algebraically closed. Slightly later, Reineke proved every strongly minimal group is Abelian [Reineke 1975]. An *algebraic group* is a variety G (page 100) over a field k equipped with a group operation from $G \times G \to G$ that is a morphism (in the sense of algebraic geometry[8]). The definition of an algebraic group (over an algebraically closed field) yields

---

[6] Page 58 of [Borovik & Nesin 1994].

[7] I am not giving a detailed historical survey here so many attributions and references are omitted. Poizat [Poizat 1987, Poizat 2001] provides the general setting as in the late 1980s. [Borovik & Nesin 1994] gives a good overview of the finite rank case in the mid 1990s. Cherlin's webpage [Cherlin 2004] lays out the Borovik program in broad strokes with references. A more recent summary is [Altinel et al. 2008].

[8] The *Hrushovski–Weil* theorem establishes ([Poizat 2001], Chapter 4.5) that an equivalent definition is: a group defined in an algebraically closed field.

immediately that it is interpretable[9] in the field and so has finite Morley rank. The fundamental geometric definitions of the dimension (Krull/Weil dimension) of a definable subset over an ACF, in particular an algebraic group, yield the same value as the Morley rank (page 130). Since every algebraic group is interpretable in an algebraically closed field, it is FMR.

The driving *Cherlin–Zilber conjecture*, 'A simple group of finite Morley rank is an algebraic group over an algebraically closed field', links model theory with algebraic geometry. The more than 30 year project to solve the conjecture has developed as an amalgam of stability theoretic tools with many different tools from algebraic, finite, and combinatorial group theory.

The basic scheme for understanding the structure of groups relies on the Jordan–Hölder theorem: each finite group can be written (uniquely up to the order of the decomposition) as $G = G_0 \rhd G_1 \rhd \ldots \rhd G_n = 1$, where G_{i+1} is normal in G_i and the quotient groups G_i/G_{i+1} are simple. Thus identifying the finite simple groups is a key step to understanding all finite groups. (We also need to know the nature of the extension at each level.) The analysis of FMR groups replaces induction on the cardinality of a group by induction on Morley rank.

Work in the late 1970s showed similar properties of algebraic groups over an algebraically closed field and groups of low finite Morley rank. Cherlin extended from algebraic groups to FMR groups the propositions: rank 1 implies abelian; rank 2 implies solvable. Zilber showed that a solvable[10] connected[11] FMR group which is not nilpotent[12] interprets an algebraically closed field. This implies that every FMR group 'involves' an algebraically closed field; the issue is, 'how close is the involvement?' Although algebraic groups over algebraically closed fields have FMR, groups of finite Morley rank are clearly more general. The Prüfer group $\mathbb{Z}_{p^\infty}$ is FMR but not algebraic and FMR groups are closed under direct sum while algebraic groups are not. But the role of rank/dimension in each of the cases and the identification of the field in the group led to the idea that groups of finite Morley rank were some kind of natural closure of the algebraic groups. In particular, it led to the conjecture above, that the basic building blocks are the same.

The classification of the finite simple groups identifies most of them as falling into families of algebraic groups over finite fields, the Chevalley groups. As we discussed in Example 1.4.2, families such as the Chevalley groups are a natural notion in model theory. They are the solutions of the same formal definition of a matrix group as the field changes.

---

[9] See Definition 4.5.2.
[10] A group is *solvable* if the quotients in the Jordan–Hölder decomposition are abelian.
[11] A group is definably connected if it has no definable subgroup of finite index.
[12] Nilpotent imposes a stronger condition on the decomposition series.

Chain conditions illustrate the analogy between finite groups and FMR groups. The descending chain conditions on all subgroups for finite groups is replaced by the descending chain condition on *definable* subgroups. This allows a model theoretic definition of 'connected' to replace the topological definition in the study of algebraic groups. Induction on the cardinality of the group is replaced by induction on its Morley rank. The use of definability now provides a common framework for the study of algebraic and finite groups.[13] The quest for the conjecture led to intricate analysis of all three generations of the proof of the classification of finite simple groups. The main strategy of the proof is an induction on the 'minimal counterexample' and the possibilities for this counterexample are sorted as in the finite case.

We quoted in Chapter 4.5 a passage from [Borovik & Nesin 1994] highlighting the role of the logical notion of interpretation in many areas of mathematics. Again, the ability to recognize and codify these uses depends on having formalized the areas involved. Formalization enables one to focus on the key idea of a proposition. The standard statement of the Borel–Tits theorem takes half a page and gives a laundry list of the possible kinds of maps (albeit considering the fields of definition of the groups). Zilber (see Theorem 4.17 [Poizat 1987, Poizat 2001]) gives the following elegant statement.

Theorem 6.1.1 (Borel–Tits à la Zilber/Poizat) *Every pure group isomorphism between two simple algebraic groups over algebraically closed fields K and L respectively can be written as the composition of a map induced by a field isomorphism between K and L followed by a quasi-rational function over L.*

The maps are constructed as compositions of *definable* and abstract maps. The statement and proof of the Borel–Tits theorem illustrate the role of formalization in providing context and clarity to mathematical results. Zilber replaces the laundry list by a three line conceptual description of a decomposition of the isomorphism into two more clearly understood types of map.

6.2 Formal Methods as a Tool in Mathematics

In 'A path to the epistemology of mathematics: Homotopy theory', Marquis [Marquis 2008] argues that mathematicians deal not only with 'objects' but with 'tools' and writes, 'Our main objective is to show that a typical

---

[13] Borovik had independently introduced the notion of ranked group (one which admits a collection of subsets containing the finite sets and closed under Boolean operations, projection, quotient). Poizat showed the class of such groups is exactly the FMR groups.

component of twentieth century mathematics is the emergence, proliferation, and establishment of *systematic mathematical technologies* within mathematics.'

We described in Chapter 1.1 formalization as a process of clarifying the understanding of an area of mathematics and emphasized in Chapter 1.2 the importance of choosing the correct set of primitives. In this section we discuss first the ordinary mathematical development of two subjects and then the role that *full* formalization played in the solution of major problems.

Thus we support the claims of Theses 1 and 2 that model theory increasingly contributes to a wide range of mathematics. These are essentially empirical claims. We demonstrate this impact by examining how formalization and in particular virtuous properties function as tools for doing and organizing mathematics in three fields of mathematics.

Example 6.2.1 (Diophantine Geometry) Even though arithmetic is the quintessential wild theory (page 10), we will see how it can be non-definably embedded into tame structures. This is a standard technique in number theory that goes under that name of *Diophantine geometry* or in Grothendieck style *arithmetic algebraic geometry*. Model theory provides tools to identify tame targets for such an embedding.

The problem of finding natural number or *integer* solutions to an equation such as $x^n + y^n = z^n$ is commonly called after Diophantus. As a problem in formal number theory the Yuri Matiyasevich, Julia Robinson, Martin Davis, and Hilary Putnam negative solution of Hilbert's 10th problem establishes the lack of an algorithm to determine the solvability (in integers) of such equations. A variant asks whether the equation has infinitely many *integral* solutions. This problem is easily equivalent to asking for infinitely many *rational* solutions of an equation in one less variable.

The twentieth century[14] approaches the revised problem by studying the equation as first defining a variety V in $\mathbb{C}^n$ and then asking about its integer solutions. We restrict here to curves, equations $f(x,y) = 0$ with f a polynomial with rational coefficients. The first reduction is to note that any curve can be replaced by an irreducible smooth curve while preserving finite/infinite number of solutions. The key invariant for this problem becomes the *genus*[15] g of the curve: it is significant for both

---

[14] The transformations described below merit the kind of detailed account given for Dirchelet's notion of character in [Avigad & Morris 2016, Avigad & Morris 2014]. Note that number theory exhibits a long sequence of context-changing domain extensions (page 125): $\mathbb{N}, \mathbb{Z}, \mathbb{Q}, \mathbb{C}$, the p-adic numbers and topology on $\mathbb{C}$.

[15] A curve is smooth if it has no self-intersections or cusps. The genus of a smooth curve can be computed by its degree d as $\binom{d-1}{2}$ but for non-smooth curves the degree is not a fine enough invariant.

number theory and topology. The set of solutions in $\mathbb{C}^2$ of an equation of genus g is a Riemann surface that looks like g donuts joined together.[16] It is comparatively straightforward to show that for genus 0, there are always infinitely many solutions and, for genus 1, there are infinitely many if there are any.[17]

Conjecture 6.2.2 (Mordell, 1922) *A curve of genus greater than 1 over the field $\mathbb{Q}$ of rational numbers (generalized to over any finite extension of $\mathbb{Q}$) has only finitely many rational points.*

Note for example that this conjecture immediately implies a weak form of Fermat's conjecture; that the Fermat equation has only finitely many solutions. Some experts thought this conjecture fanciful until Faltings solved a more general version in 1984. We need some more concepts.

An *abelian variety* is a variety equipped (by another variant on the geometric definition of multiplication) with the structure of an abelian group (equivalently a group definable in an algebraically closed field). We will not discuss the embedding of any (smooth projective) curve into its Jacobian J_C, which by the Mordell–Weil theorem is a finitely generated (as a group) Abelian variety that is the minimal definable group containing C and that properly contains C if the genus of C is at least 2.

Theorem 6.2.3 (Faltings' Theorem) *Let G be an abelian variety defined over the field of complex numbers $\mathbb{C}$ and $\Gamma \subset G(\mathbb{C})$ be a finitely generated subgroup of the group of $\mathbb{C}$-points on G.*
Then $X(\mathbb{C}) \cap \Gamma$ is a finite union of cosets of subgroups[18] of Γ.

The transformations required to state this conjecture represent several context-changing domain changes (Chapter 5.2). The problem is transformed from the original wild environment of the natural numbers to the slightly tamer rationals and then to a domestic expansion of an algebraically closed field. Conceptually, analytic and topological tools came into play in defining genus and the Jacobian variety.[19] But all of this work was carried out informally. Recent works show the relevance of a full formalization. But the particular formalization depends on prior work in abstract stability theory that provides a fruitful framework. Pillay explains the connection in a volume dedicated to Hrushovski's proof of 'functional Mordell-Lang'.

---

[16] There is a typical picture at https://en.wikipedia.org/wiki/Genus_(mathematics).

[17] See [Scanlon 2001] or Ellenberg's article on arithmetic algebraic geometry in [Gowers 2008].

[18] This condition, which implies the finiteness conjectured by Mordell, is Lang's contribution to the *Mordell–Lang* conjecture.

[19] After the fact, algebraic definitions were given.

The purpose of this note is to point out that the use of model-theoretic and stability-theoretic methods should not be so surprising, as the *full* Lang conjecture itself is equivalent to a purely model-theoretic statement. The structure $(\mathbb{Q}, +, \cdot)$ is wild (undecidable, definable sets have no structure, etc.), as is the structure $(\mathbb{C}, +, \cdot)$ with a predicate for the rationals. What comes out of the diophantine type conjectures however is that *certain* enrichments of the structure $(\mathbb{C}, +, \cdot)$ (more specifically expansions obtained by adding a predicate, not for $\mathbb{Q}$ itself, but rather for the $\mathbb{Q}$-points of certain algebraic groups) are not wild, in particular are stable. [Pillay 1999]

Here is a later precise statement of the model theoretic version.[20]

Fact 6.2.4 (Faltings as Rephrased by Pillay (Moosa–Scanlon)) *Let G be an abelian variety defined over the field of complex numbers $\mathbb{C}$ and $\Gamma \subset G(\mathbb{C})$ be a finitely generated subgroup of the group of $\mathbb{C}$-points on G. Then the induced structure on Γ is stable and weakly normal.*[21]

After defining weakly normal, the connection to Mordell–Lang is clear.

Definition 6.2.5 *A definable set X is weakly normal if the intersection of every infinite family of distinct conjugates (i.e. $a^{-1}Xa$) of X is empty.*

A theory is weakly normal *if each formula is a Boolean combination of weakly normal formulas.*

Following ([Marker 2002], 312), we sketch the proof of Mordell's conjecture[22] 6.2.2 from Fact 6.2.4. Let C be an irreducible smooth curve of genus at least 1 with Jacobian J_C. We show $C(\mathbb{Q})$ is finite; let $J_C(\mathbb{Q}) = \Gamma$. By Fact 6.2.4, $C \cap \Gamma$ is a finite union of cosets of subgroups (of J_C and thus of Γ). If some coset aH is infinite then its Zariski closure $\overline{aH}(\mathbb{C})$ must equal $C(\mathbb{C})$ (since C is irreducible and has Morley rank equal Weil dimension equal 1). But then there is a definable group structure[23] on C. This contradicts that $J_C \supsetneq C$ is the smallest group containing G.

The direct translation of Mordell–Lang to characteristic p is blatantly false. Hrushovski proved a conjectured relativized version known as Mordell–Lang for function fields. His proof integrates Shelah's notions of orthogonality and p-regularity and such notions from geometric stability theory as *one-based* with tools of arithmetic algebraic geometry. The proof used the model theoretic analysis of differentially closed fields (ω-stable) in characteristic 0 and separably closed fields (strictly stable)

---

[20] The proof remains primarily algebraic; but Pillay's formulation inspires generalization.

[21] Weakly normal is equivalent to 1-based (Theorem 5.4.7).

[22] An easier variant, since the torsion group is locally finite, shows the Manin–Mumford conjecture that C contains only finitely many torsion (finite order) points of J_C.

[23] Just transfer the group structure on $\overline{H}$ to $\overline{aH}(\mathbb{C})$ by a push-through construction.

[Marker et al. 1996] in characteristic p (page 101). See [Bouscaren 1999], by geometers and model theorists, for a proof with background in both model theory and algebraic geometry, [Hrushovski 1997] for an overview and connections to stability theory, and [Scanlon 2001, Scanlon 2012] for a more recent account of work in this area showing other applications of the model theoretic methods.[24]

From one standpoint, Hrushovski gives one of several later proofs of Faltings' theorem. However, not only does it yield extensions, it provides a powerful new twist on formalization. Find inside a wild structure such as $(\mathbb{C}, +, \mathbb{Q})$ a substructure $(\mathbb{C}, +, (X(\mathbb{C}) \cap \mathbb{Q}))$ whose theory is stable and, in this case, weakly normal, so tame.

Example 6.2.6 (Valued Fields) Model theory has played a heavy role in the study of valued fields since Ax–Kochen–Ershov (page 9). Two examples growing out of valued fields yield two further ways to use first order formalization. In the first case, rather than finding a complete theory, the use of an incomplete theory is essential. In the second, more modern extensions into neo-stability theory are the tools for using the formalization.

Motivic integration: The mathematical analysis of functions of p-adic numbers, p-adic analysis, has played a major role in number theory for the last 100 years. In what Scanlon [Scanlon 2009] calls the prehistory of motivic integration, Denef's groundbreaking result showed the rationality[25] of a certain Poincaré series; the geometers' tool of desingularization is avoided by an induction on quantifier rank using p-adic cell decomposition. Kontesevich introduced the important tool of motivic integration in 1995 to solve a question about Calabi–Yau manifolds. Model theorists and geometers (including Denef/Cluckers/Hales/Hrushovski/Kazhdan/Loeser) jointly developed motivic integration theory in the twenty-first century. Unlike much of our earlier discussion, it deals with incomplete theories in an essential way because the goal is to establish uniform results for fields of different characteristics. This tool draws on the ideas of cell decomposition arising in the study of o-minimality and with issues arising from the study of p-adically closed fields as NIP theories (page 130) and [Haskell et al. 2007], our next topic.

---

[24] The proof uses the extension from finding the field in the first order case to the Zariski geometry case [Hrushovski & Zilber 1993, Hrushovski 1996]. A crucial step involves the *Manin kernel* which was investigated using the (formal) methods of differential algebra by Buium and Hrushovski.

[25] Number theorists assign infinite series to algebraic situations. It is important to prove such a series is actually a rational function.

Algebraically closed valued fields: The study of valued fields intrinsically splits into a stable (field) sort and an unstable sort (ordered value group). The right choice of vocabulary varies with the problem; it is nontrivial to achieve elimination of quantifiers and even more complicated to obtain the elimination of imaginaries[26] proved in [Haskell et al. 2007], whose authors write, 'The purpose of this paper is to give the foundations of a study of structures which are first-order interpretable in an algebraically closed valued field; that is, which live on a quotient of a power of the field by a definable equivalence relation.' In [Haskell et al. 2007], the authors introduce the concept of *stable domination*: the behavior of a stably dominated type is determined by a small part of it, that lies in the stable sort of the model. That is, the authors develop the model theory to study many of the algebraic topics arising in both valued fields and motivic integration. In [Hrushovski & Loeser 2016], the use of model theory extends into the study of Berkovich spaces (Google this!), an important early twenty-first century topic.

We saw in this section that the search for appropriate primitives to find a framework for investigating a problem is an important epistemological concern in both informal mathematics and crucially when constructing a full formalization. But we see also that recently, finding frameworks that satisfy specific virtuous properties that were identified in pure model theory has become an important research tool. These virtuous properties include the place in the (neo)-stability hierarchy, the classification of definable geometries, and, as discussed in the next section, o-minimality.

6.3   First Order Analysis

Our earlier examples illustrating Thesis 3, the effectiveness of first order model theory in mathematics, have come from algebra. Classical analysis is built on the Dedekind complete ordered field (page 253) of real numbers. Thus, it naturally takes place in second order logic ([Shapiro 2005], 5.3) or in set theory. There is little question that these (Notation 1.3.5, Chapter 3) are the frameworks in which most analysts work. Nevertheless contemporary model theory provides two successful first order frameworks for formalizing analysis. We call these *axiomatic analysis* and *definable analysis*. The underlying motif of the first is sounded by Ritt, who founded

---

[26] Johnson [Johnson 2014] 'hopes to provide a more conceptual proof'; in particular he eliminates the use of five technical notions.

differential algebra in the 1930s. After describing the advances of algebraic geometry in the early twentieth century, he wrote:

To bring to the theory of systems of differential equations, which are algebraic in their unknowns and their derivatives, some of the completeness enjoyed by the theory of systems of algebraic equations is the aim of the present monograph. ([Ritt 1950], iii–iv)

Axiomatic analysis studies behavior of fields of functions with operators but *without* explicit attention in the formalism to continuity but rather to the algebraic properties of the functions. The function symbols of the vocabulary act on the functions being studied; the functions are elements of the domain of the model. As reported in [Birkhoff & Kreyszig 1984], Hadamard in 1928 described functional analysis as being on a higher level of abstraction than the theory of functions because it regards the functions as variables.[27] Thus, *differential algebra* is axiomatized by first order sentences in the vocabulary $(+, \times, 0, 1, \partial)$ (where ∂f is interpreted as the derivative[28] of the function f); the key axiom is the Leibniz rule: $\partial(fg) = f\partial g + g\partial f$. The first order formulation is particularly appropriate because many of the fields involved are non-Archimedean.

We discussed in Chapter 5.2 the epistemological advantages of finding a universal domain for a particular area of mathematics. Model theoretic methods play an important role in the search for a universal domain appropriate to different areas of analysis. Kolchin incorporated Robinson's notion of *differentially closed field* (Definition 4.4.5) in his 'bible' of differential algebra [Kolchin 1973] and it is one of the tools of differential algebraists. Blum [Blum 1968] provided the first of several axiomatizations (page 211) and she showed the characteristic 0 theory was ω-stable. By Shelah's uniqueness theorem for prime models over sets for ω-stable theories, differential closures are unique up to isomorphism. But Kolchin, Rosenlicht, and Shelah independently showed they are not minimal (by constructing in different ways a strongly minimal set with trivial geometry). Hrushovski and Itai [Hrushovski & Itai 2003] lay out as an application of 'Shelah's philosophy' the following model theoretic fact (based on Buechler's dichotomy) fundamental to the study of differential fields: 'an *algebraically* closed differential

---

[27] Invoking Quine, the functions are now objects as 'to be is to be the value of variable'. Unlike functional analysis, there is no way to discuss the evaluation of a function in the differential algebra setting and no metric on the space. Continuous logic (page 97, Chapter 6.4) with its intrinsic metric space provides a better framework for functional analysis where convergence is a central topic.

[28] The theory extends to consider partial derivatives.

field K is *differentially* closed if every strongly minimal formula over K has a solution in K.' Even more, by the general theory of superstability, their study reduces to the study of strongly minimal sets and definable simple FMR groups that are associated with strongly minimal sets.

In the next few paragraphs we sketch the interactions of modern model theory with classical analysis to solve two problems in the century old study of *Painlevé equations*. The formal setting here is the theory of differentially closed fields with constant field the complex numbers $\mathbb{C}$. In 1900 Painlevé began the study of nonlinear second order ordinary differential equations (ODE) satisfying the Painlevé property (no movable singularities). In general such an equation has the form

$$y'' = f(y, y')$$

with f a rational function (i.e. in $\mathbb{C}(t)$). He classified such equations into 50 canonical forms and showed that 44 of these were solvable in terms of 'previously known' functions. Here is a canonical form for the third of the remaining classes; the Greek letters are the constant coefficients; t is the independent variable satisfying $y' = 1$ and the goal is to solve for y.

$$P_{III}(\alpha, \beta, \gamma, \delta): \quad \frac{d^2y}{dt^2} = \frac{1}{y}\left(\frac{dy}{dt}\right)^2 - \frac{1}{t}\frac{dy}{dt} + \frac{1}{t}(\alpha y^2 + \beta) + \gamma y^3 + \frac{\delta}{y}$$

The first problem is to show that a generic equation (i.e. the constant coefficients are algebraically independent) of each of the six forms is irreducible. For this, one must take on the logician's task: 'What does *not reducible* mean?' By reducible Painlevé meant solvable from 'known functions'. The Japanese school clarified 'solvable' to mean, roughly speaking, generated from solutions to order one ordinary differential equations (ODE) and algebraic functions through a fixed family of constructions (integration, exponentiation, etc.). In the formal setting, this is equivalent to showing that if the order two differential equation is strongly minimal; then there can be no *classical solutions*. This problem was solved (without the formalization) in each of the six cases by the Japanese school (led by Umemura) in the late 1980s.

The second problem is the conjecture that if $y_1, \ldots y_n$ are n independent solutions of a generic strongly minimal Painlevé equation then that set along with its first derivatives is also algebraically independent. The new step is to invoke the Zilber trichotomy (page 83), which holds for differentially closed fields,[29] to reduce to a trivial strongly minimal set.

---

[29] The first proof of the trichotomy was in [Hrushovski & Sokolović 1993], which depends on the analysis in [Hrushovski & Zilber 1993]; a more direct proof is given in [Pillay & Ziegler 2003].

Pillay and Nagloo [Nagloo & Pillay 2016] show the other alternatives are impossible in this situation and indeed that the strongly minimal set is $\aleph_0$-categorical. Then using the geometric triviality (from the Zilber trichotomy) heavily, along with tools from the Japanese analysts, they show the required algebraic independence. In contrast, Freitag and Scanlon [Freitag & Scanlon 2015] show the order three algebraic differential equation over $\mathbb{Q}$ satisfied by the analytic j-function defines a *non-$\aleph_0$*-categorical strongly minimal set with trivial forking geometry.

Marker and Pillay [Marker et al. 1996, Pillay 1995] provide an overall summary of the interaction of model theory and differential algebra, especially differential Galois theory. Freitag [Freitag 2014] illustrates the continued interplay between various ranks on (differential) groups and rings and stability theoretic ranks (Definition 5.3.1). [Sànchez & Pillay 2016] show how model theoretic ideas unify and extend results on differential equations obtained by more traditional mathematical methods.

Definable analysis has a lower level of abstraction; the domain of the functions remains the universe of the model. The functions being studied are the compositions of the functions named in the vocabulary; one cannot quantify over them. This study builds on the already major interaction between real algebraic geometers and *o*-minimality. The refinement in real algebraic geometry of Tarski's elimination of quantifiers to the method of *cylindric algebraic decomposition* [Collins 1975, Cohen 1969], which represents each *n*-dimensional set as a union of fewer-dimensional connected cells, advances the epistemological objectives of Manders (Chapter 2.1) and has major mathematical consequences.

The extension from real algebraic geometry to analysis began by studying real exponentiation: add to the field vocabulary a unary function symbol exp and make it behave as an exponential function by adding the axiom $\exp(a + b) = \exp(a) \times \exp(b)$ to the axioms for real closed fields.

Building on van den Dries' insights concerning Tarski's quantifier elimination for the real field, Pillay and Steinhorn defined o-minimality as an analog of strong minimality. Strong minimality characterizes the definable subsets (no matter how extensive the ambient vocabulary) as easily describable using only equality (finite or cofinite).

Definition 6.3.1 (o-Minimal Theory) *The vocabulary τ of an o-minimal theory T must contain (among other predicates) a symbol $<$ which linearly orders each model. T is o-minimal (order minimal) if every τ-formula is equivalent to a Boolean combination of intervals.*

Thus, an o-minimal T admits a very strong form of quantifier elimination. Every subset definable in the large language is described in a very simple way using only the order. Even though the definition concerns only subsets of the universe, o-minimality implies the existence of *cell decomposition* generalizing the clear understanding of definable sets in all dimensions from real algebraic geometry to definable analysis. On the one hand, this notion illustrates Manders' accessibility criterion (page 60); on the other, its immense consequences reinforce the significance of that requirement. Key examples are the real field and the real field with exponentiation [Marker 1996]. The geometer Teissier[30] examines at length the connections of o-minimality with Grothendieck's notion of 'tame topology'. Alex Wilkie (Seminaire Bourbaki, November 2007) motivated the notion:

The notion of an o-minimal expansion of the ordered field of real numbers was invented by Lou van den Dries as a framework for investigating the model theory of the real exponential function $\exp : \mathfrak{R} \mapsto \mathfrak{R} : x \mapsto e^x$, and thereby settle an old problem of Tarski. More on this later, but for the moment it is best motivated as being a candidate for Grothendieck's idea of tame topology as expounded in his Esquisse d'un Programme. It seems to me that such a candidate should satisfy (at least) the following criteria. (A) It should be a framework that is flexible enough to carry out many geometrical and topological constructions on real functions and on subsets of real Euclidean spaces. (B) But at the same time it should have built in restrictions so that we are a priori guaranteed that pathological phenomena can never arise. In particular, there should be a meaningful notion of dimension for all sets under consideration and any that can be constructed from these by use of the operations allowed under (A). (C) One must be able to prove finiteness theorems that are uniform over fibred collections.

None of the standard restrictions on functions that arise in elementary real analysis satisfy both (A) and (B). For example, there exists a continuous function $G : (0, 1) \mapsto (0, 1)^2$ which is surjective, thereby destroying any hope of a dimension theory for a framework that admits all continuous functions. ...

Rather than enumerate analytic conditions on sets and functions sufficient to guarantee the criteria (A), (B) and (C) however, we shall give one succinct axiom, the o-minimality axiom, which implies them. Of course, this is a rather open-ended (and currently flourishing) project because of the large number of questions that one can ask under (C). One must also provide concrete examples of collections of sets and functions that satisfy the axiom and this too is an active area of research. [Wilkie 2007]

Crucially, the order topology in an o-minimal theory is definable so continuity and limits can be discussed. Every o-minimal theory is unstable but this definition illustrates Thesis 1, by isolating a robust family of applicable

---

[30] See ([Teissier 1997], 232–236).

first order theories. Wilkie [Wilkie 1996] proved that the real exponential field is o-minimal[31] and model complete. In addition to the exponential, adding the Γ function (defined on positive reals) or any analytic function on a bounded domain maintains o-minimality [Marker 2000, Dries 1999]. The limitation is that adding additional functions to the vocabulary may destroy o-minimality. The major such obstacle is *global* oscillatory functions such as $\sin(x)$ as the theory T_{an} obtained by adding each restriction of an analytic function to a compact set remains o-minimal [Dries & Miller 1994].

Often the objects definable in o-minimal structures are proved classical. Peterzil and Starchenko [Peterzil & Starchenko 2000] study the foundations of calculus in this setting. Pillay [Pillay 1988] showed that if G is a group definable in an o-minimal structure M, G can be equipped with a definable manifold structure making G a topological group. In particular, if M is an expansion of the reals then G can be given the structure of a Lie group.

In axiomatic analysis, the study of *ordered* differential fields[32] provides an approach to *asymptotic analysis*. Du Bois-Reymond, followed by Hardy, initiated the deep study of rates of growth of real functions over a century ago. In their study of real analysis, Bourbaki introduced the notion of Hardy field: a subfield K of the ring of germs at $+\infty$ of differentiable functions $f \colon (a, +\infty) \to \Re$ with $a \in \Re$ that is closed under taking derivatives. Van den Dries, Macintyre, and Marker [Dries et al. 1997] solved an old problem of Hardy by showing the functional inverse of the function $(\log x)(\log \log x)$ is not in a field of logarithmic-exponential series. Seeking a universal domain to include Hardy fields was one motivation to study the field of transseries $\mathbb{T}$. Van der Hoeven [van der Hoeven 2006] defines, 'A transseries is a formal object, constructed from the real numbers and an infinitely large variable $x > 1$, using infinite summation, exponentiation and logarithm.' He notes the double significance of 'transseries'; the infinite series can be transfinite (have ordinal length beyond ω) and they can model asymptotic behavior of transcendental functions. The monumental [Aschenbrenner et al. 2017] provides a *first order model complete theory* in a vocabulary for an ordered, valued differential field, whose models include these transseries. They show that the complete theory of certain naturally described models of this theory are NIP (page 130) and 'o-minimal at infinity'. In [Aschenbrenner et al. 2016], they show that the natural embedding

---

[31] This earned him a Karp prize. The Karp prize is awarded by the ASL every five years for 'a connected body of research'.

[32] See a more detailed survey in the introduction to [Aschenbrenner et al. 2017].

of the differential field of transseries into Conway's field of surreal numbers with the Berarducci–Mantova derivation is an elementary embedding. Since the theory is unstable and hence creative (page 64), new models continually occur which cannot be decomposed into countable models as classifiable theories can; this lends some support to Ehrlich's thesis that the 'correct' universal model is universal for arbitrarily large models and so a proper class (e.g. the surreals [Ehrlich 2012]).

Hafner and Mancosu [Hafner & Mancosu 2008] critiqued Kitcher's unification model of explanation using examples from real algebraic geometry to assess the effect of formalization via real closed fields. The extensions to analysis described here suggest this area is ripe for further investigation[33] to evaluate the explanatory value of the concept of o-minimality.

Axiomatic analysis, definable analysis, and continuous logic[34] support Thesis 1 by providing three frameworks to solve new and old problems in mathematical analysis. Further, research using $L_{\omega_1,\omega}(Q)$ has opened new vistas around complex exponentiation (page 170). These different frameworks highlight the importance of the various choices required in our definition of formalization (Definition 1.0.1). *Axiomatic analysis* uses first order quantification over functions but has no means for evaluating functions. In *definable analysis*, the functions under study are terms in the language and so can be evaluated as can any further functions first order definable from them. These distinctions are enforced by the vocabulary and axioms. In continuous logic, a metric space is built into the logic and so one can study convergence. Thus, by providing local foundations, model theory provides a more faithful formalization of mathematical practice.

6.4   What Are the Central Notions of Model Theory?

In the provocative [Macintyre 2003b], Macintyre writes,

It seems to me now uncontroversial to see the fine structure of definitions as becoming the central concern of model theory, to the extent that one could easily imagine the subject being called definability theory in the near future.

While it is simply true that most 'structures' of ordinary mathematics can be construed as Tarskian structures, few model theorists can have failed to notice how unappealing the formulation is to other mathematicians. ... However, in those parts

---

[33] In fact the third approach discussed in ([Hafner & Mancosu 2008], 162) is directly generalized by o-minimality.

[34] Continuous logic (page 97) supports continuing research on C^*-algebras, von Neumann algebras, and other topics (Chapter 6.4).

of model theory with more relevance for algebra and geometry, the set-theoretical, 'rigorous' foundation seems to me to have given practically nothing, and arguably to be currently inhibiting.

On the other hand Scanlon writes,

Model theory is the systematic study of models. Of course, this answer invites the question: What is a model? While in common parlance, a model is a mathematical abstraction of some real system, problem or event; to a logician a model is the real object itself and models in the sense that it is a concrete realization of some abstract theory. More formally, a model M is a nonempty set M given together with some distinguished elements, functions defined on certain powers of M, and relations on certain powers of M. [Scanlon 2002]

These are in fact two aspects of the same idea. Certainly, *definability* is a central notion of model theory. But a collection of *sentences* of the language of fields defines a certain collection of fields, while, having fixed a field, say, the complex numbers, a collection of *formulas* (say conjunctions of equations) describes a definable set[35] (an algebraic variety). And in fact there is a deep connection between the two that arises from Shelah's work. Van den Dries [Dries 2005] remarks, 'It may be surprising that how many models a theory has of a given size can be relevant for the structure of the definable sets in a given model.' A simple example of this is a special case of the contrapositive of the main gap theorem: if a theory has fewer than the maximal number of models in some uncountable cardinality, it does *not* interpret either an infinite linear order or the ring of integers.

However, definability has to be considered in an even broader sense to encompass the standpoint of contemporary model theory. The goal is not just the understanding of a specific theory or the definable subsets of models of such theory, although that is often a happy consequence. The goal (Chapter 5.3) is to understand the relations amongst theories and to discover common syntactically definable relations that illumine concepts that appear in different areas of mathematics.

One unexpected application of stability theory, witnessing Thesis 2, is to the classification of finite structures. Cherlin summarizes the connection:

---

[35] Perhaps Macintyre intended something a bit more abstract than the second: ignore the field and just consider the structure of the definable sets. This notion generalizes into sheaf and topos theory. But he doesn't mention the model theoretic developments in that direction discussed here on page 303 in his paper. Macintyre's notes [Macintyre 2003a] give a less polemic and much more detailed account of how the abstract model theoretic ideas are integrated into number theoretic studies.

Shelah's notion of stability is one of the fundamental concepts of pure model theory; Lachlan realized that in the context of homogeneous relational systems, it is equivalent to 'smooth limit of finite' (cf. §4). This has led to a fruitful interaction of model theory and the theory of permutation groups, which involves an interplay between the group theoretic analysis of large finite structures, and the combinatorial analysis of their infinite limits. All of this depends ultimately on the classification of the finite simple groups. [Cherlin 2000]

The interaction between finite and infinite structures, described by Cherlin, has a striking success in the exploitation by Malliaris and Shelah of stability theoretic ideas to improve the famous Szemerédi lemma in finite graph theory [Malliaris & Shelah 2014]. Even more, the same circle of ideas led to the proof in ZFC of a 70 year old conjecture concerning topological invariants of the continuum.[36]

Most of the discussion here is of ideas developed in the last millennium. Given the chapter title of 'contemporary model theory', I owe something on recent developments. Much of Chapter 14 is devoted to the theory of abstract elementary classes; we focus here on compact logics. Peterzil and Starchenko [Peterzil & Starchenko 2010] developed much of the theory of complex analytic spaces in o-minimal expansions of $\mathfrak{R}$ even though there is no formal notion of convergence of power series or integrals. Half of the 2013 Karp prize[37] was awarded to Kobi Peterzil, Jonathan Pila, Sergei Starchenko, and Alex Wilkie for 'their efforts in turning the theory of o-minimality into a sharp tool for attacking conjectures in number theory, which culminated in the solution of important special cases of the André-Oort Conjecture by Pila.' This conjecture was a major topic of the spring 2014 program at the Mathematical Science Research Institute in *Model theory, arithmetic geometry and number theory*. Many of these developments spring from Hrushovski's interpreting groups and fields from abstract model theoretic constructions (Chapter 5.6); these methods rise almost to a third stage of the paradigm shift, embedding a wild problem into a tame context. We only sketched some of these contributions in Chapter 6.3. They require not only profound knowledge of the particular subject area but specific results and techniques from the general development of classification theory that preclude more detailed exposition here. Zilber has proposed an $L_{\omega_1,\omega}(Q)$ axiomatization for the theory of the complex exponential function (Chapter 7.1). A schema expressing Schanuel's conjecture

---

[36] Most such conjectures had been settled by forcing to show the conjecture is independent; unusually, in this case [Malliaris & Shelah 2013], it was proved.

[37] For award details see http://vsl2014.at/2014/07/awards-at-the-logic-colloquium/.

(Definition 7.1.3) is an important one of the axioms. The Pink–Zilber conjectures are far-reaching assertions about algebraic subgroups of semi-abelian varieties. An initial motivation for these conjectures is that they yield a first order sentence expressing Schanuel's conjecture. Active work on the more theoretical aspects of infinitary logic continues in the study of abstract elementary classes (Chapter 14.)

Hrushovski [Hrushovski 2012] introduced stability theory methods into the study of approximate groups; Breuillard's [Breuillard 2016] survey takes the model theory work as analogy; this may be an example of a formalism-free approach (Chapter 14).

Here are two major recent results in continuous logic (Chapter 4.2). The fine structure of ultrafilters also arises in the model theory of Banach spaces. In a series of papers beginning with [Farah et al. 2013, Farah et al. 2014] Farah, Hart, and Sherman used the fact that infinite-dimensional $C*$ algebras and II_1-factors[38] (two-one) are model theoretically unstable to prove that, under the negation of the continuum hypothesis, these objects always have at least two non-isomorphic ultrapowers. McDuff had asked if ultraproducts of II_1-factors were necessarily isomorphic which turns out to be equivalent to asking if they are all stable; they answer no. Goldbring and Sinclair use the omitting types theorem and model theoretic forcing[39] (for continuous model theory) in 'On Kirchberg's embedding problem' [Goldbring & Sinclair 2015].

In another direction, several conferences at Banff and Oaxaca have focused on 'neo-stability theory', the development of a finer classification of unstable theories, motivated both by abstract considerations [Shelah 2004] and by work with more applied motivations (e.g. [Hrushovski & Pillay 2011, Haskell et al. 2007]). Strikingly, this work has also contributed to combinatorics. One subclass of NIP theories is called *distal theories*. Chernikov and Starchenko [Chernikov & Starchenko 2016] strengthened Szemerédi's lemma for distal theories and showed that the 'common denominator' of the stable improvement and theirs holds for all NIP theories. Simon (2012) and Chernikov (2013) earned the Sacks prize[40] for work in this area. Simon[41] summed up the connections between Chapters 5 and 6.

---

[38] A factor is a simple von Neumann algebra, that is, a von Neumann algebra with no ideals. A II_1-factor is a factor satisfying further specialized conditions. www.math.ku.dk/english/research/conferences/what_is/a_II_1_factor.pdf.

[39] Model theoretic forcing is a technology invented by A. Robinson for constructing models, usually related to lowering quantifier complexity [Cherlin 1976].

[40] The ASL awards the Sacks prize for the most outstanding dissertation in mathematical logic.

[41] I freely translated from the French introduction to Simon's thesis [Simon 2011].

This domain is composed traditionally of pure model theory (study of the combinatorial properties of the class of definable sets) and applied model theory (study of particular structures, coming from algebra and geometry). The two parts nourish one another; the study of classical structures motivates the development of an abstract theory which in return then is applied to the concrete case. The principal applications are found in algebraic geometry, in arithmetic[42], and in real geometry.

---

[42] In our terminology, he means Diophantine geometry.

7 | Infinitary Logic

Infinitary logic occupies a peculiar place between first and second order logic. Since infinite conjunctions are manifest in the very definition it seems less secure than second order logic which after all has formation rules much like first order. Only after a little thought does one see that second order logic is immensely more powerful and has much stronger metatheoretic commitments, although incomparable in definability strength.

Recall that $L_{\omega_1,\omega}$ allows countable Boolean combinations of formulas but only finite strings of first order quantifiers; $L_{\infty,\omega}$ strengthens the logic by allowing conjunctions of any cardinality (Chapter 1.3.5). Moore [Moore 1997] describes the 'continuous development of infinitary logic' as beginning with the work of Henkin, Karp, Scott, and Tarski in the middle 1950s. Moore also reports a substantial prehistory including Boole, Peirce, Schroeder, Löwenheim, Skolem, Hilbert, Zermelo, and others. The key contentual distinction seems to be that the work up to Zermelo is entirely semantic. Only what Moore calls the 'continuous development' culminating in Karp's completeness theorem and the active school of the 1960s–70s fully incorporates the Tarskian syntax–semantics distinction.

We first consider the role of infinitary logic in formalizing mainstream mathematics by reviewing attempts to formalize the Lefschetz principle. This analysis relies on the completeness of the theory of the algebraically closed field of a fixed characteristic. In Chapter 7.1 we describe the role of categoricity in power for the logic $L_{\omega_1,\omega}$ and investigations of the theory of complex exponentiation. Chapter 7.2 surveys the connections among many areas of logic arising from Vaught's famous question, 'Does the continuum hypothesis hold for the countable models of a first order sentence?' In Chapter 7.3 we return to second order logic with a slight twist. We consider *second* order logic with infinite conjunctions of various lengths.

Why Infinitary Logic? The Lefschetz Principle: We discuss an example of logicians carefully choosing an appropriate logic to formalize a well-known heuristic principle. These local foundations are tested against a goal set by André Weil. The Lefschetz principle was long known informally by algebraic geometers and appeared in Lefschetz's 1953 text, *Algebraic Geometry*.

Barwise and Eklof describe the issues around formalizing the Lefschetz principle as follows.

What we call Lefschetz principle has been stated by Weil ([Weil 1962], 306) as follows : 'for a given value of the characteristic p, every result, involving only a finite number of points and varieties, which has been proved for some choice of universal domain remains valid without restriction; there is but one algebraic geometry of characteristic p for each value of p; not one algebraic geometry for each choice of universal domain.' Weil says that a formal proof of this principle would require 'a "formal metamathematical" characterization of the type of proposition' to which it applies; 'this would have to depend upon the "metamathematical" i.e. logical analysis of all our definitions.' [Barwise & Eklof 1969]

Seidenberg [Seidenberg 1958] argues that Weil's formulation is weaker than Lefschetz intended. He gives a formulation (the minor principle: Theorem 4.3.3) in first order logic, which encompasses Weil's version. But Seidenberg argues that this does not really reflect mathematical practice and conjectures that the Lefschetz principle should be formulated in terms of what he call almost-elementary sentences, a fragment of $L_{\omega_1,\omega}$. Barwise and Eklof [Barwise & Eklof 1969] address Weil's question head-on. 'Thus, in contrast to previous mathematical formulations of the Lefschetz principle which arose from general logical considerations, ... our starting point has been an analysis of the definitions of algebraic geometry.' They extend Theorem 4.3.3 to a transfer principle in an infinitary version of finite type theory to encompass such notions as integers, affine and abstract varieties, polynomial ideals, and finitely generated extensions of the prime field. In [Eklof 1973], Eklof builds on work of Feferman to construct a simpler logic than that in [Barwise & Eklof 1969], a many-sorted language for $L_{\infty,\omega}$.

7.1 Categoricity in Uncountable Power for $L_{\omega_1,\omega}$

In this section we survey the status of categoricity and categoricity in power for sentences of $L_{\omega_1,\omega}$. The main results are in [Shelah 1983a, Shelah 1983b], expounded in [Baldwin 2009]. Since, for a countable vocabulary, there are $2^{\aleph_0}$ inequivalent sentences and $2^{2^{\aleph_0}}$ theories but a proper class of structures, some theories must fail to be categorical. By Scott's theorem, every *complete* sentence of $L_{\omega_1,\omega}$ is categorical in $\aleph_0$. In contrast to the first order case, there are countable structures that are categorical (have exactly one model) for $L_{\omega_1,\omega}$. By the downward Löwenheim–Skolem theorem (page 90), no *sentence* of $L_{\omega_1,\omega}$ can have a unique model which is uncountable. But the $L_{\omega_1,\omega}$-*theory* of the reals is categorical.

A countable structure is categorical if and only if it has no proper $L_{\omega_1,\omega}$-elementary submodel. For sentences in $L_{\omega_1,\omega}$, categoricity in power $\aleph_1$ *implies* the existence of a complete sentence satisfied by the model of cardinality $\aleph_1$. It is open whether this implication holds for $\aleph_2$-categoricity in $L_{\omega_1,\omega}$.

The best generalization of Morley's theorem to $L_{\omega_1,\omega}$ is due to Shelah [Shelah 1983a, Shelah 1983b]. Shelah shows that one can more profitably study this subject by focusing on classes of the form $EC(T, Atomic)$, the class of atomic[1] models of a complete countable first order theory. The class of models of a complete sentence ϕ of $L_{\omega_1,\omega}$ is in 1–1 correspondence with an $EC(T, Atomic)$-class (Chapter 6 of [Baldwin 2009]). Making this translation is a key simplification. An $EC(T, Atomic)$-class is *n-excellent* if it is possible to find a unique minimal amalgamation[2] of n independent countable models in the class and excellent if n-excellent for every finite n.

Theorem 7.1.1 (Shelah 1983) *Assume only ZFC. If K is an* excellent $EC(T, Atomic)$-*class then if it is categorical in one uncountable cardinal, it is categorical in all uncountable cardinals.*

Theorem 7.1.2 (Shelah 1983) *Assume ZFC*$+2^{\aleph_n} < 2^{\aleph_{n+1}}$ *for finite n. If an* $EC(T, Atomic)$-*class K is categorical in $\aleph_n$, for all $n < \omega$, then it is excellent.*

The necessity to assume categoricity in $\aleph_n$ for *all* finite n was shown in [Hart & Shelah 1990] and refined in [Baldwin & Kolesnikov 2009]. Thus for $L_{\omega_1,\omega}$ the study of categoricity in power has made a substantial start but much remains open; an outstanding question from a foundational standpoint is whether the *very weak generalized continuum hypothesis* (VWGCH: for all n, $2^{\aleph_n} < 2^{\aleph_{n+1}}$) is actually needed. A few papers aim to find an analog of the stability hierarchy in this framework [Grossberg & Hart 1989, Baldwin & Shelah 2012, Vasey 2016b].

The connection between these theorems and the solution of the main gap conjecture (Chapter 5.5) illustrate the relation between studies in first order and infinitary logic. Shelah began the study of n-dimensional amalgamation (or rather the stronger requirement of n-excellence) in his study of categoricity in $L_{\omega_1,\omega}$ in the papers, first submitted in 1976, which

---

[1] M is *atomic* if each finite sequence from M realizes a complete type over the empty set.

[2] Let $X = \{x_1, x_2, x_3\}$ be independent elements in an infinite-dimensional vector space. Each of the seven proper subsets Y of X generates a ≤ 2-dimensional subspace V_Y and together they generate a 3-space. More generally, such a (unique minimal) amalgam exists in each model with an abstract dependence relation; we say the class (is n-excellent) has n-amalgamation.

eventually appeared[3] as [Shelah 1983a, Shelah 1983b]. During 1981–82, Shelah realized that in the first order case n-excellence reduced to 2-excellence. He then named the first order version of failure of 2-excellence, the dimensional order property (DOP). With this property he could solve the main gap problem for $\aleph_1$-saturated models, using the slightly weaker $\aleph_\epsilon$-saturation. In 1982, he realized he could study arbitrary models using the omitting types order property; this is a special case of the $L_{\omega_1,\omega}$-order property, which itself was introduced in [Shelah 1972] and continued in [Grossberg & Shelah 1983, Grossberg & Shelah 1986]. The first order result [Shelah 1982] beat into print the infinitary results that inspired it.

There are important mathematical structures, e.g. complex exponentiation,[4] which exhibit the Gödel phenomena and so cannot, at least a priori, be analyzed by stability techniques *in first order logic*. However, Zilber [Zilber 2005b, Zilber 2004, Baldwin 2009] has conjectured a means for such an analysis in the logic $L_{\omega_1,\omega}(Q)$. The following number theoretic conjecture plays a central role.

Definition 7.1.3 (Schanuel's Conjecture) *If n elements $a_1, \ldots a_n$ of the complex field $\mathbb{C}$ are linearly independent over the rational field Q, the transcendence degree of the field $Q(a_1, \ldots a_n; e^{a_1}, \ldots e^{a_n})$ is at least n.*

Zilber's proposed axioms for complex axiomatization $(C, +, \cdot, 0, 1, \exp)$ are:

(1) the first order theory of algebraically closed fields plus $\exp(x + y) = \exp(x) \cdot \exp(y)$;
(2) specify in $L_{\omega_1,\omega}$ that the kernel of the exponential map is isomorphic to $(Z, +)$ and in $L_{\omega_1,\omega}(Q)$ that the closure of a countable set is countable;
(3) Schanuel's conjecture;
(4) A schema asserting that all 'reasonable' systems of exponential equations are solvable;
(5) A variant[5] of Shelah's excellence condition.

He proved in [Zilber 2005b] an abstract theorem establishing the consistency and categoricity in power of so-called quasiminimal excellent classes. In [Zilber 2004], an ingenious variant on the Hrushovski construction (page 83) shows the axioms above define such a class: the *Zilber fields*. Amazingly, [Bays et al. 2014] showed that the excellence

---

[3] Grossberg provided the history in this paragraph.
[4] That is, the structure, $(\mathbb{C}, +, \times, 0, 1, e^x)$.
[5] Zilber conceived his notion independently but at least ten years later.

hypothesis was redundant. Whether these axioms are in fact true for complex exponentiation remains completely open. Little progress has been made in 50 years on the Schanuel conjecture and Zilber's fourth schema led to the Pink–Zilber conjectures (page 165) which are a lively topic in algebraic number theory.

Already in the 1980s, Shelah shifted much of his emphasis from the syntactic notion of $L_{\omega_1,\omega}$ to abstract elementary classes; Zilber's quasiminimal class is also given in a formalism-free manner; see Chapter 14.

7.2 The Vaught Conjecture

In the seminal paper, Vaught asked

Can it be proved, without the use of the continuum hypothesis, that there exists a complete theory having exactly $\aleph_1$ non-isomorphic denumerable models? [Vaught 1961]

The problem continues to excite interest in many areas of logic after more than 50 years including Vaught Conjecture Conferences in 2005 (Notre Dame) and 2015 (Berkeley). We expound various approaches to this problem from the standpoints of model theory, computability theory, and both descriptive and axiomatic set theory. At first sight, this question might appear to be independent of ZFC: it is, after all, the continuum hypothesis for the set of countable models of a complete first order theory. But the problem[6] can be rephrased as:

First Order Vaught Conjecture Every complete first order theory has either countably many or a perfect (page 123) set of countable models.

Morley's proof [Morley 1970] that the number of countable models of any T is finite, $\aleph_0, \aleph_1$, or $2^{\aleph_0}$ (via a perfect set) elevated interest in Vaught's conjecture. I say the proof rather than the result because the key step is to analyze T by building an uncountable sequence of countable fragments of $L_{\omega_1,\omega}$, L_α such that for arbitrarily large α there may be a proper extension $T_{\alpha+1}$ of T_α to a complete theory in L_α. Vaught's conjecture then becomes that there is such a sequence of length $\aleph_1$. Morley shows that if the sequence stops, there are either countably many or $2^{\aleph_0}$ models. Morley's analysis works the same way even if we begin with a sentence of $L_{\omega_1,\omega}$. Thus Vaught's conjecture was reformulated as:

---

[6] If the original version of Vaught's conjecture is provable in ZFC then so is the perfect set version and that version is absolute (page 46).

$L_{\omega_1,\omega}$ **Vaught Conjecture** Every sentence of $L_{\omega_1,\omega}$ has either countably many or a perfect set of countable models.

Thus, we are discussing this problem in the infinitary logic chapter. However, Vaught and others further reformulated the problem into one in descriptive set theory [Gao 1996]. First, one can view the collection of countable models as a standard Borel space on 2^ω. Then isomorphism is a Σ_1^1-equivalence relation on this space. Morley's theorem was generalized to Burgess' theorem: Every Σ_1^1-equivalence relation on a standard Borel space has either $\leq \aleph_1$-classes or a perfect set of classes. Still further, the equivalence relation can be seen as the *orbit space* of the *logic action*, the action of the group S_ω of all permutations of ω on 2^ω. Then one generalizes the conjecture still more.

Topological Vaught Conjecture Whenever a Polish group acts continuously on a Polish space, there are either countably many orbits or continuum many orbits.

Morley's result applies equally well to pseudo-elementary classes (page 39) in $L_{\omega_1,\omega}$. But Vaught's conjecture fails in this generality. One first order pseudo-elementary counterexample, pointed out by Ken Kunen,[7] is the class of 'groupable orders', linear orders which admit a group operation compatible with the order (so the group operation is in the expanded vocabulary). Such countable linear orderings are of the form $Z^\gamma \times \mathbb{Q}$ (where Z^γ is a specific notion of γth power of an ordered group; see [Rosenstein 1982]). There are $\aleph_1$ such countable orderings.

From a model theoretic perspective, the stability hierarchy provides a partial solution in the first order case. The Vaught conjecture is true for ω-stable theories [Shelah et al. 1984] and for superstable theories of finite U-rank [Buechler 2008]. The difficulty in actually computing invariants is illustrated by a surprising twelve year gap between knowing the conjecture held for differentially closed fields of characteristic 0 and knowing which way the answer went. Shelah proved the conjecture for ω-stable theories in 1980–81; so it holds for differentially closed fields of characteristic 0. The proof that there are continuum many models came in 1992 by Hrushovski and Sokolovic. (See [Marker 2007] and [Pillay 1996] for details.) The argument uses substantial work in geometric stability theory and an argument of Manin on elliptic curves to show the theory has the *essentially non-isolated dimensional order property*, eni-dop.[8]

---

[7] Harvey Friedman gave an earlier more complicated example.
[8] See Example 8.3.4.

Another ostensibly model theoretic approach to Vaught's conjecture aims to approximate the countable by the uncountable and take advantage of the analysis of the number of uncountable models made possible by the use of stationary sets and arguments concerning the spectrum of first order theories. Thus, Baldwin observed [Baldwin 2007] that any *first order* counterexample to Vaught's conjecture has $2^{\aleph_1}$ models in $\aleph_1$. This follows by propositional logic from two earlier theorems of Shelah. One, with distinct descriptive set theoretic overtones [Shelah 1978, Baldwin 1989], asserts that a theory which is not ω-stable has $2^{\aleph_1}$ non-isomorphic models in $\aleph_1$; the second is Vaught's conjecture for ω-stable theories. In the same article, Baldwin asked whether a counterexample to Vaught's conjecture must have a model in $\aleph_2$. Hjorth replied immediately. He applied his ongoing work on the Hanf number of complete sentences of $L_{\omega_1,\omega}$ and descriptive set theory, to show that if there is a counterexample to Vaught's conjecture then there is an $L_{\omega_1,\omega}$-counterexample with no model in $\aleph_2$. A reworking in [Baldwin et al. 2016a] of Hjorth's argument shows it has little to do with either descriptive set theory or $\aleph_2$. They define the notion of a *receptive model*. A receptive model M is a countable model that contains an infinite definable set P of *absolute indiscernibles*.[9] Suppose there is an M which has a Scott sentence[10] ϕ_M and has no model in $\aleph_2$, and P^M is receptive. Then one gets Hjorth's theorem by relativizing an arbitrary counterexample to Vaught's conjecture to P^M. They construct such an M by a modification of the Fraïssé method (similar to Hjorth); moreover every model of ϕ_M in $\aleph_1$ is maximal, strengthening, 'no model in $\aleph_2$'.

After Hjorth's result, Sacks and others approached Vaught's conjecture by trying to show any counterexample to it has a model of cardinality $\aleph_2$. This possibility was given credence since Harrington [Marker 2011] had in the early 1970s proved (but not published), using admissible set theory and Levy absoluteness between $H(\aleph_1)$ and $H(\aleph_2)$, that every Vaught counterexample has models in $\aleph_1$ of arbitrarily high Scott rank. In the last few years there have been three further proofs of this result [Baldwin et al. 2016a, Larson 2017, Knight et al. 2016]; models of high Scott rank are found, but none have proper extensions in $\aleph_1$. The real difficulty [Baldwin et al. 2016a] is to show every counterexample has a pair $M \precnsim N$ in $\aleph_1$; but high Scott rank says nothing about this issue.

---

[9] P is a set of absolute indiscernibles in M if every permutation of P extends to an automorphism of M.

[10] A *Scott sentence* is a sentence of $L_{\omega_1,\omega}$ that has a unique countable model. Every countable model, indeed, every model that realizes only countably many $L_{\omega_1,\omega}$-types over the empty set, satisfies a Scott sentence.

By focusing on the orbit equivalence relations the spotlight can be turned from the number of classes to analyzing the models in a single isomorphism class. The most striking result in this area turns the problem to one in computability theory. One intuition about Vaught's conjecture is that easily describable classes of countable models that have $\aleph_1$ members up to isomorphism are closely related to ordinals. A natural example is the ordinals themselves. Spector proved that any ordinal that is hyperarithmetic in a real x is recursive in x. Montalban [Montalban 2013] shows the analog for any theory with fewer than the maximal number of countable models happens almost everywhere (i.e. on a cone of Turing degrees).

Theorem 7.2.1 (Montalban, ZFC + PD) *An $L_{\omega_1,\omega}$ sentence, T, has less than $2^{\aleph_0}$ many countable models if and only if we have that, for every $X \in 2^{\aleph_0}$ on a cone of Turing degrees, every X-hyperarithmetic model of T has an X-computable copy.*

A standard connection with descriptive set theory is to analyze the complexity of the isomorphism relation on models of a given theory. Shelah's counting of models is too crude; refine it by analyzing the Borel complexity of isomorphism ($\approx_T$) on models of a first order theory T. Here are two surprising results [Laskowski et al. 2017].

Theorem 7.2.2 (Koerwien+Ulrich) *There is an ω-stable, depth 2 theory T for which $\approx_T$ is properly Σ^1_1. But $\approx_T$ is not Borel complete.*

For the second, we must establish some notation.

Definition 7.2.3 (Refining equivalence relations:) *Let $L = \langle E_n : n < \omega \rangle$ and consider L-theories T that say: Each E_n is an equivalence relation; E_0 consists of a single class; Each E_{n+1} refines E_n, i.e., $E_{n+1}(a,b)$ implies $E_n(a,b)$. In order to make T complete, one need only say how many classes E_{n+1} partitions each E_n-class into:*
REF_ω says: Each E_{n+1}-class partitions each E_n-class into infinitely many classes.
REF_2 says: Each E_{n+1}-class partitions each E_n-class into two classes.

Theorem 7.2.4 (Laskowski–Rast–Ulrich)

(1) $\approx_{REF_\omega}$ *is Borel complete so properly Σ^1_1.*
(2) $\approx_{REF_2}$ *is superstable and properly Σ^1_1 but not Borel complete.*

Note REF_ω is strictly stable while REF_2 is superstable. The study of Borel invariants is a level deeper that of the model theoretic version. The model theorist seeks sets of cardinal invariants while the descriptive set theorist approach seeks the more refined analysis of the level in the hierarchy of Borel reductions of the isomorphism equivalence relation on the models.

After connecting Vaught's conjecture with the spectrum of Scott heights of models of the theory, Lascar [Lascar 1985] formulates the interest in the conjecture as:

I hope that I have convinced the reader that Vaught's conjecture is not merely a question of counting models of a theory. It is rather a way of expressing in precise mathematical terms our intuition concerning the existence of absolute invariants for structures in a countable language.

Thus, the interest of Vaught's conjecture stems not merely from its age but from its ability as a test question to stimulate research in three areas of logic. A positive solution to the topological Vaught's conjecture would provide more evidence for the generality of the use of Polish groups to analyze phenomena in numerous areas of mathematics. A counterexample would validate the importance of model theoretic methods to assign invariants, and stimulate even further questions about exactly where in the stability hierarchy a positive solution to the Vaught conjecture lies.

7.3   Déja vu: Categoricity in Infinitary Second Order Logic

We give here an example of Thesis 3, the importance of choice of logic in determining an appropriate formalization. We rehearse here some results of Hyttinen, Kangas, and Väänänen [Hyttinen et al. 2013] that identify in a systematic way a proper class of categorical structures.[11] Consider the logic $L^2_{\kappa,\omega}$ which allows first and second order quantification and conjunctions of length $< \kappa$. In this family of logics, as κ varies there are a class of sentences so the cardinality argument for the existence of non-categorical structures fails. In fact, every structure of cardinality κ is categorical in $L^2_{\kappa^+,\omega}$. The goal is to identify those structures of cardinality κ that are categorical in $L^2_{\kappa,\omega}$ (κ not $\kappa+$). As this work draws on the first order stability hierarchy discussed in Chapter 2.4, we give more detail on the main gap theorem.

Any text in stability theory shows that if T is stable then via the notion of non-forking an independence relation can be defined on all models of T which generalizes the independence notion in combinatorial geometries discussed in Chapter 3.3. In general the closure relation is not a geometry because $cl(cl(X)) \neq cl(X)$. But on the set of realizations of a regular

---

[11]   The authors use 'M is characterizable in $\mathcal{L}$' to mean M is the unique model of an $\mathcal{L}$-sentence; in $\mathcal{L}$ section we will write '$\mathcal{L}$-categorical'.

(page 136) type it is. Thus in a model M and for any regular type p with domain in M, we can define the dimension of the realizations of p in M.

If T is not stable or even not superstable, T has 2^κ models in every uncountable κ (Theorem 8.3.3). We recounted in Chapter 5.5 that if either DOP or OTOP holds T has the maximal number of models in each uncountable cardinal. While if the dimensional order property and the omitting types order property do not hold (NDOP and NOTOP), T is classifiable.

For a classifiable theory (Chapter 5.5) the number of models in $\aleph_\alpha$ is bounded by $\beth_\beta(\alpha)$ where $\beta = \beta(T)$ is a bound on the rank of the decomposing tree for all models of T (independently of cardinality). [Shelah 1990] claims that each model of such a theory is characterized by a sentence in a certain 'dimension logic', quantifying over the dimension of a set. Unfortunately there are technical difficulties in the definition of this logic.[12] However, [Hyttinen et al. 2013] shows how to find such a categorical sentence in $L^2_{\kappa,\omega}$. Thus they obtain (we state a slightly less general form):

Theorem 7.3.1 (Hyttinen, Kangas, and Väänänen) *Suppose that κ is a regular cardinal such that $\kappa = \aleph_\alpha$, $\beth_{\omega_1}(|\alpha| + \omega) \leq \kappa$, and $2^\lambda < 2^\kappa$ for all $\lambda < \kappa$. The countable complete theory T is classifiable if and only if for every model M of T with $|M| \geq \beth_{\omega_1}$, the $L^2_{\kappa,\omega}$ theory of M is categorical.*

A highly technical argument from classifiability yields that the decomposition of the models (and the dimensions of the types involved) can be defined in $L^2_{\kappa,\omega}$. Conversely, if a theory lies on the chaotic side of the main gap, it has 2^κ models in κ. But there are only $2^{<\kappa}$ sentences in $L^2_{\kappa,\omega}$ so there must be a sentence which is not categorical in the logic $L^2_{\kappa,\omega}$.

So using the virtuous properties developed in first order logic, the authors are able to uniformly identify a large family of structures with cardinality κ that are categorical in $L^2_{\kappa,\omega}$. But categoricity is used in Huntington's role[13] of 'sufficiency'. It is again a test of an axiomatization. In contrast to the ad hoc search for the axioms of the fundamental structure, Shelah's structure theorems provide a strategy for obtaining the axiomatization. Categoricity of a theory is an informative property in this situation. But while the axiomatizations of the fundamental structures informed us about the principles underlying proofs in the underlying number theory and real analysis, these axiomatizations inform us about the structure of the models of the underlying classifiable first order theory.

---

[12] These issues are circumnavigated in [Bouscaren & Hrushovski 2006].
[13] Footnote 3.1.

8 | Model Theory and Set Theory

We chronicle here the divorce of first order model theory from set theory. Recalling Maddy's [Maddy 2007] injunction (page 13) to 'adjudicate the methodological questions of mathematics – what makes for a good definition,' we explain this major aspect of the paradigm shift as the decision to choose definitions of model theoretic concepts that reduce the set theoretic overhead. To explore this development we follow the prescription of Thesis 2: connect various local formalizations. That is, we look at formalized set theory ZFC and analyze its use in model theory.

These investigations expand on recent work studying the entanglement of mathematics with set theory [Kennedy 2015, Kennedy 2013, Parsons 2013, Väänänen 2012]. These works, as on page 46, find that the amount of entanglement depends crucially on the logic chosen for formalizing mathematics.[1] Paraphrasing Väänänen [Väänänen 2012], it is very difficult to tell the difference between the *second order view* (mathematics is the study of higher order properties of structures) and *the set theory view* (formalizing mathematics in set theory); see Chapters 7.3 and 11. The high entanglement of second order logic with set theory is expressed more precisely by it being *symbiotic* with the power set [Kennedy 2015]. Parsons' general theme in [Parsons 2013] is closer to the approach in this chapter. He considers, a bit more generally, logic and mathematics, and argues that the ontological commitments of first order logic are relatively weak. But his argument is not a general metaphysical one. Similarly to Maddy's Second Philosopher he precisely analyzes the commitments within second order arithmetic for specific results in basic model theory. For example, he points to the fact[2] [Simpson 2009] that the completeness theorem is equivalent to weak König's lemma over RCA_0. Analogously we investigate the necessity in contemporary first order and infinitary model theory of specific ZFC techniques or specific axioms. To what extent should, could, or must a model theorist go beyond ZFC?

---

[1] We leave the notion of entanglement vague and refer to more precise expositions. E.g. Kennedy discusses the entanglement with set theory of logics with various Lindstrom quantifiers in [Kennedy 2015].

[2] This point becomes essential in Fact 10.3.4.

What are the specific commitments for doing model theory (as opposed to mathematics in general)? We consider six aspects of the interaction of set theory and model theory. In the first section, we describe an apparent close relation between model theory and set theory that was dissolved by the paradigm shift. Chapter 8.2 expounds three combinatorial results, provable in ZFC, that are widely used in model theory and their connection with indiscernibility. Chapter 8.3 explores the question, *why is $\aleph_0$ exceptional?*, while introducing a combinatorial technique that is part of the answer. Our discussion in Chapter 8.4 of *the entanglement of model theory with cardinality* begins by noting that the basic model theoretic tenet interpreting the equality relation as identity builds in the centrality of cardinality. Then we explain the connection between two of the most important counting functions in model theory. We turn in Chapter 8.5 to our theme of the importance of *local* foundations by pointing to some results that require the axiom of replacement as theorems about arbitrary first order theories but are provable in Zermelo set theory for stable or simple theories. Finally Chapter 8.6 considers how axiomatic set theory continues to play a role in model theory, even of the most applied sort. Chapter 8.7 emphasizes the complementary arguments that dominate the chapter: the absoluteness of the concepts of first order model theory and the increasing acceptance across mathematics of the type of infinitary combinatorics it uses.

8.1   Is There Model Theory without Axiomatic Set Theory?

The generally accepted answer around 1970 was a resounding no! Reflecting in 1997 ([Shelah 2000b], 41), Shelah recalls, 'in '69 Morley and Keisler told me that model theory of first order logic is essentially done and the future is the development of model theory of infinitary logics (particularly fragments of $L_{\omega_1,\omega}$). By the eighties it was clearly not the case and attention was withdrawn from infinitary logic (and generalized quantifiers, etc.) back to first order logic.' He had begun his account with some evidence for this claim.

During the 1960s, two-cardinal theorems were popular among model theo-rists. ... Later the subject becomes less popular; Jensen complained 'when I start to deal with gap n two-cardinal theorems, they were the epitome of model theory and as I finished, it stopped to be of interest to model theorists.' I sympathize, though model theorists have reasonable excuses: one is that they want ZFC-provable theorems or at least semi-ZFC ones (see 1.20t of [Shelah 2000a]) the second is that it has not been clear if there were any more. ([Shelah 2000b], 41)

What Shelah modestly omits from this description is how it came about during the 1970s that model theorists had the option of ZFC-provable theorems. We now refine our overview in Chapters 1.3 and 2.4 of the crucial point: the shift from the study of logics to the study of theories.

We begin with some background on 2-cardinal theorems. We discussed in Chapter 4.2 the Löwenheim–Skolem–Tarski theorem, which asserts that a first order theory in a countable vocabulary that has an infinite model has a model in every infinite cardinality. Vaught generalized the situation by asking not just about the cardinality of the whole model but about the cardinality of a definable subset. We say that M is a *two-cardinal model* if there is an infinite definable (Definition 1.3.4) subset of M with cardinality strictly less than $|M|$. A simple example would be a vocabulary which contains two unary predicates P, Q; the theory T requires that both are infinite and they partition the universe.[3] Then it is clear that for any pair of infinite cardinals $\kappa \geq \lambda$ there is a model M of T such that $|M| = \kappa$ and $|Q^M| = \lambda$. Is it the case that for every model and smaller definable set that every pair of cardinals is possible? We need to fix some notation.

(1) A *two-cardinal model* is a structure M with a definable subset D with
$\aleph_0 \leq |D| < |M|$.

(2) We say a first order theory T in a vocabulary with a definable subset D
admits (κ, λ) if there is a model M of T with $|M| = \kappa$ and $|D^M| = \lambda$.
So T has a two-cardinal model just if it admits (κ, λ) for some κ, λ with
$\kappa > \lambda$. We write $(\kappa, \lambda) \to (\kappa', \lambda')$ if *every theory* that admits (κ, λ) also
admits (κ', λ').

The first two-cardinal theorem was tied closely to categoricity in power. Vaught proved:

Theorem 8.1.1 *An $\aleph_1$-categorical theory T in a countable vocabulary τ has no two-cardinal model.*

This establishes the property which along with ω-stability characterizes (Chapter 3.3) countable theories that are categorical in an uncountable power. Since it is easy to see that an $\aleph_1$-categorical theory has a model in $\aleph_1$ where every infinite definable set is uncountable,[4] Theorem 8.1.1 follows immediately from[5]

---

[3] The axioms include a schema of sentences saying for each n, 'there exist at least n elements in P' and similarly for Q. We introduced the predicate P to avoid writing negation symbols.

[4] Take an $\aleph_1$-chain of models M_α, systematically realizing each formula with parameters from M_α in $M_{\alpha+1}$.

[5] Theorem 8.1.2 appears in [Morley & Vaught 1962]; it generalized a slightly weaker result than Theorem 8.1.1 in [Vaught 1961].

Theorem 8.1.2 (Vaught) *For any $\kappa > \lambda$, $(\kappa, \lambda) \to (\omega_1, \omega)$.*

The proof of Theorem 8.1.2 illustrates both the importance of formalization and the mathematical importance of the push-through principle, Fact 3.2.3. First note that if T has a two-cardinal model, witnessed by a τ-predicate $D(x)$, this implies the consistency[6] of the new theory T', in the extended vocabulary obtained by adding a new predicate P to τ, that satisfies: (i) The τ-reduct (page 39) of both the model M of T' and the substructure $P(M)$ are models of T; and (ii) $D(M)$ is contained in $P(M)$, which is a *proper* substructure of M.

Secondly, working in τ', we can assume that $P(M)$ and M are both homogeneous as τ-structures and realize the same τ-types over $\emptyset$; thus they are isomorphic τ-structures (page 128). Construct a continuous increasing chain of models $\langle M_\alpha : \alpha < \omega_1 \rangle$, which are all isomorphic and with $D(M_\alpha) = D(M_0)$ for all α. The successor stage is guaranteed by the push-through construction with $P(M_{\alpha+1} \restriction \tau) = M_\alpha \restriction \tau$, Fact 3.2.3, while the limit stage follows since the union of homogeneous models is homogeneous and all models in the sequence realize the same τ-types over $\emptyset$.

Later, Vaught asked ([Vaught 1965], 390),

Question 8.1.3 *For which $\kappa > \lambda$, $\kappa' \geq \lambda'$, does (κ, λ) imply (κ', λ')?*

Vaught answered with his Löwenheim–Skolem theorem for cardinals far apart. Before stating the theorem, let's see that it is the best possible. Consider predicates P_0 and P_1 and an extensional relation E (page 113) so that we can think of an $a \in P_1$ coding a subset of P_0, namely, $\{b \in P_0 : E(b, a)\}$. Then in any model M of the theory T axiomatizing this conception, $|P_1^M| < 2^{|P_0^M|}$. So if λ is $\aleph_0$, κ is at most $2^{\aleph_0}$. It is easy to iterate[7] this to get gaps with n iterations of power set. For each theory T_n, every model has at most $\beth_n(\aleph_0)$ elements.[8] The next Vaught theorem tells us that this is the worst that can happen.

Theorem 8.1.4 (Vaught) $(\beth_\omega(\lambda), \lambda) \to (\mu_1, \mu_2)$ *when* $\mu_1 \geq \mu_2$.

This evolved into a 'big question', 'For what quadruples of cardinals does $(\kappa, \lambda) \to (\kappa', \lambda')$ hold?' In the 1960s a small industry of proving such theorems developed, a small industry which turned out to be intimately connected with axiomatic set theory. Further progress on this question involved a massive intertwining of model theory and set theory. The choice

---

[6] The consistency is guaranteed by applying the Löwenheim–Skolem theorem to interpret $P(M)$ as a *proper* elementary submodel N of a (κ, λ) that contains all realizations of D.

[7] In T_2 there will be a new predicate P_3 whose elements code subsets of P_1.

[8] See page 142 for the $\beth$-notation.

of the cardinals evolved through more complicated two-cardinal theorems/ conjectures of Chang whose proofs used the GCH and then Jensen's abstruse notion of a morass. Foreman [Foreman 2010] lists as tools for classical two-cardinal transfer results from the 1960s: morasses, Erdős cardinals, $V = L$, etc. A later result of Foreman showed the equivalence between such a transfer theorem and a partition theorem[9] on cardinal numbers.

The introduction of the stability hierarchy revolutionized the situation. Independently, Shelah [Shelah 1969]and Lachlan [Lachlan 1972] revised the problem and solved it in ZFC.

Theorem 8.1.5 (Shelah, Lachlan) *If T is stable then*
$\forall(\kappa > \lambda, \kappa' \geq \lambda')$ *if T admits* (κ, λ) *then T also admits* (κ', λ').

Much later Timothy Bays [Bays 1998] established the analogous result for o-minimal theories (see Chapter 2.4).

Theorem 8.1.6 (Bays) *If T is o-minimal then* $\forall(\kappa > \lambda, \kappa' \geq \lambda')$ *if T admits* (κ, λ) *then T also admits* (κ', λ').

By reversing the question, the reliance on axioms beyond ZFC is eliminated.

The *set* theorist asks, 'For which *cardinals* (κ, λ) does T have a (κ, λ)-model for every *theory T?*'

The *model* theorist asks, 'For which *theories T* does T have a (κ, λ)-model for every T for all *cardinals* (κ, λ)?'

The first question is far beyond ZFC. The second can be answered for many important cases in ZFC; and we can explain why it fails in some other cases. In particular, we will see that ZFC itself is one of the 'complicated' theories. This switch of the question exemplifies the *paradigm shift*.

Here is an another example of the importance of the hierarchy. In the seminal paper of Morley and Vaught [Morley & Vaught 1962] the question of in which cardinalities there are saturated models appeared complicated. Assuming the GCH, it has a simple answer: all regular cardinals. The notion of *special model* has to be introduced to deal with singular cardinals. In ZFC, classification theory yields:

Theorem 8.1.7 *Let T be a countable theory.*

(1) *If T is ω-stable then it has a saturated model in every cardinality.*
(2) *If T is strictly[10] superstable then it has a saturated model in every cardinality greater than or equal to the continuum.*

---

[9] Partition theorems are a family of combinatorial results explained around Fact 8.2.5.
[10] In discussing the hierarchy we say a theory is strictly in class X if it satisfies X but fails the lower conditions in the hierarchy. So here we could have said superstable but not ω-stable.

(3) *If T is strictly stable then it has a saturated model in every cardinality λ satisfying $\lambda^{\omega} = \lambda$.*

(4) *If T is unstable then the cardinals in which it has a saturated model is a more complicated issue and may depend on set theory.*

We studied both the importance of saturation and the nature of the stability hierarchy in Chapter 5.2. As noted in Chapter 2.4, the stability hierarchy is given by absolute syntactic properties. That is, the computation of when a *stable* theory has a saturated model is impervious to changes in the background set theory. The crucial point is that instead of asking that *every theory satisfies a property* (e.g. two-cardinal transfer or existence of saturated models), one asks, *for which theories does the property hold?*

Thus we have seen, first order model theory thrives without extending the set theory beyond ZFC. This frees model theory for the applications to traditional mathematics discussed in Chapter 6. We will see in Chapter 8.6 that for infinitary logic axiomatic set theory again plays an important role; but even that role is perhaps more limited than what I described for first order logic around 1970. Thus while even first order logic is deeply entangled with *cardinality*, infinitary model theory (Chapter 7) is entangled with *cardinal arithmetic* but first order model theory is not.

8.2   Is There Model Theory without Combinatorial Set Theory?

In this chapter we investigate the notion of indiscernability from two standpoints. On the one hand, we show how the requirements on vocabulary set down in Chapter 1.2 illuminate issues concerning the identity and indiscernibility dating from at least Leibniz. On the other, we show how the existence of sets of distinct elements that are in a technical sense indiscernible is a powerful tool in model theory.

Leibniz's law sometimes is used to name the conjunction of two principles:

(1) the indiscernibility of identicals
(2) the identity of indiscernibles.

The first is a truism of model theory; for any relation R the schema $\wedge_i(x_i = y_i) \rightarrow [R(\mathbf{x}) \leftrightarrow R(\mathbf{y})]$ is a basic assumption of any logic considered by model theorists.

On the other hand the second principle requires some clarification and then, as we see below, is rejected. I was drawn to the role of these principles by the PhD thesis of Ken Martin [Martin 2015]. He argues that the axiom of choice preserves the 'identity of indiscernibles'. For, if we are given a set of apparently indistinguishable distinct objects, thus violating the principle, the axiom of choice (most strikingly in the form of the well-ordering principle) allows us to define a well order on the set. The objects are now distinguishable; each is the αth element for some appropriate α. While the elements are now discernible from the standpoint of set theory, they remain indiscernible[11] in the model theoretic context in which they arose.

Definition 8.2.1 *For an ordered set* $(I, <)$, *a sequence* $\langle a_i : i \in I \rangle$ *is order-indiscernible in a* τ*-structure M if for any* τ*-formula* $\phi(\mathbf{x})$ *with* $\mathbf{x} = \langle x_1, \ldots x_k \rangle$ *and properly ordered sequences* $i_1 < i_2 < \ldots i_k$ *and* $i'_1 < i'_2 < \ldots i'_k$ *from I:*

$$\phi(a_{i_1} \ldots a_{i_k}) \leftrightarrow \phi(a_{i'_1} \ldots a_{i'_k}).$$

A dense linear order is a sequence of order indiscernibles for $\tau = \langle < \rangle$. Starting with infinitely many (linearly ordered by $(Q, <)$) copies of the ordered set of integers Z, one finds a sequence of indiscernibles by choosing one element from each copy of Z.

There are several crucial parameters to this definition. First, we deal with ordered sequences. We will shortly see a stronger notion of indiscernibility; the difference between the two highlights the classification of theories. But first notice the reason that the model theorist rejects the 'identity of indiscernibles'. Model theoretic indiscernibility is *only* with respect to the family of relations specified by τ (and the ambient logic) in a fixed structure. Our requirement that one specifies the vocabulary for a particular topic recognizes that any particular study can focus on only a certain set of characteristics. It is then natural to accept that there may be elements that are indiscernible for these characteristics but nevertheless distinct. In fact, this is a theorem.

Theorem 8.2.2 [Ehrenfeucht & Mostowski 1956] *Let T be a theory with an infinite model and* $(I, <)$ *an infinite linear order. Then there is a model M of T that contains a set of order-indiscernibles (for the vocabulary of T) indexed by I.*

We now describe three basic combinatorial principles: the pigeon-hole principle, Ramsey's theorem, and the Erdős–Rado theorem. They are infinitary generalizations of the principle: given a finite set of properties

---

[11] See http://modeltheory.wikia.com/wiki/Indiscernible_sequence for basic introduction.

and enough objects two of the objects cannot be distinguished by those properties.

The pigeon-hole principle concerns unary predicates. If $n + 1$ letters are filed in n pigeon holes two letters must be in the same box and those letters cannot be distinguished by our predicates. For a slightly more complex example suppose that all Boolean combinations of the predicates are allowed. Since there are only 2^n possible such combinations,[12] if we have $2^n + 1$ objects two of them agree on every property and are indiscernible.

Ramsey's theorem[13] extends this result to n-ary relations. For simplicity we state the version for one binary relation. It says that given any k there is an n such that any edge coloring of the complete (each pair of vertices is connected) graph on n vertices contains a complete subgraph of size k where all edges have the same color. Ramsey proved both this version and the often used one, where the choice of n is replaced by 'infinite'. Theorem 8.2.2 is an easy consequence of Ramsey's theorem and compactness. (See e.g. page 179 of [Marker 2002].)

Many applications of Theorem 8.2.2 use a further tool from the original paper. First Skolemize (page 91) the theory T. Then an *Ehrenfeucht–Mostowski model* is the Skolem hull of a set of indiscernibles created by Theorem 8.2.2.

But now suppose we demand a stronger notion of indiscernibility.

Definition 8.2.3 *For a set (I), a sequence $\langle a_i : i \in I \rangle$ is* fully indiscernible *in a τ-structure M if for any τ-formula $\phi(\mathbf{x})$ with $\mathbf{x} = \langle x_1, \ldots x_k \rangle$ and sets of distinct elements $i_1, i_2, \ldots, i_k$ and $i'_1, i'_2, \ldots, i'_k$:*

$$\phi(a_{i_1} \ldots a_{i_k}) \leftrightarrow \phi(a_{i'_1} \ldots a_{i'_k}).$$

Here is the amazing connection between the stability hierarchy and indiscernibility (Theorem 2.13 of [Shelah 1978]).

Theorem 8.2.4 *A first order theory T is stable if and only if every sequence of order-indiscernibles is a fully indiscernible set.*

A straightforward example of a fully indiscernible set is the basis of a vector space. Developing this analogy, indiscernible sets become the building blocks for the classification of models (Chapter 5.5).

The last few paragraphs wandered away from combinatorics into pure model theory. But there is one more essential combinatorial idea used in

---

[12] E.g. if there are two properties A and B, either neither, just A, just B, or both hold.

[13] Ramsey [Ramsey 1930] established this combinatorial result as a lemma in proving the decidabilitiy of the class of universal first order sentences.

model theory. Ramsey's theorem depends essentially on the hypothesis that there are only finitely many colors. The Erdős–Rado theorem requires some technical notation: $\mu \to (\lambda)^{r+1}_\kappa$ means if f is a coloring of the $r + 1$-element subsets of a set of cardinality μ, in κ many colors, then there is a homogeneous set X of cardinality λ (that is, all $r + 1$-element subsets of X are mapped to same value by f). Marker [Marker 2002] gives a clear proof.

Fact 8.2.5 (Erdős–Rado) *The following implications hold.*

(1) $\beth_n(\kappa)^+ \to (\kappa^+)^{n+1}_\kappa$.

(2) $\beth_{\alpha+n}(\kappa) \to (\beth_\alpha(\kappa)^+)^n_{\beth_\alpha(\kappa)}$.

Using this combinatorial fact, Morley [Morley 1965b] shows the following two-cardinal theorem. The two-cardinal notion is extended by comparing the size of the model not with the size of a set definable by a predicate U but with the number of realizations of a type p (Example 5.1.2). More simply, the question now becomes, not 'is the set of realizations infinite?' but 'is the type realized at all?'

Theorem 8.2.6 (Morley) *If T is a first order theory in a countable vocabulary and p is a type over the empty set which is omitted in a model of cardinality $\beth_{\omega_1}$ then there are arbitrarily large models of T that omit p.*

Using the Erdős–Rado theorem iteratively, Morley shows that there is a model 'generated' by a countable set of order indiscernibles that omits p. Compactness then gives arbitrarily large models. A very accessible proof is in [Marker 2002]. Significantly, this result is easily rephrased[14] as the following 'upward Löwenheim–Skolem theorem'. (See chapters 4 and 6 of [Baldwin 2009].)

Theorem 8.2.7 (Morley) *Let ϕ be a sentence in $L_{\omega_1,\omega}(\tau)$ for a countable vocabulary τ. If ϕ has a model of cardinality $\beth_{\omega_1}$ then it has arbitrarily large models.*

Our conclusion: there is no model theory without combinatorial set theory – but a little goes a long way. Using only Ramsey's theorem, we obtain infinite sets of *order indiscernibles*; for stable theories the set is of *full indiscernibles*. With the Erdős–Rado theorem we obtain a still stronger existence theorem for indiscernibles, which is explored in Chapter 8.5. In fact only one more combinatorial idea is essential[15] and we discuss it in the next section.

---

[14] In the more general language of Definition 8.5.2, Theorem 8.2.6 and Theorem 8.2.7 say that the Hanf number for omitting types and the Hanf number for existence for $L_{\omega_1,\omega}$ are both $\beth_{\omega_1}$.

[15] This is a slight exaggeration; the Halpern–Lauchli and Erdős–Makkai theorems are used.

8.3   Why Is $\aleph_0$ Exceptional for Model Theory?

Shelah writes,

Some people think this [Vaught's conjecture] is the most important question in model theory as its solution will give us an understanding of countable models which is the most important kind of models. We disagree with all those three statements. ([Shelah 1990], xxi)

We discussed the first two assertions, focused specifically on Vaught's conjecture, in Chapter 7.2; here we focus on the importance of countable models. A priori as a student of mathematics Shelah sees no particular distinction for countable models. Certainly, one could justify concentrating on countable models if they were in some sense typical. But they are not. We already in the Introduction pointed out that categoricity in $\aleph_0$ is a very special case.[16] To answer the question that titles this section, we identify some unusual model theoretic properties[17] of $\aleph_0$. We first describe some combinatorial properties of uncountable sets which behave rather uniformly in all uncountable regular cardinals but fail miserably in $\aleph_0$.

The Ramsey and Erdős–Rado theorems are two combinatorial tools used extensively in model theory. Here is a further somewhat more technical combinatorial method. The closed unbounded filter and stationary sets are basic tools, to understand $\aleph_1$ and indeed any uncountable regular cardinal; the tool has a vast number of applications. We now explain why in model theory this filter is fundamental to understand the *exceptional nature* of $\aleph_0$. One of the fundamental structures, the rational order, is categorical in $\aleph_0$ and has the maximal number of models in $\aleph_1$. Shelah explains that in the second respect, dense linear orders are prototypical of unstable theories. Via coding stationary sets, every unstable theory has 2^κ non-isomorphic models in every uncountable cardinal κ, as we will now see.

A collection $\mathcal{F}$ of subsets of a set X is called a *filter* if $\mathcal{F} \neq \emptyset$, any superset of an element of $\mathcal{F}$ is in $\mathcal{F}$, and the intersection of two elements of $\mathcal{F}$ is in $\mathcal{F}$.

---

[16] For one thing, $\aleph_0$-categoricity implies via the Ryll-Nardzewski theorem that T is locally finite, a model generated by finitely many elements is finite.

[17] Boney has suggested in correspondence some deeper reasons why the cardinal $\aleph_0$ is exceptional. Three related issues that have arisen are: (1) Any cardinal λ satisfying $\lambda^{<\omega} = \lambda$ has nice cardinal arithmetic. But even with large cardinals, generalizations of these properties make requirements on the cofinality of λ. (2) ω satisfies many large cardinal definitions (e.g., compactness) if you remove uncountability, although it doesn't satisfy supercompactness! (3) ω is the unique infinite cardinal that has no limit cardinals below it. This makes inductive constructions much much easier. Thus, $L_{\infty,\omega}$-equivalence gives isomorphism only for countable structures, hiding properties which become visible in the uncountable.

Metaphorically, the closed unbounded (cub) sets are 'big', analogs to sets that have probability (equivalently measure) 1, while the stationary sets are analogs to sets with positive probability and small sets, those contained in the complement of a cub, are similar to those of measure 0.

Definition 8.3.1 *Let κ be a regular uncountable cardinal.*

(1) *A subset C of κ is*

    (a) closed *if the limit β of a sequence from C satisfies $\beta < \kappa$, then $\beta \in C$.*

    (b) unbounded *if for every $\alpha < \kappa$, there is a $\beta \in C$ with $\beta > \alpha$.*

(2) *The* club *or* cub *filter on κ, $CUB(\kappa)$ is the collection of all subsets of κ that contain a cub-set (one that is both closed and unbounded).*

(3) *A set $S \subset \kappa$ is* stationary *if S has a non-empty intersection with every closed unbounded set.*

Fix a set, $F = \langle f_n : n < \omega \rangle$, of countably many finitary operations on $\kappa = \aleph_1$. Observe that the set of $\alpha \in \kappa$ such that α is closed under all the f_i is a cub. (That is, if κ is the domain of an algebra, there is a cub C^F of proper subalgebras.)

Moreover tracing through the definition you will see that if S is stationary, there is an $\alpha \in C^F \cap S$. That is, if κ is the universe of a model, each stationary set contains an α which is the domain of a subalgebra (of an elementary submodel if we have Skolemized).

If we start with a first order theory and Skolemize (page 91), we can prove the Löwenheim–Skolem theorem in the form every model of size $\aleph_1$ has a countable elementary submodel. Taking the Skolem functions as the f_n, we see there is not just one but a closed unbounded set of countable elementary submodels.

Here is proved a second useful combinatorial fact about stationary sets.[18] Continuing our metaphor, there are many disjoint sets of positive measure.

Fact 8.3.2 *For any regular $\kappa > \omega$.*

(1) *(Ulam) There is family of κ stationary subsets of κ that are pairwise disjoint.*

(2) *(Sulokay) Thus, there exist a family of stationary sets S_i for $i < 2^\kappa$ such if $i \neq j$, $S_i \triangle S_j$ is stationary.*[19]

Crucially both the fact that the non-stationary ideal is well behaved and the existence of many stationary sets depend on κ not having cofinality[20] ω. This much of the many-models argument is combinatorial set theory.

---

[18] See e.g. 6.12 of [Kunen 1980] or II.4.12 of [Eklof & Mekler 2002].

[19] The *symmetric difference* of X and Y, $X \triangle Y$, is $(X - Y) \cup (Y - X)$

[20] If κ has countable cofinality, closed becomes vacuous and stationary becomes 'contains a tail'.

Using the classes of orders guaranteed by the failure of superstability or of stability Shelah manages to code[21] families of 2^κ non-isomorphic linear orders into models of size κ to prove the following theorem.

Theorem 8.3.3 *Let T be a countable complete first order theory. If T is either unstable or unsuperstable[22] then for any* uncountable κ, T *has 2^κ models of size κ.*

The following easy example of the coding (taken from [Hodges 1987]) indicates the gap between the countable and the uncountable.

Example 8.3.4 Let T be the theory of a bipartite graph,[23] say with red and green vertices, such that each green vertex is connected to exactly two red vertices, while each pair of red vertices is connected to infinitely many green vertices. Then T is $\aleph_0$-categorical and ω-stable. Nevertheless, T has $2^{\aleph_1}$ models of power $\aleph_1$. Indeed we can code any symmetric graph (G, E) on $\aleph_1$ vertices into a model M of T; identify the vertices of G with the green vertices of M. Then if $E(a, b)$ connect countably many elements to both a and b and uncountably many if $\neg E(a, b)$. A theory which admits such a coding (i.e. by the dimension of the set attached to pairs) is said to have the *dimensional order property* and this property[24] implies T has the maximal number of models in every uncountable cardinal. In order to get the maximal number of models in $\aleph_0$, one must have the *essentially non-isolated dimensional order property*. In the example this would require making the type of the green point over the pair determining it non-principal (so we would have a choice of the 'dimension' of realizations even in the countable). See the discussion of differentially closed fields in Chapter 7.2.

One intriguing result in Vaught's seminal paper [Vaught 1961] asserts there is no complete first order theory with exactly two countable models. This has spurred a lively line of research into theories with finitely many countable models. They cannot be superstable. For uncountable cardinalities, the answer is clearer. If a theory has finitely many models but more

---

[21] He first codes stationary sets into linear orders to get 2^κ linear orders and then proves that the coding extends to Ehrenfeucht–Mostowski models over the orderings. The proof is significantly easier when κ is regular. The unsuperstable case replaces the linear order by trees of width κ and height $\omega + 1$.

[22] Of course, proving the result for unsuperstable theories implies it for unstable. But the combinatorics is significantly easier for the stable case. Further consequences of stability can be used to prove the unsuperstable case.

[23] The graph is partitioned into sets X and Y such that all edges connect an element of X with an element of Y.

[24] See Chapters 7.1 and 8.4.

than one of cardinality $\aleph_1$ then it is $\aleph_0$ categorical and has finitely many models in each $\aleph_n$. So again the condition in $\aleph_1$ has more consequences. A similar curiosity is that Lachlan [Lachlan 1975] proved there is no complete theory with exactly four models in $\aleph_1$.

Vaught's conjecture (in the original form, Chapter 7.2) is that a countable complete first order theory has either countably many or $2^{\aleph_0}$ countable models. We have just seen that for unsuperstable theories the problem is easy for every uncountable κ. While, as we discuss in Chapter 7.2, the stability hierarchy allows one to solve some special cases, the missing combinatorics of stationary sets prevents a simple transfer of Theorem 8.3.3 to countable models.

Theorem 8.3.3 is sometimes misunderstood as being simply a counting problem with a not very interesting result. But the deeper meaning is this: it is not possible for any of these theories to fix a cardinal γ depending only on T such that every model of T is determined by the assignment of γ cardinals, to the nodes of a tree with bounded height, as specific invariants. In contrast, Chapter 5.5 showed all models of a classifiable theory have such a prescribed system of invariants.

This discussion has distinguished $\aleph_0$ by the inapplicability of the methods of stationary sets and has given us the chance to explain those methods. Perhaps even more telling is that model theory[25] concerns itself with finitary relations and functions and often with countable vocabulary. In this situation and in the absence of $\aleph_0$-categoricity, there will be countable models of different finite dimensions. This is the crux of the exceptional nature of $\aleph_0$ in Morley's categoricity theorem.

8.4 Entanglement of Model Theory and Cardinality

Most mathematical theorems are either specifically about structures of cardinality at most the continuum or are true uniformly without regard to cardinality. On the one hand, there is unique complete linear order with a countable dense subset. On the other, one can show that a group such that every element has order two is abelian, or discuss the solvabilty of a group with no consideration of the cardinality of the group. By the entanglement of model theory and cardinality I mean that this dichotomy is false for model theory.

This entanglement springs from the treatment of the equality symbol in logic and the *exact* definition of truth in a structure. Axiomatically, any text

---

[25] The study of abstract elementary classes (Chapter 14) violates this stricture.

in logic will posit that the symbol '=' is an equivalence relation: satisfies the transitive, symmetric, and reflexive properties. Further, it satisfies the axiom scheme which defines what universal algebraists call a congruence. For any natural number n and any n-ary function symbol f and any n-ary predicate symbol P:

$$(x_1 = y_1 \wedge \ldots \wedge x_n = y_n) \to f(x_1, \ldots x_n) = f(y_1, \ldots y_n)$$
$$(x_1 = y_1 \wedge \ldots \wedge x_n = y_n) \to P(x_1, \ldots x_n) = P(y_1, \ldots y_n)$$

Thus for any structure M, the quotient obtained by dividing out the equivalence relation induced by equality gives a structure whose universe is the equivalence classes in M and where the predicate and function symbols are well-defined.

But *model theory makes a stronger convention.* The inductive definition of truth in a structure[26] demands that the equality symbol be interpreted as identity.

The entanglement of model theory with cardinality is now ordained! This is easy to see for finite cardinalities.

$$(\exists x_1 \ldots x_n) \bigwedge_{1 \le i < j \le n} x_i \ne x_j \wedge (\forall y) \bigvee_{1 \le i \le n} y = x_i$$

is true exactly for structures of cardinality n. But the effect is also clear for infinite cardinals. Again, a relatively simple example is the proof in Chapter 4.1 of the upwards Löwenheim–Skolem theorem by adding constants *and demanding* that they are distinct. We explore more intriguing entanglements in the remainder of this section.

Morley's categoricity theorem (Theorem 3.3.4) says there is little cardinal dependence for categoricity in power for first order logic; all uncountable cardinals behave the same way. There is more dependence for categoricity in power of $L_{\omega_1, \omega}$-sentences; even assuming the VWGCH (page 23), one needs categoricity in every $\aleph_n$ to conclude categoricity in all uncountable cardinalities. We discussed in Chapter 7 examples of categoricity up to $\aleph_n$ [Baldwin & Kolesnikov 2009, Hart & Shelah 1990] but no further. Boney [Boney 2014b] proves, under large cardinal hypotheses, that there is eventually little cardinal dependence for categoricity of an abstract elementary class K; either for all sufficiently large cardinals, K is categorical; or beyond some point it is never categorical in a successor cardinal.

---

[26] There is a valuation v mapping terms $\mathbf{t}$ in the language to elements of the structure M and the base step in the definition of truth is $M \models t_1 = t_2$ if and only if $v(t_1)$ **is** $v(t_2)$. In [Tarski 1965], Tarski doesn't actually give the formal definition but argues informally on page 61 for the interpretation of equality as identity. See the last couple of paragraphs of Chapter 4.7 for more on the role of 'extensional' equality.

In Chapter 5.3, we emphasized that the stability classification is defined by syntactic conditions on *countable* objects that are absolute and make sense for any first order theory. Here we emphasize that these syntactical properties echo through all cardinalities. This echoing takes place in two ways. In Chapter 5.3 we partially described the possible values of the stability spectrum function and in Chapter 5.5 we calculated the spectrum function.

Keisler [Keisler 1976] showed that the unstable theories split syntactically into three classes: simply ordered, multiply ordered and having the independence property whose associated stability spectrum functions are $\mathrm{Ded}(\lambda)$, $\mathrm{Ded}(\lambda)^\omega$, and 2^ω, where $\mathrm{Ded}(\lambda)$ is the supremum of the cardinalities of linear orders with dense subsets of size λ. We describe the use of the absoluteness of NIP (not the independence property) to prove a theory has NIP in Chapter 8.6.

The syntactically given stability hierarchy gives as in Chapter 8.3 many models for unsuperstable theories. This hierarchy is further modulated by conditions beyond first order (OTOP, DOP, deep) on the models of a theory that determine the function $I(T, \kappa)$. (See Chapter 5.5 for details.) There is little entanglement with cardinal arithmetic because while the various functions might depend on cardinal arithmetic (e.g. many collapse under GCH), their non-decreasing character is absolute. Morley's conjecture that the spectra functions are non-decreasing was a *test question*. The positive answer shows the power of the stability theory framework.

Cardinal arithmetic enters the situation if we make an (over)-simplification of the main gap theorem and read it to say that the other spectra are everywhere strictly smaller that the maximal spectra. The schema giving these spectrum functions are absolute, e.g. $f(x) = \min(2^{\aleph_\alpha}, \beth_{d-1}(|\alpha + \omega| + \beth_2))$. We showed in Theorem 5.5.1 that each of the non-maximal spectra functions was bounded by $\beth_{\omega_1}(\alpha)$.

While on the surface, $2^{\aleph_\alpha}$ appears to grow much faster than $\beth_\beta(\alpha)$, there are anomalies. If α is a fixed point of the $\aleph$-function[27] and GCH holds then our sample f agrees with $2^{\aleph_\alpha}$ at $\aleph_\alpha$. So we need a bit more set theory to *ensure* the two functions are distinct, starting with the following standard [Jech 1978] result.

Easton's theorem: Let G be a non-decreasing function from (some subset of the) ordinals to ordinals such that $\aleph_\alpha$ is regular for each α and $\mathrm{cf}(\aleph_{G(\alpha)}) > \aleph_\alpha$ then there is a model of ZFC in which

$$2^{\aleph_\alpha} = \aleph_{G(\alpha)}.$$

---

[27] There is a proper class of cardinals such that $\aleph_\alpha = \alpha$. The first such cardinal is below the first inaccessible.

Corollary: It is consistent with ZFC that on every sufficiently large regular cardinal $\aleph_\alpha$, for every spectrum function g that is not maximal, $g(\aleph_\alpha) < 2^{\aleph_\alpha}$.

Proof. Define a function G as in Easton's theorem for every regular $\aleph_\alpha$ by setting: $\aleph_{G(\alpha)} = (\beth_{\omega_1}(\alpha))^+$. Clearly G satisfies the hypotheses of Easton's theorem. In the model of ZFC obtained by Easton forcing, $g(\aleph_\alpha) < 2^{\aleph_\alpha}$ on every regular cardinal so consistently f is almost always less than $2^{\aleph_\alpha}$.

But stability theory is not just about counting models; it is about establishing a dimension theory. In Chapters 5.6 and 5.5 we described such notions as regular types, weight, and domination as a means to regulate the dimensions of types in the same model. These refinements play a crucial role in the argument for our second thesis, the applicability of the model theoretic classification and techniques in the study of traditional mathematics. The examples here have already demonstrated the deep entanglement of model theory with cardinality. The strategy to calculate the spectra function led to the study of the geometries of regular types and the analysis of the relationship among the dimensions of different geometries and that led to finding definable groups. While order relationship among various spectra might depend on set theory, the underlying definition of the functions and the model theoretic distinction concerning the existence of invariants is immune to the vagaries of cardinal arithmetic.

8.5   Entanglement of Model Theory and the Replacement Axiom

We consider briefly the issue of whether in assuming ZFC, I have taken an unnecessarily strong metatheory. We do not pursue this in detail. A few papers study the use of the axiom of choice, most decisively, [Shelah 2009b]. Recall that the replacement schema is the axiom which guarantees (with power set) that the cumulative hierarchy is unbounded. Investigating the role of the replacement axiom sharpens the question, 'What set theory is used in model theory?' Friedman proved [Friedman 1971] that Borel determinacy implies the consistency[28] of the existence of $\beth_{\omega_1}$. Are there such examples of necessary uses of replacement in first order model theory? We show here the status of a possible example depends on stability hierarchy.

---

[28] https://gowers.wordpress.com/2013/08/23/determinacy-of-borel-games-i/#comment-42098.

We denote by ZC (Zermelo Set Theory) the axiom system ZFC with-out either replacement or foundation,[29] but with the full axiom of sepa-ration. Hanf [Hanf 1960] introduced the following extremely general and soft argument. Let $P(K, \lambda)$ range over such properties as: a class of models K has a model in cardinality λ; K is categorical in λ; or the type q is omitted in some model of K of cardinality λ.

Theorem 8.5.1 (Hanf) *Fix a* set *of classes K of a given kind (e.g. the classes of models of a given similarity type defined by sentences of the logic $L_{\mu,\nu}$ for some fixed μ, ν). For any property $P(K, \lambda)$ there is a cardinal κ such that if $P(K, \lambda)$ holds for some $\lambda > \kappa$ then $P(K, \lambda)$ holds for arbitrarily large λ.*

The easy argument for this from replacement appears as Theorem 4.18 of [Baldwin 2009] where there is a more detailed discussion. This general principle manifests itself in the notion of Hanf number.

Definition 8.5.2 (Hanf Numbers) *The* Hanf number *for a property P that is downward closed[30] is the least μ such that if there is a model in K with cardinality $> \mu$ that has property P, then there is a model with property P in* all *cardinals greater than μ.*

This argument yields the existence of a Hanf number for the existence of models of any reasonable logic (compare Chapter 1.3). The calculation of this number (see Theorem 8.2.7) relies on a fundamental tool of model theory: constructing indiscernibles realizing types from a *prescribed set.*

Theorem 8.5.3 (Morley) *Let M be a big saturated model of a first order theory T. For* every *large enough set $I \subset M$, there exists an infinite sequence of order indiscernibles $J \subset M$ such that for every finite $\mathbf{b} \in J$ there is an $\mathbf{a} \in I$ with $\mathrm{tp}(\mathbf{b}/\emptyset) = \mathrm{tp}(\mathbf{a}/\emptyset)$.*

The crux here is the requirement that for each $\mathbf{b} \in J$, $\mathrm{tp}(\mathbf{b}/\emptyset)$ is actually realized in I. In the standard Ehrenfeucht–Mostowski theorem (Theorem 8.2.2), which uses only Ramsey's theorem and compactness, it is only demanded that every finite subset of each $\mathrm{tp}(\mathbf{b}/\emptyset)$ is realized in I.

---

[29] In fact, the discussion here is only about replacement; there is no use of foundation in any of the proofs. (Kunen remarks in III.4 of [Kunen 1980], 'Unlike the other axioms, Foundation has no application in ordinary mathematics.') We chose ZC and thus omitted foundation in [Baldwin 2015] for comparison with McLarty's discussion of the logical strength of Fermat's last theorem [McLarty 2010].

[30] P is *downward closed* if there is a κ_0 such that if $P(K, \lambda)$ holds with $\lambda > \kappa_0$, then $P(K, \mu)$ holds if $\kappa_0 < \mu \leq \lambda$.

We can guarantee Theorem 8.5.3 for an arbitrary theory T only if $|M| \geq \beth_{\omega_1}$ and M is required to be saturated.[31] Morley [Morley 1965b] showed both that $\beth_{\omega_1}$ sufficed for the cardinality of I and that it was necessary. But this necessity argument itself uses replacement[32] to show $\beth_{\omega_1}$ exists.

So Theorem 8.5.3 and Hanf numbers for omitting types require the existence of $\beth_{\omega_1}$ even to be stated. Those notions are about size or about 'logics'. But there are theorems whose statements do not rely on replacement but their standard proofs do. We show that by restricting the theories, this reliance is eliminated. Here the interesting class is that of *simple* theories (Definition 8.5.6). In the extended stability hierarchy described in Chapter 2.4, simple theories include stable theories.

Shelah first developed the notion that a type $p \in S(B)$ divides over $A \subseteq B$ to indicate that any realization of p 'depended' on B more than on A. But he couldn't prove non-dividing types extended to complete non-dividing types. He invented the notion of forking for types that satisfied such an extension property.[33] He later proved the equivalence of forking and dividing for stable theories without any reliance on replacement. The strength of stationary types in stable theories allows the construction of the required indiscernibles using only Ramsey's theorem (as in Ehrenfreucht-Mostowski). See Section V.3, in particular Theorem V.3.9, of [Baldwin 1988a].

Here is a theorem clearly stated in ZC, but for which all proofs known before 2014 used replacement. Byunghan Kim [Kim 1998] proved:

Theorem 8.5.4 (Kim) *For a simple first order theory non-forking is equivalent to non-dividing.*

The usual easily applicable descriptions of simple theories involve uncountable objects. But the definitions below of simple, non-forking, and non-dividing are equivalent in ZC to statements about countable sets of formulas [Casanovas 1999]. Nevertheless, the argument for Kim's theorem employs Morley's technique for omitting types; that standard argument uses the Erdős–Rado theorem up to $\beth_{\omega_1}$ and thus the replacement axiom.

We first show that this proposition is properly formulated without any use of replacement. For this, we simply repeat the basic definitions from [Casanovas 1999] where the exact result we are after is given a short

---

[31] For arbitrary M, Morley proves there exist N and $J \subseteq N$ that satisfy the condition.

[32] Roughly, replacement says that the image of a set under a functional relation is a set. Thus the map $n \to \beth_n$ yields the existence of $\aleph_\omega$.

[33] See the end of Chapter 13.1 for the nomenclature.

complete proof. We work in a complete first order theory in a countable vocabulary.

k-inconsistency: Let $\langle a_i : i < \omega \rangle$ be a sequence of finite tuples in a model of a first order theory T. A set of formulas $X = \{\phi(\mathbf{x}, a_i) : i < \omega\}$ is k-inconsistent if every k element subset of X is inconsistent. With this notion in hand we can define forking and dividing. The basic notion is dividing; intuitively if a formula $\phi(x, \mathbf{b})$ divides over A any element c satisfying $\phi(c, \mathbf{b})$ is more constrained by $A\mathbf{b}$ than by A. For example, suppose $\phi(x, y, a)$ (with $a \in A$) defines an equivalence relation with infinitely many classes. Then for any $b \notin \mathrm{acl}(A)$, $\phi(x, b)$ constrains x over A by picking a particular equivalence class. The formula $\phi(x, b)$ 2-divides over A in the following sense.

Definition 8.5.5 (Forking and Dividing) *Let $A \cup \{a\} \cup \{a_j : j < n\}$ be a subset of a model of T.*

(1) *The formula $\phi(\mathbf{x}, \mathbf{b})$ k-divides over A if for some set $I = \{\mathbf{b}_i : i < \omega\}$ such that $\{\phi(\mathbf{x}, \mathbf{b}_i) : i < \omega\}$ is k-inconsistent and all the $\mathbf{b}_i$ realize $\mathrm{tp}(\mathbf{b}/A)$. ϕ divides over A if it k-divides over A for some k.*
(2) *The formula $\phi(\mathbf{x}, \mathbf{b})$ forks over A if for some finite set of formulas $\psi_j(\mathbf{x}, \mathbf{b}_j)$ with $j < n$, $\phi(\mathbf{x}, \mathbf{b}) \vdash \bigvee_{j<n} \psi_j(\mathbf{x}, \mathbf{b}_j)$ and each $\psi_j(\mathbf{x}, \mathbf{b}_j)$ divides over A.*

The formula $\phi(\mathbf{x}, \mathbf{y})$ has the *tree property* with respect to $k < \omega$ if there is a tree $(a_s : s \in \omega^{<\omega})$ (in some model of T) such that for all $\eta \in \omega^\omega$, the branch $\{\phi(\mathbf{x}, a_{\eta \restriction n}) : n < \omega\}$ is consistent and for all $s \in \omega^{<\omega}$, the family of siblings $\{\phi(\mathbf{x}, a_{s\widehat{\ }i}) : i < \omega\}$ is k-inconsistent.

Definition 8.5.6 *T is simple if there is no formula $\phi(\mathbf{x}, \mathbf{y})$ which has the tree property in T.*

There is a direct proof of the following result in [Casanovas 1999].

Theorem 8.5.7 *Let T be a simple theory. A partial type $\pi(\mathbf{x}, a)$ divides over A if and only it forks over A.*

None of the arguments given in [Casanovas 1999] directly invoke replacement. Crucially, Lemma 1.1 of that paper is a variant of Theorem 8.5.3. Grossberg, Iovino, and Lessmann asked for an analog of Theorem 8.5.3 in a weak set theory and I asked for a proof of Kim's theorem in [Baldwin 2015]. Building on work of Adler, [Vasey 2017a] answered both questions.[34]

---

[34] Tsuboi [Tsuboi 2014] independently answered the second.

Theorem 8.5.8 (Vasey: Existence of Morley sequences in simple theories (ZC)) *Assume T is simple. Let $A \subset B$ be sets. Let $p \in S(B)$ be a type that does not fork over A. Let I be a linearly ordered set. Then there is a Morley sequence[35] $I := \{b_i | i \in I\}$ for p over A.*

Corollary 8.5.9 *Theorem 8.5.7 is provable in ZC.*

This section illustrates the role of classification theory in addressing a more specifically foundational issue. When a certain useful result is formulated for arbitrary theories it requires the replacement axiom. But it is provable in ZC if one restricts to appropriate levels of the stability hierarchy. This illustrates Thesis 2 in two ways. First, the restriction improves the results in pure model theory by weakening the set theoretic hypothesis. Since the applications are often to theories that are easily demonstrated to be low in stability hierarchy and are mathematically important, the proof strength of results about specific important mathematical theories is also reduced.

8.6   Entanglement of Model Theory with Extensions of ZFC

In Chapter 8.1 we argued that, contrary to received opinion around 1970, most interesting and applicable first order model theory takes place in ZFC. In Chapter 8.5, we briefly discussed weaker metatheories. However, there are several methodological motives for using extensions of ZFC as a tool in model theory. The general principle is: *A theorem under additional hypotheses is better than no theorem at all.* We explain in the next few pages three ways that 'better' can be interpreted:

(1) *Oracular*: The result may guide intuition towards a ZFC result.
(2) *Metatheoretic*: The set theoretic hypothesis is automatically eliminable.
(3) *Entangled*: There is a deep connection between the model theory and the set theory.

Oracular: The result may guide intuition towards a ZFC result. One of many noteworthy examples of this phenomena is the theorem.

Theorem 8.6.1 (Shelah) *Any complete sentence in $L_{\omega_1,\omega}(Q)$ that is categorical in $\aleph_1$ has a model in $\aleph_2$.*

---

[35] A *Morley sequence* for a type $p \in S(A)$ is now defined as a sequence $I = \langle a_i : i < \delta \rangle$ of realization of p such that the type of a_i over its predecessors does not fork over A and I is order indiscernible.

In particular this theorem implies that categoricity (exactly one model) is impossible[36] for $L_{\omega_1,\omega}(Q)$. Shelah's [Shelah 1975] complicated proof of Theorem 8.6.1 developed stability theory in the infinitary context and assumed $V = L$. But in [Shelah 1983a] he removed the set theoretic hypothesis and gave a beautiful short (stability-free) proof ([Baldwin 2009], Chapter 8).

Such a result can have important secondary consequences. Keisler proved, under the generalized continuum hypothesis, any two elementarily equivalent models have isomorphic ultrapowers. The original argument for the Ax–Kochen–Ershov theorem in number theory (page 9) relied on Keisler's theorem and absoluteness. Shelah improved the combinatorial methods to prove Keisler's isomorphism of ultrapowers result and thus Ax–Kochen directly in ZFC. Thus the Ax–Kochen proof lost its metatheoretic content (appeal to forcing and absoluteness).

Metatheoretic: The set theoretic hypothesis is automatically eliminable. For example, suppose *the conclusion is absolute*: the form of the conclusion guarantees that if it is consistent with ZFC it is provable in ZFC (page 46). While the proof at hand uses a proper extension of ZFC, it shows that there is a proof in ZFC. As we just noted, one of many such arguments is the original Ax–Kochen proof which relied on the absoluteness of the number theoretic conclusion to eliminate the GCH.

Shelah gave metatheoretic proofs of two results concerning NIP. We use two facts about $\mathrm{Ded}(\kappa)$ (page 191): (i) $\kappa < \mathrm{Ded}(\kappa) \leq \mathrm{Ded}(\kappa)^{\aleph_0} \leq 2^\kappa$ and (ii) which inequalities can be strict is independent of ZFC. By II.1 of [Shelah 1978], $g_T^m(\kappa) = g_T^1(\kappa) = g_T(\kappa)$ (page 131). Easily, if a formula $\phi(\mathbf{x},\mathbf{y})$ has the independence property ($m = \lg(\mathbf{x})$), for some A, $|S_m(A)| = 2^{|A|}$. Immediately (a) if for every κ, $g_T^1(\kappa) \leq \mathrm{Ded}(\kappa)^{\aleph_0} < 2^\kappa$ then T has NIP and (b) no formula[37] $\phi(x,\mathbf{y})$ has the independence property. The last inequality in the hypothesis is only consistent; but the absoluteness of NIP gives the result in ZFC. Laskowski [Laskowski 1992] and Poizat [Poizat 1985] made (b) oracular with combinatorial arguments in ZFC. Laskowski constructs from a formula $\phi(\mathbf{x},\mathbf{y})$ with the independence property, a formula $\psi(x,\mathbf{z})$ with the independence property.

Gehret applied the argument a half century later to axiomatic analysis. An *asymptotic couple*, with theory $T_{\log}$, is the value group of a certain

---

[36] I asked whether such a sentence with an uncountable model was possible for $L_{\omega,\omega}(Q)$ and Friedman included it in his 102 problems in mathematical logic [Friedman 1975].

[37] Theorems 4.9–4.11 of [Shelah 1978]. Note x is a singleton.

kind of valued field with an associated operator that arises in the study of transseries (page 161). As in the study of valued fields [Haskell et al. 2007], generalizing o-minimal to NIP is desirable. Gehret [Gehret 2017] showed that in the theory $T_{\log}$ there are at most $\mathrm{Ded}(\lambda)^{\aleph_0}$ 1-types over a model of size λ; by arguments above $T_{\log}$ has NIP. One hopes for an algebraic proof of Gehret's result. But for now, Shelah's (a) and this application remain metatheoretic.

Recently, a number of papers in infinitary logic have used a more subtle version of the absoluteness argument with a motto *Consistency implies Truth*. Here the actual proposition Φ being proved is not absolute. But given a model M of set theory in which Φ is true one constructs (often by an ultralimit) a model[38] M^* of set theory such that Φ is absolute between M^* and V. For example, the aim might be to construct uncountable models of an $L_{\omega_1,\omega}$ sentence; this property is certainly not absolute between V and all models of set theory. However, in special situations one can construct an M^* with $\aleph_1^{M^*} = \aleph_1^V$.

Note that our classification of results or rather of proofs is time-dependent. The original Ax–Kochen proof was metatheoretic; after the Shelah–Keisler isomorphism theorem, it became oracular.

Full Entanglement of Model Theory with Extensions of ZFC. By full entanglement we mean results that provably depend on extensions of ZFC. We examine three types of examples in the book. We discussed in Chapter 7 a few theorems about infinitary logic that currently use hypotheses beyond ZFC for model theoretic results. In Chapter 1.3 we expound the entanglement of various model theoretic logics with set theory. And in Chapter 14, we consider the greater entanglement of axiomatic set theory including large cardinal axioms with abstract elementary classes.

8.7   Moral

Model theory studies structures for all vocabularies and in various logics. There is (at least as yet) no pretense of an axiomatic formulation.[39] Thus, the natural setting for doing model theory is the same as any mathematical study, ZFC. We have seen that there were serious doubts a half century ago

---

[38] A more sophisticated version constructs a family of models [Baldwin et al. 2016a, Baldwin & Larson 2016, Baldwin et al. 2015, Baldwin et al. 2016b].

[39] The axiomatic framework provided by abstract elementary classes has so far been incapable of expressing the distinctions relevant to traditional mathematics imposed by full formalization.

whether ZFC provided an adequate foundation, even for first order logic. Supporting Thesis 3, these doubts have been banished for first order logic, but not for infinitary logic.

However, even when working in ZFC, some are uneasy about the use of infinite combinatorics. Chapters 8.2 and 8.3 show that only a few combinatorial notions are involved: local finiteness, stationary sets, and Ramsey and Erdős–Rado theorems. These are no longer exotic tools: certainly not for logicians and increasingly not in mathematics at large. Gowers [Gowers 2000] writes, 'subjects that appeal to theory-builders are, at the moment, much more fashionable than the ones that appeal to problem-solvers. Moreover, mathematicians in the theory-building areas often regard what they are doing as the central core (Atiyah uses this exact phrase) of mathematics, with subjects such as combinatorics thought of as peripheral and not particularly relevant to the main aims of mathematics.' He then defends the problem-solvers, in particular the work of Erdős and E. Szemerédi, pointing to the use of probabilistic combinatorics in Banach space theory, which won Gowers a Fields medal. (If one thinks this defense self-interested note that Michael Harris (fully in the theory-building camp) writes (footnote 23, page 364 of [Harris 2015]), 'Ten years after the [Gowers] article's publication, the case no longer needs to be made.') Gower's work can be closely connected with the model theory (including stability) of Banach spaces [Iovino 1999]. Nevertheless, some combinatorial proofs, e.g., the use of *ded*(λ) as in some of the metatheoretic examples, may eventually be seen to be unnecessary.

Gerald Sacks presciently summarized more than 50 years ago the ability to disentangle first order model theory from set theory.

Sacks Dicta

The central notions of model theory are absolute and absoluteness, unlike cardinality, is a logical concept. That is why model theory does not founder on that rock of undecidability, the generalized continuum hypothesis, and why the Łos conjecture is decidable. [Sacks 1972]

There remains a purity issue. How much of set theory is actually used? We saw in Chapter 8.5 that the stability classification calibrates those theories for which the replacement axiom is necessary for important model theoretic results. We classified proofs which rely on extensions of ZFC into three classes. Clearly, an oracular proof is impure; it is shown to be oracular by finding a proof avoiding the stronger axiom. Metatheoretic proofs are candidates to become oracular. Set theoretic legerdemain with no apparent connection to the topic is used to prove a result. Thus the search for a second proof that exposes the *real*[40] reason for the result and becomes pure is certainly justified. If a result is fully entangled, then we embed the set theoretic hypothesis in the statement and have a prima facie pure statement.

---

[40] That is, a reason more closely connected to the particular hypotheses. Thus, Laskowski's proof (page 197) replaces an argument using absoluteness and infinite cardinal arithmetic by finite combinatorics.

PART III
....

Geometry

The more modern interpretation:- Geometry treats of entities which are denoted by the words straight line, point, etc. These entities do not take for granted any knowledge or intuition whatever, but they presuppose only the validity of the axioms, such as the one stated above, which are to be taken in a purely formal sense, i.e. as void of all content of intuition or experience. These axioms are free creations of the human mind. All other propositions of geometry are logical inferences from the axioms (which are to be taken in the nominalistic sense only). The matter of which geometry treats is first defined by the axioms. Schlick in his book on epistemology has therefore characterized axioms very aptly as 'implicit definitions.' [Einstein 2002]

We have identified Einstein's 'modern interpretation' with Hilbert. But we take Einstein's 'free creations' in a limited sense. The axioms represent and sharpen prior intuitions. In this part we examine the historical relationship between certain intuitions, often formed by earlier axiomatizations, and new sets of axioms. We aim to evaluate axiomatizations of the geometric continuum.

In Chapter 9.1, we consider several accounts of the purpose of axiomatization and adjust Detlefsen's notion of *descriptive completeness* by fixing a criterion for evaluating axiom systems: *modest descriptively complete axiomatization*.

We lay out in Chapter 9.3 various sets of axioms, crucially formulated in different logics, for geometry and correlate them with the specific sets of propositions from Euclid that they justify. We emphasize those propositions of Euclidean, Cartesian, and Hilbertian geometry which might be thought to require the Archimedean or Dedekind axiom but do not; Hilbert's proof that the first order axioms suffice to define a field yields these geometric propositions. In particular, the notions of similarity and area of polygons are so grounded. This leads to the conclusion argued in Chapter 11 that Hilbert's full axiomatization is immodest. Such a *formula* as $A = \pi r^2$ is not justified on the basis of Hilbert's first order axioms (even with Archimedes); but in Chapter 10, we expand the first order theory of Euclidean geometry EG, by adding a constant π which allows us to compute the area and circumference of a circle. Invoking o-minimality we do the same for the Descartes/Tarski geometry.

In Chapter 11 we analyze various meanings for 'complete', both logical and 'topological'. We explore the relations between the Archimedean and completeness axioms and argue that Hilbert's completeness axiom is immodest even with respect to Descartes' conceptions.

9 | Axiomatization of Geometry

We now see that the careful delineation of vocabulary, structure, and emphasis on the choice of logic in Part I both draw from Hilbert's *Grundlagen* and in retrospect illuminate the distinctions made there. By the *geometric continuum* we mean the line situated in the context of the plane. The following two propositions[1] represent changing conceptions of geometry over the last two and one half millennia.

(*) Euclid VI.1: Triangles and parallelograms which are under the same height are to one another as their bases.

Hilbert[2] gives the area of a triangle by the following formula.

(**) Hilbert: Consider a triangle ABC having a right angle at A. The measure of the area of this triangle is expressed by the formula

$$F(ABC) = \frac{1}{2}AB \cdot AC.$$

At first glance each describes the familiar method to calculate the area of a triangle. But clearly they are not identical. Euclid tells us that the two-dimensional area of two triangles 'under the same height' is *proportional* to their 1-dimensional bases. Hilbert's result is not a statement of proportionality; it tells us the *2-dimensional measure* of a triangle is computed as a product of the *1-dimensional measures* of its base and height. Hilbert's rule looks like a statement of basic analytic geometry, but it wasn't. He derived it from an axiomatic geometry similar to Euclid's, which in no way builds on Cartesian analytic geometry.

One can see several challenges that Hilbert faced in formulating a new axiom set in the late nineteenth century:

---

[1] Diagrams illustrating the Euclidean propositions about area appear with Theorem 9.5.4 and Remark 9.5.6.

[2] Hilbert doesn't state this result as a theorem; and I have excerpted the statement below from an application on page 66 of [Hilbert 1962]. Hilbert defines proportionality in terms of segment multiplication on page 50. 'Negative' segments are introduced in Section 17 on page 53.

(1) Delineate the relations among the principles underlying Euclidean geometry. In particular, identify and fill 'gaps' or remove 'extraneous hypotheses' in Euclid's reasoning.

(2) Reformulate propositions such as VI.1 to reflect the nineteenth century understanding of real numbers as measuring both length and area.

(3) Ground the geometry of Descartes, nineteenth century analytic geometry, and rigorous foundations for mathematical analysis.

The third aspect of the third challenge is not obviously explicit in Hilbert. We will argue Hilbert's completeness axiom is unnecessary for the first two challenges and at least for the Cartesian aspect of the third. The gain is that it grounds mathematical analysis (provides a rigorous basis for calculus); that Hilbert desired this is more plausible than that he thoughtlessly assumed too much. For such a judgement we need some idea of the goals of axiomatization and when such goals are met or even exceeded. We frame this discussion in terms of the notion of *descriptive axiomatization* from [Detlefsen 2014], which is discussed above.

But the axiomatization of a theory of geometry that had been developing for over two millennia leads to further considerations. How does one correlate distinct statements such as (*) and (**), which, in some sense, express the same proposition? We lay out the relations among three perspectives on a mathematical topic.

(1) *A data set* [Detlefsen 2014], a collection of propositions about the topic.

(2) A system of axioms and theorems for the topic.

(3) The different conceptions of various terms used in the topic at various times.

Previous descriptions of complete descriptive axiomatization omit the possibility that the axioms might be too strong and so obscure the 'cause' for a proposition to hold. We introduce the term 'modest' descriptive axiomatization to denote one which avoids this defect. We give several explicit lists of propositions from Euclid and draw from [Hartshorne 2000] for an explicit linking of subsets of Hilbert's axioms as justifications for these lists. We conclude that Hilbert's first order axioms provide a modest complete descriptive axiomatization for most of Euclid's geometry. In Chapter 11.2 we argue that the second order axioms aim at results that are beyond (and even in some cases antithetical to) the Greek and even the Cartesian view of geometry. So Hilbert's axioms are immodest as an axiomatization of traditional geometry. This conclusion is no surprise to Hilbert[3] although it may be to many readers.[4] Indeed, Hilbert writes:

This thought [coordinatization] with *one blow* renders every *geometrical problem accessible to analysis.* So Descartes became the creator of analytic geometry. The theorems of the Greeks were *proved anew,* and then *generalised.* So there appeared on the scene through Cartesius a *sudden turn, a means, a unified method – the formula and calculation.*[5]

Even more, with Dedekind the transcendental numbers are set firmly in this universe. Hilbert groups his axioms for geometry into five classes. The first four are first order. Group V, Continuity, contains Archimedes' axiom, which can be stated in the logic $L_{\omega_1,\omega}$, and a second order completeness axiom equivalent (over the other axioms) to Dedekind's completeness line in the plane. Hilbert[6] closes the discussion of continuity with 'However, in what is to follow, no use will be made of the "axiom of completeness".' Why then did he include the axiom? Earlier in the same paragraph,[7] he writes that 'it allows the introduction of limiting points' and enables one 'to establish a one-one correspondence between the points of a segment and the system of real numbers.' Implicitly, he is grounding geometrically the rigorous mathematical analysis of Cantor, Dedekind, Weierstrass, et al. But an *explicit important motivation for Hilbert* opens the next section: to bring out the significance of the various groups of axioms.

9.1 The Goals of Axiomatization

Here, we place our analysis in the context of recent philosophical work on the purposes of axiomatization. We explicate Detlefsen's notion of

---

[3] In the preface to [Hilbert 1962] the translator Townsend writes, 'it is shown that the whole of the Euclidean geometry may be developed without the use of the axiom of continuity.' Hilbert lectured on geometry several summers in the 1890s and his notes (German) with extremely helpful introductions (English) appear in [Hallett & Majer 2004]. The first *Festschrift* version of the *Grundlagen* does not contain the continuity axioms. I draw primarily on the 2nd (Townsend) edition of Hilbert and on the 10th [Hilbert 1971].

[4] The first 10 URLs from a Google search for 'Hilbert's axioms for Euclidean geometry' contained 8 with no clear distinction between the geometries of Hilbert and Euclid and two links to Hartshorne, who carefully distinguishes them.

[5] This translation by Michael Hallett is from Hilbert's Lecture Notes on Projective Geometry on ([Hallett & Majer 2004], 24). In the valuable article [Giovannini 2016], Giovannini points a similar comment to Hilbert's 1891 lecture notes in ([Hallett & Majer 2004], 22) emphasizing that Descartes established analytic geometry as a method of calculation.

[6] See ([Hilbert 1971], 26).

[7] The section entitled *The Vollständigkeitsaxiom* in ([Hallett & Majer 2004], 426–435) contains a thorough historical description.

'data set' and investigate the connection between axiom sets and data sets of sentences for an area of mathematics. Hilbert begins the *Grundlagen* with:

> The following investigation is a new attempt to choose for geometry a *simple* and *complete* set of *independent axioms* and to deduce from them the most important geometrical theorems in such a manner as to bring out as clearly as possible the significance of the groups of axioms and the scope of the conclusions to be derived from the individual axioms. [Hilbert 1971]

Hilbert described the general axiomatization project in 1918.

> When we assemble the facts of a definite, more or less comprehensive field of knowledge, we soon notice these facts are capable of being ordered. This ordering always comes about with the help of a certain *framework of concepts* [*Fachwerk von Begriffen*] in the following way: a concept of this framework corresponds to each individual object of the field of knowledge, a logical relation between concepts corresponds to every fact within the field of knowledge. The framework of concepts is nothing other than the *theory* of the field of knowledge. ([Hilbert 1918a], 1107)

Detlefsen [Detlefsen 2014] describes such a project as a *descriptive axiom-atization*, motivating the notion with Huntington's remark (Huntington's emphasis):

> [A] miscellaneous collection of facts …does not constitute a *science*. In order to reduce it to a science the first step is to do what *Euclid* did in geometry, namely, to *select a small number of the given facts as axioms and then to show that all other facts can be deduced from these axioms by the methods of formal logic.* [Huntington 1911]

Hallett ([Hallett 2008], 204) delineates the meaning of facts in this con-text, 'simply what over time has come to be accepted, for example, from an accumulation of proofs or observations. Geometry, of course, is the central example …' He presaged the emphasis on data set (Hilbert's facts) that pervades our discussion:

> Thus completeness appears to mean [for Hilbert] 'deductive completeness with respect to the geometrical facts.' …In the case of Euclidean geometry there are various ways in which 'the facts before us' can be presented. If interpreted as 'the facts presented in school geometry' (or the initial stages of Euclid's geometry), then arguably the system of the original Festschrift [i.e. 1899 French version] is adequate. If, however, the facts are those given by geometrical intuition, then matters are less clear. ([Hallett & Majer 2004], 434)

Detlefsen introduces the term *data set* (i.e. facts[8]) and describes a local descriptive axiomatization as an attempt to deductively organize a data set for some topic. The axioms are *descriptively complete* if all elements of the data set are deducible from them. This raises two questions. What is a sentence? Who commonly accepts?

From the standpoint of modern logic, a natural answer to the first question would be to specify a logic and a vocabulary and consider all sentences in that language. Detlefsen argues ([Detlefsen 2014], 5–7) that this is the wrong answer. He thinks Gödel errs in seeing the problem as completeness in the now standard sense of a first order theory.[9] Rather, Detlefsen presents us with an *empirical* question. We (at some point in time) look at the received mathematical knowledge in some area and want to construct a set of axioms from which it can all be deduced. In our case we want to compare the commonly accepted sentences from 300 BC with a twentieth century axiomatization. As we see below, new interpretations for terms arise. Nevertheless, a specific data set delineates a certain area of inquiry. The data set is inherently flexible; conjectures are proven (or refuted) from time to time. The study here tries to distinguish propositions that are simply later deductions about the same intuitions and those which invoke radically different assumptions. By analyzing the interpretations in particular cases, we can specify a data set. Comparing geometry at various times opens a deep question worthy of more serious exploration than there is space for here. In what sense do (*) and (**) opening this chapter express the same thought, concept, etc.? Rather than address the issue of what is expressed, we will simply show how to interpret (*) (and other propositions of Euclid) as propositions in Hilbert's system. See Chapter 9.2 for this issue.

Geometry is an example of what Detlefsen calls a *local* as opposed to a *foundational* (global) descriptive axiomatization. Beyond the obvious difference in scope, Detlefsen points out several other distinctions. He asserts in ([Detlefsen 2014], 5) that the axioms of a local axiomatization are generally among the given facts while those of a foundational axiomatization are found by (paraphrasing Detlefsen) tracing each truth in a data set back to the deepest level where it can be properly traced. Hilbert's geometric axioms have a hybrid flavor. Through the analysis of the concepts involved,

---

[8] There is an interesting subtlety here (perhaps analogous to Shapiro's algebraic and non-algebraic theories). In studying Euclidean geometry, we want to find the axioms to describe an intuition of a structure. Suppose, however, that our body of mathematics is group theory. One might think the data set was the sentences in the vocabulary of group theory true in all groups. (The axioms are evident.) But these sentences are not in fact the data set of 'group theory' as studied; that subject is concerned about the properties and relations between groups.

[9] We dispute some of his points in Remark 10.2.2.

Dedekind arrived at a second order axiom[10] that is not in the data set but formed the capstone of the axiomatization: Dedekind completeness for geometry.

We return to our question, 'What is a sentence?' The first four groups of Hilbert's axioms are sentences of first order logic: quantification is over individuals and only finite conjunctions are allowed. Archimedes' axiom[11] can be formulated in $L_{\omega_1,\omega}$. But the Dedekind postulate[12] in any of its variants is a sentence of a kind of second order logic. Adopting this syntactic view, there is a striking contrast between the Greek data set for such subjects as number theory and geometry and the axiom systems advanced near the turn of the twentieth century. Except for the Archimedean axiom, the earlier data sets are expressed in first order logic.

An aspect of choosing axioms seems to be missing from the account so far. Hilbert provides the following insight into how axioms are chosen:

> If we consider a particular theory more closely, we always see that a few distinguished propositions of the field of knowledge underlie the construction of the framework of concepts, and these propositions then suffice by themselves for the construction, in accordance with logical principles, of the entire framework. ...
>
> These underlying propositions may from an initial point of view be regarded as the axioms of the respective field of knowledge ... [Hilbert 1918a]

By a *modest* axiomatization of a given data set,[13] we mean one that implies all the data and not too much more.[14] Of course, 'not too much

---

[10] We discuss the 'equivalence' of Dedekind's and Hilbert's formulation of completeness in Chapter 11.1.

[11] The Archimedean axiom is a property of an ordered group (or field). In the logic, $L_{\omega_1,\omega}$, quantification is still over individuals but now countable conjunctions are permitted so it is easy to formulate Archimedes' axiom: $\forall x, y (\bigvee_{m\in\omega} mx > y)$. By switching the roles of x and y we see each is reached by a finite multiple of the other. Robinson defined this logic, formulated Archimedes' axiom in it, and used it to prove the existence of non-Archimedean fields with specified properties [Moore 1997, Robinson 1951].

[12] See the caveats on 'second order' in Chapter 11.1. Dedekind defines the notion of a cut in a linearly ordered set I (a partition of $\mathbb{Q}$ into two intervals (L, U) with all elements of L less than all elements of U). He postulates that each cut has unique realization, a point above all members of L and below all members U – it may be in either L or U ([Dedekind 1963], 20). If either the L contains a least upper bound or the upper interval U contains a greatest lower bound, the cut is called 'rational' and no new element is introduced. Each of the other (irrational) cuts introduces a new number. It is easy to see that the *unique* realization requirement implies the Archimedean axiom for an ordered group. By Dedekind completeness of a line, I mean the Dedekind postulate holds for the linear ordering of that line.

[13] We considered replacing 'modest' by 'precise' or 'safe' or 'adequate'. We chose 'modest' rather than one of the other words to stress that we want a sufficient set and one that is as necessary as possible. As the examples show, 'necessary' is too strong. Later work finds consequences of the original data set undreamed by the earlier mathematicians. Thus just as, 'descriptively complete', 'modest' is a description, not a formal definition.

[14] This concept describes normal work for a mathematician. 'I have a proof; what are the actual hypotheses so I can convert it to a theorem?'

more' is a rather imprecise term. One cannot expect a list of known mathematical propositions to be deductively complete. By more, we mean the axioms introduce essentially new concepts and concerns or add additional hypotheses proving a result that contradicts the explicit understandings of the authors of the data set. (See the end of Chapter 10.3 and Chapter 12.4.)

We are investigating modern axiomatizations for an ancient data set. As we'll see below, using Notation 9.3.3, Hilbert's first order axioms (HP5) are a modest axiomatization of the data (Euclid I): the theorems in Euclid about polygons (not circles) in the plane. We give an example later showing that HP5 + CCP (circle–circle intersection), while modest for Euclid II, is an immodest *first order* axiomatization of polygonal geometry. In the twentieth century the process of formalization is usually attentive to modesty so it is a bit hard to find non-artificial examples. However, to study complex exponentiation, [Zilber 2005b] defined a *quasiminimal excellent class* (page 306). The axioms asserted models were combinatorial geometries satisfying extra conditions, most importantly *excellence*. Although his axioms were informal, they can be formalized in $L_{\omega_1,\omega}(Q)$. Quite unexpectedly, [Bays et al. 2014] showed 'excellence' was not needed for the main result. Thus, the original axiomatization was immodest.

Are the second order Peano axioms axiom immodest for (Greek) arithmetic? While in some ways this restates the programs for studying weak arithmetics and reverse mathematics, a historical study would involve the type and length of analysis done here.

In our view, *modesty* and *purity* are distinct, though related, notions. We just saw that formalization is useful to check modesty. [Hallett 2008, Arana & Mancosu 2012, Detlefsen & Arana 2011, Baldwin 2013a] argue that purity asks about *specific arguments* for a proposition. [Baldwin 2013a] emphasized that the same theorem can have both pure and impure proofs. In contrast, modesty concerns the *existence of proofs from appropriate hypotheses*.

No single axiom is modest or immodest; the relation has two arguments: a set of sentences is a modest axiomatization of a given data set. If the axioms are contained in the data set the axiomatization is manifestly modest and this is just a *mathematical fact* that can be clarified by formalization. But some proper subset of the axioms might imply the whole set. This might just happen by a clever proof. But, the cases studied here are more subtle. New interpretations of the basic concepts (multiplication and number) developed over time so that the sentences attained essentially new meanings. As (**) illustrates, such is the case with Euclid's VI.1. Distinct *philosophical issues* arise in checking modesty and immodesty. For modesty, a historical investigation, as in this chapter, can explore how changing conceptions are reflected in new proofs and whether such arguments are pure. For immodesty, the conceptual content of the new axioms (not in the data set) must be compared with that of the data set.

In the quotation above, Hilbert takes the axioms to come from the data set. But this raises a subtle issue about what comprises the data set. For examples such as geometry and number theory, it was taken for granted that there was a unique model. Even Hilbert adds his completeness axiom to guarantee categoricity and to connect with the real numbers. So one could argue that the early twentieth century axiomatizers took categoricity as part of the data.[15] As a metatheoretic statement it is certainly not in the data set of the Greeks for whom 'categoricity' is meaningless.

Hallett ([Hallett & Majer 2004], 429) formulates the completeness issue in words that fit strikingly well in the 'descriptive axiomatization' framework, 'Hilbert's system with the *Vollständigkeitsaxiom* is complete with respect to "Cartesian" geometry.' But by no means is Hilbert's 'real' geometry[16] a part of Euclid's or even the Cartesian data set.

Beyond Descriptive Completeness: The notion of descriptive completeness as enunciated by Detlefsen and discussed in Chapter 9.1 and its application to geometry in this book is inherently historical. A certain data set is given; find axioms that (modestly) describe it. Schlimm provides a more flexible view of axiomatization as a tool that can serve several purposes.

The practical usefulness of axioms goes well beyond the context of justification and the aim of clarifying and providing foundations for mathematical theories; they are also engines for discovery in mathematics. [Schlimm 2013]

The next example illustrates that even *formal* axiomatization can play the role Schlimm describes. In trying to clarify the Ritt–Kolchin theory of differential algebra, Robinson [Robinson 1959a] introduced the powerful notion of a differentially closed field (Chapter 6.3). He developed the concept in analogy to that of algebraically closed fields. He did not provide explicit axioms for the theory but proved that the formal theory of differential fields (of characteristic 0) had a model completion (Chapter 4.4)

---

[15] In fact Huntington invokes Dedekind's postulate in his axiomatization of the complex field in the article quoted above [Huntington 1911].

[16] As we clarify our understanding of Cartesian geometry, stated in Notation 9.3.2 and elaborated in Chapter 10.2, we argue that Hilbert's view (the study of the Dedekind real plane) of 'Cartesian geometry' does not agree with Descartes. This view is supported in [Bos 2001], [Crippa 2014b], [Giovannini 2016], and [Panza 2011].

with a recursive set of axioms. Blum [Blum 1968] developed an informative set of axioms in terms of the solutions of differential polynomials (indexed by degree). Thirty years later [Pierce & Pillay 1998] provided an influential set of 'geometric axioms' for the theory.

9.2 Descriptions of the Geometric Continuum

First, we distinguish the *geometric continuum* from the *set theoretic continuum*. Then we sketch the background shift from the study of various types of magnitudes by the Greeks, to the modern notion of a collection of real numbers which are available to measure any sort of magnitude.

Conceptions of the Continuum: We now explain our decision to define the *geometric* continuum as the line situated in the plane. Sylvester describes three divisions of mathematics:

There are three ruling ideas, three so to say, spheres of thought, which pervade the whole body of mathematical science, to some one or other of which, or to two or all of them combined, every mathematical truth admits of being referred; these are the three cardinal notions, of Number, Space and Order.[17]

This is a slightly unfamiliar trio. We are all accustomed to the opposition between arithmetic and geometry. While Newton famously founded the calculus on geometry [Detlefsen & Arana 2011], the 'arithmetization of analysis' in the late nineteenth century reversed the priority. From the natural numbers the rational numbers are built by taking quotients and the reals by some notion of completion. And this remains the normal approach today. We want here to consider reversing the direction again: building a firm grounding for geometry and then finding first the field and then some completion and considering incidentially the role of the natural numbers. In this process, Sylvester's third cardinal notion, order, will play a crucial role. The notion that one point lies between two others will be fundamental and an order relation will naturally follow; the properties of space will generate an ordered field and the elements of that field will be called numbers, although they are not numbers in the Greek conception.

We now argue briefly that we address a real problem: there are different conceptions of the continuum (the line); hence different axiomatizations may be necessary to reflect these different conceptions. These different conceptions are witnessed by such collections as [Ehrlich 1994,

---

[17] As quoted in [Mathias 1992].

Salanskis & Sinaceur 1992] and further publications concerned with the constructive continuum and various non-Archimdean notions of the continuum. [Feferman 2008a] lists six[18] different conceptions of the continuum: (i) the Euclidean continuum, (ii) Cantor's continuum, (iii) Dedekind's continuum, (iv) the Hilbertian continuum, (v) the set of all paths in the full binary tree, and (vi) the set of all subsets of the natural numbers. For our purposes, we will identify (ii), (v), and (vi) as essentially cardinality based since they have lost the order type imposed by the geometry; so, they are not in our purview. We want to contrast two essentially geometrically based notions of the continuum: those of Euclid and Hilbert. And we identify Dedekind's and Hilbert's conceptions for reasons described in Chapter 11.1.

We began by stipulating that by 'geometric continuum', we meant the line situated in the plane. One of the fundamental results of twentieth century geometry is that *any* plane[19] can be coordinatized by a 'ternary field'. A ternary field is a structure with one ternary function $f(x, y, z)$ such that f has the properties that $f(x, y, z) = xy + z$ would have if the right hand side were interpreted in a field. In accord with our concerns with Euclidean geometry here, we assume the axioms of congruence and the parallel postulate; this implies that the ternary field is actually a field. But these geometric hypotheses are necessary. In [Baldwin 1994], I constructed an $\aleph_1$-categorical projective plane where the ternary field is as wild as possible (in the precise sense of the Lenz–Barlotti classification [Yaqub 1967]): the ternary function cannot be decomposed into an addition and multiplication.

Ratio, Magnitude, and Number: In this section we give a short review of Greek attitudes toward magnitude and ratio. Fuller accounts of the transition to modern attitudes appear in such sources as [Mueller 2006, Euclid 1956, Stein 1990]. We by no means follow the 'geometric algebra' interpretation decried by Grattan-Guinness in [Grattan-Guinness 2009]; rather, we contrast the Greek meanings of propositions with Hilbert's understanding. When we rephrase a sentence in algebraic notation we try to make clear that this is a modern formulation and often does not express the intent of Euclid.

Euclid develops arithmetic in chapters VII–IX of *The Elements*. What we think of as the 'number' 1 was the unit: a number (Definition VII.2) is a

---

[18] Smorynski [Smorynski 2008] notes that Bradwardine already reported five in the fourteenth century.

[19] This result is usually stated in terms of a projective plane, any system of points and lines such that two points determine a line, any two lines intersect in a point, and there are four non-collinear points. This is merely for convenience (Chapter 12.2).

multitude of units. These are counting numbers. So from our standpoint (considering the unit as the number 1) Euclid's numbers (in the arithmetic) can be thought of as the 'natural numbers'. The numbers[20] are a discretely ordered collection of objects.

Following Mueller we interpret magnitudes as in *The Elements* to be 'abstractions from geometric objects which leave out of account all properties of those objects except quantity: length for lines, area of plane figures, volume of solid figures, etc.' [Mueller 2006].[21] Mueller emphasizes the distinction between the properties of proportions of magnitudes developed in Book V and those of number in Book VII. The most easily stated is implicit in Euclid's proof of Theorem V.5; for every m, every magnitude can be divided in m equal parts. This is, of course, false for the (natural) numbers.

There is a second use of 'number' in Euclid. It is possible to count unit magnitudes, to speak of, e.g. four copies of a unit magnitude. So (in modern language) Euclid speaks of multiples of magnitudes by positive integers. On page 225 we give a modern mathematical interpretation of this usage.

Magnitudes of the same type are also linearly ordered and between any two there is a third.[22] Multiplication of line segments yields rectangles. Ratios are not objects; equality of ratios is a 4-ary relation between two pairs of homogeneous magnitudes.[23]

As a background for later discussion, here are some key points from Euclid's discussion of proportion in Book V.

(1) Definition V.4 of Euclid [Euclid 1956] asserts: Magnitudes are said to *have a ratio* to one another, which are capable, when multiplied, of exceeding one another.

(2) Definition V.5 defines 'sameness of two ratios' (in modern notation): The ratio of two magnitudes x and y are proportional to the ratio of two others z, w if for each m, n, $mx > ny$ implies $mz > nw$ (and also if $>$ is replaced by $=$ or $<$).

(3) Definition V.6 says: Let magnitudes which have the same ratio be called proportional.

(4) Proposition V.9 asserts that 'same ratio' is, in modern terminology, a transitive relation. Apparently Euclid took symmetry and reflexivity for granted and treats proportional as an equivalence relation.

---

[20] More precisely, the natural numbers greater than 1.

[21] ([Mueller 2006], 121).

[22] The Greeks accepted only potential infinity. So while, from a modern perspective, the natural numbers are ordered in order type ω, and any collection of homogeneous magnitudes (e.g. areas) are in a dense linear order (which is necessarily infinite), this completed infinity is not the understanding of the Greeks.

[23] Homogeneous pairs means magnitudes of the same type. Ratios of numbers are described in Book VII, while ratios of magnitudes are discussed in Book V.

The shift to a notion of numbers as representing ratios and real numbers was a long one and not the uniform progress of adding negatives, rationals, irrationals, and transcendentals presented in (American) pre-calculus classes. Jacob Klein ([Klein 1968], 186–197) reports some of the truly novel ideas of the Flemish mathematician Simon Stevin (1548–1620):[24] Stevin's main thesis is *the unit is a number; number is not all a discontinuous quantity; every root is a number*. Further Klein asserts ([Klein 1968], 197) that Stevin is the first mathematician who understands subtraction of a 'number' as the addition of a 'negative number'.

But through the seventeenth century the notion of even 'rational numbers' has to be argued for. Sylla ([Sylla 1984], 23) traces the history of two traditions of compounding ratios. The second, less known, tradition identifies a ratio (of natural numbers?) as a rational number, *the denomination* (and three other names from the literature: denominationes, denominators, exponentes) and multiplies ratios to obtain a *compound ratio*.

Leibniz[25] (1646–1716) writes

I have always disapproved of the fact that special signs are used in ratio and proportion, on the ground that for ratio the sign of division suffices and likewise for proportion the sign of equality suffices. Accordingly I write the ratio a to b thus: $a: b$, or a/b just as is done in dividing a by b. I designate proportion or the equality of two ratios by the equality of two divisions or fractions. Thus, when I express that the ratio of a to b is the same as that c to d, it is sufficient to write $a: b = c: d$ or $a/b = c/d$.

And, again from [Sylla 1984], Wallis (1616–1703) says:

And whatever things should be said about the addition, subtraction, multiplication, division, or even ratio, etc. of fractions, I wish exactly the same to be said about ratios or rather about the denominations of ratios. In fact, fractions (whether proper, or improper, or even irrational) are nothing else but the denominations of ratios.

Proportionality in Geometry: Returning to geometric considerations, we now contrast Euclid's notion of proportionality with the distinct notions of segment multiplication held by Descartes and Hilbert. We begin with a particular sequence of theorems that illuminate the distinction. In a discussion of the foundations of geometry Bolzano discusses the 'dissimilar objects' found in Euclid and finds Euclid's approach fundamentally flawed.

---

[24] The following phrases are direct quotes from Stevin via Klein.
[25] I take the quote from [Sylla 1984].

His general position is that one must analyze conceptually prior notions (line) before more complex notions[26] (plane). In particular, he objects to Euclid's proof in book VI that similar triangles have proportionate sides:

Firstly triangles, that are already accompanied by circles which intersect in certain points, then angles, adjacent and vertically opposite angles, then the equality of triangles, and only much later their similarity, which however, is derived by an atrocious detour [*ungeheuern Umweg*], from the consideration of parallel lines, and even of the area of triangles, etc.[27] [Bolzano 1810]

We'll call this *Bolzano's challenge*; it has two aspects: (a) the evil of using two-dimensional concepts to understand the line and (b) the 'atrocious detour' to similarity. We consider the first use essential to the *geometric continuum*. Chapter 9.4 reports how Hilbert avoids Euclid's following detour: In VI.1, using the theory of proportions (so implicitly the Archimedean axiom) from Book V, Euclid determines the area of a triangle or parallelogram. The role of the theory of proportion is to show that the area of two parallelograms whose respective base and top are on the same parallel lines (and so the parallelograms have the same height) have proportionate areas *even if the bases are incommensurable*. It is then straightforward to deduce V1.2, similar triangles have proportional sides. And from VI.2, he constructs in VI.12 a fourth proportional to three segments. Using Figure 9.1 for VI.12, Descartes defines the multiplication of line segments to give another segment,[28] so he *still relies* on Euclid's theory of proportion to justify the multiplication.

Proposition VI.12: To find a fourth proportional to three given straight lines. Let A and B and C be the three given straight lines. It is required to find a fourth proportional to A, B, and C.

In the diagram, segment A is represented as DF, B as DG, and C as DH. Now if DH is taken as the unit, DE has length $DG \cdot DF$.

Hilbert's innovation (Definition 9.2.1) is to use the same diagram to define segment multiplication and then formulate proportionality from segment multiplication.

We have three different constructions for *segment multiplication*: Euclid's computation of the area of rectangle, his construction of the fourth

---

[26] See ([Bolzano & Russ 2004], 33) and ([Rusnock 2000], 53) for a summary of Bolzano's claim that the study of the line, as 'more fundamental', must precede that of the plane.

[27] This quotation is taken from [Franks 2014].

[28] He refers to the construction of the fourth proportional ('ce qui est meme que la multiplication' [Descartes 1637]). See also Section 21, page 296 of [Bos 2001].

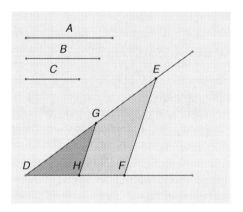

Figure 9.1 The fourth proportional

proportional, and Hilbert's definition. These three give the 'same answer'. As we now use the fourth proportional in $1 : a :: b : x$ to measure the area of the rectangle with sides of length a and b, we do not view Hilbert as introducing a new concept of multiplication – but as reinterpreting the notion as applying to line segments, which we now read as numbers, rather than magnitudes. With his multiplication, Hilbert redefines proportionality:

Definition 9.2.1 (Proportionality) *We write the ratio of CD to CA is proportional to that of CE to CB,*

$$CD : CA :: CE : CB$$

which is defined as

$$CD \times CB = CE \times CA,$$

where $\times$ is taken in the sense of segment multiplication defined by Hilbert (page 224).

Now (*) (beginning of this chapter) (VI.1) is interpreted as a variant of:

$$(**)\qquad F(ABC) = \frac{1}{2}\alpha \cdot AB \cdot AC.$$

Here $F(ABC)$ is an area function (Definition 9.5.5). But the cost is that Euclid does not (and need not) specify what we now call the proportionality constant while Hilbert must. As we'll see after Definition 9.5.5, Hilbert assigns a proportionality constant (in this case the constant α is one). Euclid depends on the Archimedean axiom to justify Eudoxus' theory of incommensurables. But as expounded in Chapter 9.5, Hilbert's treatment of area and similarity has no such dependence.

It is widely understood[29] that Dedekind's analysis is radically different from that of Eudoxus. A principal reason for this (Chapters 10.3 and 11.2) is that while Eudoxus applies his method to specific situations, Dedekind demands that *every* cut be filled. Secondly, Dedekind develops addition and multiplication on the cuts. Thus, *Dedekind's postulate should not be regarded as part of the Euclidean data set.*

9.3 Some Geometric Data Sets and Axiom Systems

We refine Hilbert's analysis of the 'significance of the groups of axioms' (introduction to [Hilbert 1962]) using the notion of a modest descriptive axiomatization. Showing a particular set of axioms is descriptively complete is inherently empirical. One must check whether each of a certain set of results is derivable from a given set of axioms. Hartshorne [Hartshorne 2000] carried out this project without using Detleftsen's terminology and we organize his results in Theorem 9.3.4.

We distinguished the Euclid–Hilbert style of formalization from the Hilbert–Gödel–Tarski style in Chapter 1.1. We take it that the Euclid–Hilbert approach includes the Hilbert insight that postulates are implicit definitions of classes of models.[30] We apply methods of modern model theory in Chapter 10 and so move to the Hilbert–Gödel–Tarski style. We will give our arguments in English; but we will be careful to specify the vocabulary and the postulates in a way that the translation to a first order theory is transparent. We will frequently switch from syntactic to semantic discussions so we stipulate precisely the vocabulary in which we take the axioms above to be formalized. We freely use defined terms such as collinear, segment, and angle in giving the reading of the relevant relation symbols.

Notation 9.3.1 *The fundamental relations of plane geometry make up the following vocabulary τ.*

---

[29] [Stekeler-Weithofer 1992] writes, 'It is just a big mistake to claim that Eudoxus's proportions were equivalent to Dedekind cuts.' Feferman [Feferman 2008a] avers, 'The main thing to be emphasized about the conception of the continuum as it appears in Euclidean geometry is that the general concept of set is not part of the basic picture, and that Dedekind style continuity considerations of the sort discussed below are at odds with that picture.' In contrast Stein [Stein 1990] argues for at least the compatibility of Dedekind's postulate with Greek thought 'reasons ... plausible, even if not conclusive – for believing the Greek geometers would have accepted Dedekind's postulate, just as they did that of Archimedes, once it had been stated.'

[30] The priority for this insight is assigned to such slightly earlier authors as Pasch, Peano, and Fano, in works such as [Freudenthal 1957] as commented on in [Bos 1993] and chapter 24 of [Gray 2011].

(1) *Two-sorted universe: points (P) and lines (L).*
(2) *Binary relation* $I(A, \ell)$: *Read: a point is incident on a line.*
(3) *Ternary relation* $B(A, B, C)$: *Read: B is between A and C (and A, B, C are collinear).*
(4) *Quaternary relation,* $C(A, B, C, D)$: *Read: two segments are congruent, in symbols* $\overline{AB} \cong \overline{CD}$.
(5) *6-ary relation* $C'(A, B, C, A', B', C')$: *Read: the two angles* $\angle ABC$ *and* $\angle A'B'C'$ *are congruent, in symbols* $\angle ABC \cong \angle A'B'C'$.

We begin by distinguishing several topics in plane geometry[31] that represent distinct data sets in Detlefsen's sense. Explicit axioms from Euclid are included in the appropriate data set. While we describe five sets here, only polygonal geometry and circle geometry are considered in this chapter; the others are treated in Chapter 10. The hierarchy is cumulative.

Notation 9.3.2 *(five data sets of geometry)*

Euclid I, polygonal geometry: Book I (except I.1 and I.22), Book II.1–II.13, Book III (except III.1 and III.17), Book VI
Euclid II, circle geometry: I.1, I.22, II.14, III.1, III.17, and Book IV
Archimedes, arc length and π: XII.2, Book IV (area of a circle proportional to the square of the diameter), approximation of π, circumference of circle proportional to radius, Archimedes' axiom
Descartes, higher degree polynomials: nth roots; coordinate geometry
Hilbert, continuity: The Dedekind plane

The division of the data sets is somewhat arbitrary and made with the subsequent axiomatizations in mind. We placed I.22, II.14, III.1, III.17, and Book IV of Euclid in our Euclid II group because they depend on circle–circle intersection. In addition to these, much of Euclid's theory of proportion in Chapters V and X follows from existence of the field.

The arguments in Euclid I go through the theory of area which depends on Eudoxus' theory of proportion and so have an implicit dependence on the Archimedean axiom; Hilbert eliminates this dependence. The role of Euclid II appears already in Proposition I of Euclid where Euclid makes the standard construction of an equilateral triangle on a given base. Why do the two circles intersect? While some[32] regard the absence

---

[31] In the first instance we draw from Euclid: Books I–IV, VI, and XII.1–2 are clearly plane geometry; XI, the rest of XII, and XIII are solid geometry; V and X deal with a general notion of proportion and with incommensurability. Thus, below we select from Books I–IV, VI, and XII.1–2 and consider certain geometrical aspects of V and X.

[32] E.g. Veblen ([Veblen 1914], 4).

of an axiom guaranteeing such intersections as a gap in Euclid, Manders ([Manders 2008], 66) asserts: 'Already the simplest observation on what the texts do infer from diagrams and do not suffices to show the intersection of two circles is completely safe.'[33] For our purposes here, we are content to accept that adopting the circle–circle intersection axiom resolves those continuity issues involving circles and lines. We separate this case as Hilbert's first order axioms do not resolve this issue;[34] he chose to resolve it (implicitly) by an appeal to Dedekind.

We explain in Chapter 10.1 why Euclid XII.2 (area of a circle is proportional to the square of the diameter) is placed in Archimedes rather than Euclid II.

Circle–Circle Intersection Postulate (CCP) If from distinct points A and B, circles with radius AC and BD are drawn such that one circle contains points both in its interior in the exterior of the other, then they intersect in two points, on opposite sides of AB.

Hilbert makes one other essential addition[35] to Euclid's axioms, by introducing Group III, the congruence axioms and in particular the

---

[33] Manders develops the use of diagrams as a coherent mathematical practice; Avigad and others [Avigad et al. 2009] have developed the idea of formalizing a deductive system which incorporates diagrams. Here is a rough idea of this program. Properties that are *not* changed by minor variations in the diagram such as subsegment, inclusion of one figure in another, whether two lines intersect or betweenness are termed *inexact*. Properties that *can be* changed by minor variations in the diagram, such as whether a curve is a straight line, congruence or a point being on a line, are termed *exact*. We can rely on reading inexact properties from the diagram. We must write exact properties in the text. The difficulty in turning this insight into a formal deductive system is that, depending on the particular diagram drawn, after a construction, the diagram may have different inexact properties. The solution is case analysis but bounding the number of cases has proven difficult.

Although I agree with the approach of Manders, Avigad et al., or Miller [Miller 2007], the goal of this chapter is comparison with the axiom systems of Hilbert and Tarski. Reformulating those systems via proof systems formally incorporating diagrams would delete CCP and so collapse Euclid I and II.

[34] Circle–circle intersection implies line–circle intersection. Hilbert Knew by 1898/99 that circle–circle intersection holds in a Euclidean plane. See ([Hallett & Majer 2004], 204–206) for a historical discussion and exact references. Hilbert is aware that his axiom groups I–IV do not suffice. Chapter 9.4 has a more detailed mathematical discussion.

[35] In [Baldwin & Mueller 2012] and [Baldwin 2013b] we give an equivalent set of postulates to EG, which returns to Euclid's construction postulates and stresses the role of Euclid's axioms (Common Notions) in interpreting the geometric postulates. See http://aleph0.clarku.edu/~djoyce/java/elements/bookI/bookI.html#cns. While not spelled out rigorously, our aim is to consider the diagram as part of the argument. For pedagogical reasons the system used SSS rather than SAS as the basic congruence postulate and made clear that the equality axioms in logic, as in Euclid's Common Notions', apply to both algebra and arithmetic. This eliminates silly six step arguments reducing subtraction of segments to the axioms of the real numbers in high school texts.

axiom SAS.[36] This replaces implicit appeals to the superposition principle. We follow Hartshorne [Hartshorne 2000] in the following nomenclature.

Notation 9.3.3 Consider the following axiom sets.[37]

(1) First order axioms:

    HP, HP5: We write HP for Hilbert's incidence, betweenness,[38] and congruence axioms. HP5 denotes HP plus the parallel postulate.

    EG: The *axioms for Euclidean geometry*, denoted EG,[39] consist of HP5 plus the CCP.

    $\mathcal{E}^2$: Tarski's axiom system for a plane over a real closed field (RCF, page 148).

    $EG_\pi, \mathcal{E}_\pi$: Two new systems (Chapter 10) that extend EG and $\mathcal{E}^2$.

(2) Hilbert's continuity axioms, infinitary and second order:

    Archimedes: The sentence in $L_{\omega_1,\omega}$ expressing the Archimedean axiom.

    Dedekind Dedekind's *second order* axiom:[40] There is a point in each irrational cut in the line.

(3) A *Hilbert plane* is any model of HP.

(4) A *Euclidean plane* is a model of EG: Euclidean geometry.

The next theorem aligns the axioms sets above with their consequences in Euclid as categorized in Notation 9.3.2 and as spelled out in Section 12 and Sections 20–23 of [Hartshorne 2000].

Theorem 9.3.4

(1) *The sentences of Euclid I are provable in HP5.*

(2) *The sentences of Euclid II are provable in EG.*

---

[36] Triangles with two pairs of sides and the included angles congruent are congruent.

[37] The names HP, HP5, and EG come from [Hartshorne 2000] and $\mathcal{E}^2$ from [Tarski 1959]. In fact, Tarski also studies EG under the name $\mathcal{E}_2''$.

[38] As we axiomatize *plane* geometry we include Pasch's axiom ([Hartshorne 2000], B4): any line intersecting one side of triangle must intersect one of the other two.

[39] In the vocabulary here, there is a natural translation of Euclid's axioms into first order statements. The construction axioms have to be viewed as 'for all – there exist' sentences. The axiom of Archimedes is of course not first order. We write 'Euclid's axioms' for those in the original as opposed to the modernized (first order) axioms for Euclidean geometry, EG. Note that EG is equivalent to (i.e. has the same models as) the system laid out in [Avigad et al. 2009], namely, planes over fields where every positive element has a square root. The latter system builds the use of diagrams into the proof rules.

[40] Hilbert added his *Vollständigkeitsaxiom* to the French translation and it appears from the 2nd edition on. In Chapter 11.1 we explore the connections between various formulations of completeness. We take Dedekind's formulation as emblematic.

(3) *The sentences of Archimedes,*[41] *arc length and* π: *Euclid XII.2, area of circle are provable in Hilbert's first order axioms plus Archimedes and also in the first order theories* EG_π *and* $\mathcal{E}_\pi$.

(4) *The sentences of Descartes: (nth roots) are provable in RCF* ($\mathcal{E}^2$).

(5) *The nineteenth century analysis of Weierstrass is provable in Hilbert's full system.*

Proof. For (1) and (2) see [Hartshorne 2000, Sections 20–23]. For (3) see Chapters 10.1–10.2 and for (4) Chapter 10.2. For (5) choose an analysis text such as [Spivak 1980].

9.4 Geometry and Algebra

Geometry was long *the foundation* of mathematics. Newton wrote,

Geometry was invented that we might expeditiously avoid, by drawing Lines, the Tediousness of Computation. Therefore these two sciences [Geometry and Arithmetical Computation] ought not be confounded. The Ancients did so industriously distinguish them from one another, that they never introduced Arithmetical Terms into Geometry. And the Moderns, by confounding both, have lost the Simplicity in which all the Elegance of Geometry consists.[42]

The nineteenth century arithmetization of analysis destroyed the notion that geometry could serve as universal foundation. But does that mean that geometry itself must be founded on arithmetic?

From Arithmetic to Geometry or from Geometry to Algebra? On the first page of 'Continuity and the irrational numbers', Dedekind writes:

Even now such resort to geometric intuition in a first presentation of the differential calculus, I regard as exceedingly useful from the didactic standpoint … But that this form of introduction into the differential calculus can make no claim to being scientific, no one will deny. [Dedekind 1963]

I do not contest that claim. I quote this passage to indicate that Dedekind's motivation was to provide a basis for calculus and indeed analysis, not geometry. But I will argue that the second order Dedekind completeness axiom is not needed for the geometry of Euclid or indeed for the grounding of the algebraic numbers, although it is for Dedekind's approach.

---

[41] See the discussion after Theorem 10.1.9.

[42] As quoted in [Guicciardini 2006] from page 228 of volume 2 of [Newton 1769].

Further I will discuss in Chapter 10.2 the possibility that a kind of 'definable' continuity provides a substitute for many (certainly not all) of Dedekind's concerns.

Dedekind provides a theory of the continuum (the continuous line) by building it up in stages from the structure which is fundamental to him: the natural numbers under successor. This development draws on second order logic in several places. The well-ordering of the natural numbers is required to define addition and multiplication by recursion. Dedekind completeness is a second appeal to a second order principle.

Perhaps in response to Bolzano's insistence, Dedekind constructs the line without recourse to two-dimensional objects and from arithmetic. Thus, he succeeds in the 'arithmetization of analysis'.

We proceed in the opposite direction for several reasons. Most important is that we are seeking to ground geometry, not analysis. Further, we assert that the concept of line arises only in the perception of at least two-dimensional space. Dedekind's continuum knows nothing of being straight or breadthless. Hilbert's proof of the existence of the field is the essence of the *geometric continuum*. By virtue of its lying in a plane, the line acquires algebraic properties.

The distinction between the arithmetic and geometric intuitions of multiplication is fundamental. The basis of the first is iterated addition; the basis of the second is scaling or proportionality. The late nineteenth century developments provide a formal reduction of the second to the first but the reduction is only formal; the intuition is lost. In this section we view both intuitions as fundamental and develop the second with the understanding that development of the first through the Dedekind–Peano treatment of arithmetic is in the background. See Remark 9.4 for the connection between the two.

From Geometry to Segment Arithmetic to Numbers: We now introduce *segment arithmetic* and sketch Hilbert's definition of the (semi)-field of segments with partial subtraction and multiplication. We assume Hilbert's first order geometry, the axiom system we called HP5 in Notation 9.3.3. See e.g. [Hilbert 1971, Hartshorne 2000, Baldwin 2013b, Giovannini 2016] for more detail.

Note that congruence forms an equivalence relation on line segments. We fix a ray ℓ with one end point 0 on ℓ. For each equivalence class of segments, we consider the unique segment $0A$ on ℓ in that class as the representative of that class. We will often denote the segment $0A$ (ambiguously its congruence class) by a. We say a segment CD (on any line) has

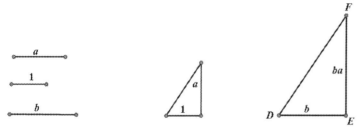

Figure 9.2 Defining multiplication

length a if $CD \cong 0A$. We define the sum of two segments as the congruence class of the result of placing one after the other on a straight line. And $AB < CD$ if AB is congruent to a segment contained in CD.

Of course there is no additive inverse if our 'numbers' are the lengths of segments which must be positive. However, this procedure can be extended to a field structure on segments on a line not a ray (so with negatives), either directly as sketched in [Baldwin & Mueller 2012] or by passing through the theory of ordered fields as in Section 19 of [Hartshorne 2000]. Following Hartshorne, here is our *official definition* of segment multiplication.[43]

Fix a unit segment class 01 and its congruence class 1. Consider two segment classes a and b (Figure 9.2). To define their product, construct a right triangle with legs of length 1 and a. Denote the angle between the hypoteneuse and the side of length a by α.

Now construct another right triangle FDE with base of length b and with the angle between the hypoteneuse and the base DE of length b congruent to α. The product ab is defined to be the length of the vertical leg EF. The parallel postulate guarantees the existence and uniqueness of the point F. It is clear from the definition that there are multiplicative inverses; use the triangle with base a and height 1. A three page proof in [Hartshorne 2000] shows multiplication is commutative, is associative, distributes over addition, and respects the order. His proof uses only the following very elementary cyclic quadrilateral theorem (Corollary 9.4.1) and connections between central and inscribed angles in a circle (Figure 9.3). Hilbert's argument uses what he calls Pascal's theorem but the closely related theorem of Desargues (Theorem 12.2.4) is the most used. In either case the domain of the algebraic structure is segments of a ray named by constants 0 and 1.

---

[43] Hilbert uses Euclid's construction of the fourth proportional. The clear association of a particular angle with right multiplication by a recommends Hartshorne's version.

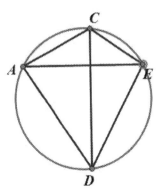

Figure 9.3 The Cyclic Quadrilateral Theorem

Corollary 9.4.1 *Let ACED be a quadrilateral. The vertices of A lie on a circle (the ordering of the name of the quadrilateral implies A and E are on opposite sides of CD) if and only if* $\angle EAC \cong \angle CDE$.

Hilbert has defined segment multiplication on the ray from 0 through 1. But to get negative numbers he must reflect through 0. Then addition and multiplication can be defined on directed segments of the line through $0, 1$[44] and thus all axioms for a field are obtained. We will discuss the historical significance of this shift just before Theorem 10.2.1. But even before that we will use the more flexible language of points, especially in Chapter 10. See Dicta 9.4.2.

The next step is to identity the points on the line and the domain of an ordered field by mapping A to OA. This naturally leads to thinking of a segment as a set of points, which is foreign to both Euclid and Descartes. Although, in the context of the *Grundlagen*, Hilbert's goal is to coordinatize the plane by the real numbers, his methods open the path to thinking of the members of any field as 'numbers' that coordinatize the associated geometries. Boyer traces the origins of numerical coordinates to 1827–1829 and writes,

It is sometimes said that Descartes arithmetized geometry but this is not strictly correct. For almost two hundred years after his time coordinates were essentially geometric. Cartesian coordinates were line segments ... The arithmetization of coordinates took place not in 1637 but in the crucial years 1827–1829. ([Boyer 1956], 242)

---

[44] Hilbert had done this in lecture notes in 1894 [Hallett & Majer 2004]. Hartshorne constructs the field algebraically from the semifield rather than in the geometry.

Boyer points to Bobillier, Möbius, Feurbach, and most critically Plücker as introducing several variants of what constitute numerical (signed distance) barycentric coordinates of a point.

Dicta 9.4.2 (Constants 1: More on 0,1) To fix the field we had to add constants 0, 1. These constants can name any pair of points in the plane.[45] But this naming induces an extension of the data set. We have in fact specified the unit. This reflects a major change in view from either the Greeks or Descartes. Here there is little effect on the data set but a major change in view.

It is easy[46] to check that the multiplication defined on the positive reals by this procedure is exactly the usual multiplication on the positive reals because they agree on the positive rational numbers.

Multiplication Is Not Repeated Addition: We now have two ways in which we can think of the product $3a$. On the one hand, we can think of laying 3 segments of length a end to end. On the other, we have given several constructions for the segment multiplication of a segment of length 3 (i.e. 3 segments of length 1 laid end to end) by the segment of length a. It is an easy exercise to show these give the same answer. But these distinct constructions make an important point. The (inductive) definition of multiplication by a natural number is indeed 'multiplication as repeated addition'. But the multiplication by another field element is based on similarity and has multiplicative inverses; so it is a very different operation. From a modern standpoint, they are quite different: no extension of natural number arithmetic is decidable but important theories of fields are.

The first notion of multiplication in the last paragraph, where the multiplier is a natural number, is a kind of *scalar multiplication* by positive integers that can be viewed mathematically as a rarely studied object: a semiring (the natural numbers) acting on a semigroup (positive reals under addition). There is no uniform definition[47] of this *scalar* multiplication within the semiring.

---

[45] The axioms of HP5 imply that the automorphism group of the plane acts two-transitively (any pair of distinct points can be mapped by an automorphism to any other such pair). This transitivity implies that a sentence $\phi(0, 1)$ holds just if either or both of $\forall x \forall y \phi(x, y)$ and $\exists x \exists y \phi(x, y)$ hold.

[46] One has to verify that segment multiplication is continuous but this follows from the density of the order since the addition respects order.

[47] Instead, there are infinitely many formulas $\phi_n(x, y)$ defining unary operations $nx = y$ for each $n > 0$.

A mathematical structure more familiar to modern eyes is obtained by adding the negative numbers to get the ring $\mathbb{Z}$, which has a well-defined notion of subtraction. The scalars are now in the ring $(Z, +, \cdot)$ and act on the module $(\mathfrak{R}, +)$. Now we can multiply by -17 but the operation is still not uniform but given by a family of unary functions.

Here are algebraic characterizations of the fields associated with models of HP5 and EG ([Hartshorne 2000], section 21).

Definition 9.4.3

(1) *A field F is* Pythagorean *if it is closed under addition, subtraction, multiplication, and division and for every $a \in F$, $\sqrt{(1 + a^2)} \in F$.*
(2) *An ordered field F is* Euclidean *if it is closed under addition, subtraction, multiplication, and division and for each positive $a \in F$, $\sqrt{a} \in F$.*

Recall that we distinguished a Hilbert plane from a Euclidean plane in Notation 9.3.3(3). As in [Hartshorne 2000], we have:

Theorem 9.4.4

(1) *HP5 is bi-interpretable with the theory of ordered Pythagorean fields.*[48]
(2) *Similarly EG is bi-interpretable with the theory of ordered Euclidean fields.*

Hartshorne [Hartshorne 2000] describes two instructive examples that connect the notions of Pythagorean and Euclidean planes.

Example 9.4.5

(1) The Cartesian plane over a Pythagorean field may fail to be closed under square root and thus the circle–circle intersection postulate also fails.[49]
(2) On page 146 of [Hartshorne 2000][50] observes that the smallest ordered field closed under addition, subtraction, multiplication, and division and square roots of positive numbers satisfies the circle–circle intersection postulate and is a Euclidean field. We denote this field by F_s for *surd field.*

---

[48] Chapter 4.5 explains the notion of interpretability. Here, Hilbert's construction interprets the field in the geometry and the *associated* Cartesian plane over a field interprets the geometry in the field.

[49] This was known to Hilbert ([Hallett & Majer 2004], 201–202); Hartshorne exhibits a hyperbolic geometry without equilateral triangles. See Exercises 39.30, 39.31 of [Hartshorne 2000].

[50] Hartshorne and Greenberg [Greenberg 2010] call this the constructible field (closed under ruler/compass constructions), but given the many meanings of constructible, we use Moise's term surd field.

Note that if HP5 + CCP were proposed as an axiom set for polygonal geometry it would be a complete descriptive but not modest axiomatization since it would prove CCP which is not in the polygonal geometry data set.

Initial Consequences for Field Arithmetic: We now investigate two sorts of statements that are true in any field associated with a geometry modeling HP5 despite prior proofs with stronger hypotheses: (1) statements of Euclid's geometry that depended in his development on the Archimedean axiom and (2) statements about the properties of real numbers that Dedekind deduces from his postulate.

We just established that one could define an ordered field in any plane satisfying HP5. The converse is routine, the ordinary notions of line and incidence in F^2 creates a geometry over any ordered field, which is easily seen to satisfy HP5. We now exploit this equivalence to show some algebraic facts using our defined operations, thus basing them on geometry. First we show square root commutes with multiplication for algebraic numbers. Dedekind ([Dedekind 1963], 22) wrote 'in this way we arrive at real proofs of theorems (as, e.g. $\sqrt{2} \cdot \sqrt{3} = \sqrt{6}$), which to the best of my knowledge have never been established before.' This is a problem for Dedekind but not for Descartes. Euclid had already, in constructing the fourth proportional, constructed from segments of length 1, a, and b, one of length ab; but he doesn't regard this operation as multiplication. When Descartes interprets this procedure as multiplication of segments, he has no problem. But Dedekind presents the problem as multiplication in his continuum and so he must prove a theorem to find the product as a real number; that is, he must show the limit operation commutes with product.

But, in an ordered field, for any positive a, if there is an element $b > 0$ with $b^2 = a$, then b is unique (and denoted $\sqrt{a}$). Moreover, for any positive a, c with square roots, $\sqrt{a} \cdot \sqrt{c} = \sqrt{ac}$, since each side of the equality squares to ac. In particular, this fact holds for any field coordinatizing a plane satisfying HP5.

Thus, the algebra of square roots in the real field is established without any appeal to limits. The usual (e.g. [Spivak 1980, Apostol 1967]) developments of the theory of complete ordered fields (following Dedekind) invoke the least upper bound principle to obtain the existence of the roots although the multiplication rule is obtained by the same algebraic argument as here. Hilbert's approach contrasts with Dedekind's.[51] The justification here for

---

[51] Dedekind objects to the introduction of irrational numbers by measuring an extensive magnitude in terms of another of the same kind ([Dedekind 1963], 9).

the existence of roots and the rules for operating on them does not invoke limits. Hilbert's treatment is based on the geometric concepts and in particular regards 'congruence' as an equally fundamental notion as 'number'.

In short, the shift here is from 'proportional segments' to 'product of numbers'. Euclid had a rigorous proof of the existence of a line segment which is the fourth proportional of $1 : a = b : x$. Dedekind demands a product of numbers; Hilbert provides this product by combining his interpretation of the field in the geometry and his geometrical definition of multiplication.

Euclid's first proof of Pythagoras' theorem (I.47) uses the properties of area that we will justify in Chapter 9.5 and his second proof (VI.31) uses the theory of similar triangles that we will develop in Chapter 9.5. Thus, in both cases Euclid depends on the theory of proportionality (and thus implicitly on Archimedes' axiom) to prove the Pythagorean theorem; Hilbert avoids this appeal.[52] Similarly, since the right-angle trigonometry in Euclid concerns the ratios of sides of triangles, the trigonometric functions on angles less than $180°$ are defined using the field multiplication, we have:

Theorem 9.4.6 *The following hold in any Euclidean plane:*

(1) *The Pythagorean theorem as well as the law of cosines (Euclid II. 12–13) and the law of sines;*
(2) *Heron's formula ($A = \sqrt{s(s-a)(s-b)(s-c)}$ where s is $1/2$ the perimeter and a, b, c are the side lengths) computes the area of a triangle from the lengths of its sides.*

Heron's formula demonstrates the hazards of the kind of organization of data sets attempted here. The geometric proof of Heron doesn't involve the square roots of the modern formula [Heath 1921]. But since in EG we have the field and we have square roots, the modern form of Heron's formula can be proved from EG. Thus as in the shift from (*) to (**) at the beginning of the chapter, the different means of expressing the geometrical property requires different proofs.

In each case we have considered in this section, the Greeks gave geometric constructions for what in modern days becomes a calculation involving the field operations and square roots. However, we still need to complete the

---

[52] However, Hilbert does not avoid the parallel postulate since he uses it to establish multiplication and thus similarity. Euclid's theory of area also depends heavily on the parallel postulate. It is a theorem in 'neutral geometry' in the metric tradition that the Pythagorean theorem is equivalent to the parallel postulate (see Theorem 9.2.8 of [Millman & Parker 1981]). But this approach basically assumes the issues dealt with here as the 'ruler postulate' (page 246) also provides a multiplication on the 'lengths' (since they are real numbers). Julien Narboux alerted me to the issues in stating the Pythagorean theorem without assuming the parallel postulate.

argument that HP5 is descriptively complete for polygonal Euclidean geometry. In particular, is our notion of proportional correct? The test question is the similar triangle theorem. We turn to this issue now.

9.5   Proportion and Area

In this section we show how the introduction of the field allows Hilbert to ground similarity and area.

Two triangles $\triangle ABC$ and $\triangle A'B'C'$ are defined to be *similar* if under some correspondence of angles, corresponding angles are congruent; e.g. $\angle A' \cong \angle A$, $\angle B' \cong \angle B$, $\angle C' \cong \angle C$.

Various texts define 'similar' as we did, or focus on corresponding sides that are proportional or require both (Euclid). We now meet Bolzano's challenge (page 215) by showing that in Euclidean geometry (without the continuity axioms) the choice doesn't matter. Defining 'proportional' in terms of segment multiplication in Definition 9.2.1, Hartshorne answers Bolzano's challenge.

Theorem 9.5.1 (EG) *Two triangles are similar if and only if corresponding sides are proportional.*

Proof of Theorem 9.5.1: If ABC and $A'B'C'$ are similar triangles then **using the segment multiplication we have defined**

$$\frac{AB}{A'B'} = \frac{AC}{A'C'} = \frac{BC}{B'C'}.$$

Proof. The point G is the incenter so $HG \cong GI \cong GJ$ (Figure 9.4). Call this segment length a. Now construct $AK \cong BL \cong MC$, all with standard unit length. Let the lengths of OL be s, NK be t, and PM be r. Let the lengths of $AI \cong AH$ be x, $BH \cong BJ$ be y, and $CI \cong CJ$ be z. By the definition of

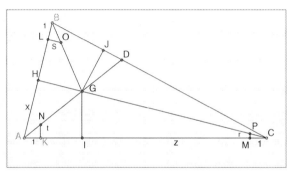

Figure 9.4 G is the center of a circle inscribed in triangle ABC

multiplication $t \cdot x = s \cdot y = r \cdot z = a$. Therefore the length of AC is
$\frac{a}{t} + \frac{a}{r} = \frac{a(r+t)}{rt}$.

Duplicate the construction on the second triangle $A'B'C'$ to get the length
of $A'C'$ is $\frac{a'}{t} + \frac{a'}{r} = \frac{a'(r+t)}{rt}$. *The crucial point is that because the angles are
congruent r, s, t are the same for both triangles.*

But then $\frac{A'C'}{AC} = \frac{a'}{a}$. Now note the same is true for the other two pairs
of sides so the sides of the triangle are proportional. The same ideas allow
one to reverse the argument and show triangles with proportional sides are
similar.

As Hilbert had shown with a somewhat different argument, in any model
M of HP5, similar triangles have proportional sides. There is no assumption
that the field is Archimedean or satisfies any sort of completeness axiom.
There is no appeal to approximation or limits.

Area of Polygonal Figures: We have just seen that Bolzano's challenge (page
215) is answered by a proof that similar triangles have proportional sides
without resorting to the concept of area. But area is itself a vital geometric
notion. We show now that, using segment multiplication, Hilbert grounds
the now familiar methods of calculating the area of polygons. As Hilbert
wrote,[53] 'We ... establish Euclid's theory of area *for the plane geometry and
that independently of the axiom of Archimedes.*'

We sketch Hartshorne's [Hartshorne 2000] exposition of this topic stressing the connections with Euclid's Common Notions. We are careful to see
how the notions defined here are expressible in first order logic[54] supporting our fifth objection to second order axiomatization of geometry in Chapter 11.2. Although these arguments are not carried out as direct deductions
from the first order axioms, in view of this expressibility, the completeness
theorem implies that the results are derivable by such a direct deduction. We
informally define those configurations whose areas are considered in this
section as follows. A *figure* is a subset of the plane that can be represented
as a finite union of disjoint triangles. There are serious issues concerning
the formalization in first order logic of the notions in this section. Notions
such as figure or polygon involve implicit quantification over integers; this
is strictly forbidden within the first order system.

We give a uniform metatheoretic definition of the relevant concepts and
prove that the theorems hold in all models of the axioms. For this, we

---

[53] Emphasis in the original: ([Hilbert 1971], 57).

[54] Poincaré ([Poincaré 1952], 15–16) uses the same example of the decomposition of polygons
into triangles to illustrate an 'advantageous' construction by recursion.

approach these notions with axiom schemes. Hilbert raised a 'pseudogap' in Euclid[55] by distinguishing area and content. In Hilbert two figures have

(1) *equal area* (equi-decomposable) if they can be decomposed into a finite number of triangles that are pairwise congruent;
(2) *equal content* (equi-complementable) if we can transform one into the other by adding and subtracting congruent triangles.

Hilbert showed: under the Archimedean axiom the two notions are equivalent, and without it they are not.

Euclid treats the equality of areas as a special case of his Common Notions. The properties of equal content, described next, are consequences for Euclid of the Common Notions and so for him need no justification. In contrast, we introduce an area function to justify that effect of the Common Notions in our development.

Fact 9.5.2 (Properties of Equal Content) *The following properties of area are used in Euclid I.35 through I.38 and beyond. They hold for equal content in HP5.*

(1) *Congruent figures have the same content.*
(2) *The content of two 'disjoint' figures (i.e. meet only in a point or along an edge) is the sum of the contents of the two polygons. The analogous statements hold for difference and half.*
(3) *(de Zolt's axiom) If one figure P is properly contained[56] in another Q then the area of the difference (which is also a figure) is positive.*

More precisely,

Definition 9.5.3 (Equal Content) *Two figures P, Q have equal content (are equi-complementable) in n steps if there are figures $P'_1 \ldots P'_n$, $Q'_1 \ldots Q'_n$ such that none of the figures overlap, each pair P'_i and Q'_i are equi-decomposable and $P = P'_1 \ldots \cup P'_n$ is equi-decomposable with $Q = Q'_1 \ldots \cup Q'_n$.*

Reading equal content for Euclid's 'equal', Euclid's I.35 (for parallelograms) and the derived I.37 (triangles) become Theorem 9.5.4 and in this formulation Hilbert accepts Euclid's proof.

---

[55] Any model with infinitesimals shows the notions are distinct and Euclid I.35 and I.36 (triangles on the same (congruent) base(s) and same height have the same area) fail for what Hilbert calls area. Since Euclid includes preservation under both addition and subtraction in his Common Notions, his term 'area' clearly refers to what Hilbert calls 'equal content', I call this a pseudogap.

[56] Hartshorne ([Hartshorne 2000], 201) gives a more precise version that $Q - P$ has non-empty interior.

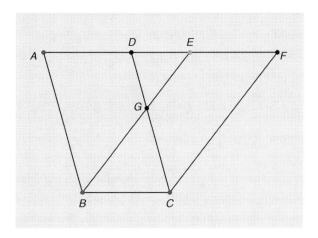

Figure 9.5 Euclid I.35

Theorem 9.5.4 *[Euclid/Hilbert] If two parallelograms (triangles) are on the same base and between parallels they have equal content in 1 step.*

Proof. $P = ADBC$ has the same content as $P' = EFBC$ (Figure 9.5) in one step as letting $Q = Q' = DGE$, $P + Q = P' + Q'$ which are decomposable as $CBG + ABE$ and $CBG + CDF$ and $ABE \approx CDF$.

Varying Hilbert, Hartshorne (Sections 19–23 of [Hartshorne 2000]) shows Fact 9.5.2 in the first order axiom system HP5 (Notation 9.3.3). The key tool is:

Definition 9.5.5 *An* area function *is a map α from the set of figures, $\mathcal{P}$, into an ordered additive abelian group with 0 such that*

(1) *For any nontrivial triangle T, $\alpha(T) > 0$.*
(2) *Congruent triangles have the same content.*
(3) *If P and Q are disjoint figures $\alpha(P \cup Q) = \alpha(P) + \alpha(Q)$.*

Semiformally, the idea is straightforward. Argue (a) that every n-gon is triangulated into a collection of disjoint triangles and (b) establish an area function on such collections. The complication for formalizing is describing the area function by first order formulas. For this,[57] add to the vocabulary for geometry a 3-ary function α_1 from points into line segments that satisfies (1) and (2). For $n > 1$, add a $3n$-ary function α_n and assert that α_n maps n-disjoint triangles into the field of line segments by summing α_1 on the individual triangles; this will satisfy (3). This handles disjoint unions

---

[57] See Definition 10.1.7 for a more detailed discussion in a slightly different context.

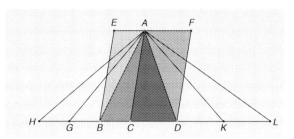

Figure 9.6 Euclid VI.1

of triangles. Hartshorne [Hartshorne 2000] (23.2–23.5) gives an inductive proof showing every polygon can be decomposed into a finite number of disjoint triangles and the area does not depend on the decomposition.

It is evident that if a plane admits an area function then the conclusions of Fact 9.5.2 hold. This obviates the need for positing separately de Zolt's axiom[58] (Fact 9.5.2(3)). In particular this implies Common Notion 4 for 'area'. Using the segment multiplication, Hilbert (compare the exposition in Hartshorne) establishes the existence of an area function for any plane satisfying HP5 and thus can compute the area of any polygon. The key point is to show that the formula $A = \frac{bh}{2}$ does not depend on the choice of the base and height. In particular, Hilbert proves (**) without recourse to the axiom of Archimedes.

Remark 9.5.6 In contrast, recall the diagram to show triangles and parallelograms which are under the same height are to one another as their bases (Figure 9.6).

If, for example, BC, GB and HG are congruent segments then the area of ACH is triple that of ABC. But without assuming BC and BD are commensurable, Euclid calls on Definition V.5 of the proportionality chapter to assert that

$$ABD : ABC :: BD : BC.$$

Thus Euclid makes a disguised appeal to Eudoxus' theory of limits while Hilbert's algebraic proof avoids both Eudoxus and the Archimedean axiom. We have now shown that the axioms for Euclidean planes suffice to prove data set Euclid II and establish proportionality, similarity, and area.

---

[58] We discuss the existence of 'pure proofs' avoiding de Zolt's axiom on page 279.

10 | π, Area, and Circumference of Circles

The geometry over a Euclidean field (every positive number has a square root) may have no straight line segment of length π. For example, the model containing only the constructible real numbers does not contain π. We want to find a theory which proves the circumference and area formulas for circles. Our approach is to extend the theory EG so as to guarantee that there is a point in every model which behaves as π does. In this chapter we will show that in this extended theory there is a mapping assigning a straight line segment to the circumference of each circle. We first introduce π to the Euclidean scheme by forming a theory EG_π. In a second direction (Chapter 10.2), we note Tarski's axiomatization of 'Cartesian' plane geometry, $\mathcal{E}^2$. Then we combine the two in a theory $\mathcal{E}^2_\pi$ to give the theory of real closed fields that include π. Given that the entire project is modern, we give the arguments entirely in the style of modern model theory.

For Archimedes and Euclid, sequences constructed in the study of magnitudes in the *Elements* are of geometric objects, not of numbers. In a modern account, as we saw already while discussing areas of polygons in Chapter 9.5, we must identify the proportionality constant and verify that it represents a point in any model of the theory.[1] Thus this goal diverges from a 'Greek' data set and indeed is orthogonal to the axiomatization of Cartesian geometry in Theorem 10.2.1.

This shift in interpretation drives the rest of this chapter. We search for the solution of a specific problem: is π in the underlying field?

10.1 π in Euclidean and Archimedean Geometry

We now describe the rationale for placing various facts in the Archimedean data set[2] in Notation 9.3.2. Three propositions encapsulate the issue:

---

[1] For this reason, Archimedes needs only his postulate while Hilbert would also need Dedekind's postulate to prove the circumference formula.

[2] The rationale is not in any sense chronological, as Archimedes attributes the method of exhaustion to Eudoxus, who precedes Euclid. Post-Heath scholarship by Becker, Knorr, and Menn [Menn 2018] has identified four theories of proportion in the generations just before Euclid. [Menn 2018] led us to the three geometric propositions whose relations are described here.

Euclid VI.1 (area of rectangle), Euclid XII.2 (area of a circle is proportional to the square of the diameter), and Archimedes' proof that the circumference of a circle is proportional to the diameter. Hilbert showed (Theorem 9.3.4) that VI.1 is provable already in HP5. While Euclid implicitly relies on the Archimedean axiom, Archimedes makes it explicit in a recognizably modern form. Euclid does *not* discuss the circumference of a circle. For determining the relation of the circumference to the diameter, Archimedes must develop his notion of arc length. By beginning to calculate approximations of π, Archimedes is moving towards the treatment of π as a number. Thus, we distinguish VI.1 from the Archimedean axiom and the theorems on measurement of a circle, and place the latter in the Archimedean data set.

Dedekind's or Birkhoff's postulates demand the identification of a straight line segment with the same length as the circumference of a circle. But this conflicts with the fourth century view of Eutocius, 'Even if it seemed not yet possible to produce a straight line equal to the circumference of the circle, nevertheless, the fact that there exists some straight line by nature equal to it is deemed by no one to be a matter of investigation[3].' Although Eutocius asserts the existence of a line of the same length as a curve but finds constructing it unimportant, Aristotle had a stronger view. Summarizing his discussion of Aristotle, Crippa ([Crippa 2014a], 34–35) points out that Aristotle takes the impossibility of such equality as the hypothesis of an argument on motion and also cites Averroes as holding 'that there cannot be a straight line equal to a circular arc.'

Thus, the validation below in the theories EG_π and $\mathcal{E}_\pi^2$ of the formulas $A = \pi r^2$ and $C = \pi d$ are answering questions of Hilbert and Dedekind not questions of Euclid though possibly of Archimedes. However, from our perspective the theory EG_π is closer to the Greek origins than to Hilbert's second order axioms.

Recall that closing a plane under ruler and compass constructions corresponds to closing the coordinatizing ordered field under square roots of positive numbers. As in Example 9.4.5, F_s (surd field) denotes the minimal such field. Note that having named 0, 1, each element of F_s is denoted by a term $t(x, 0, 1)$ built from the field operations and $\sqrt{\ }$. Such terms name the perimeter of regular polygons inscribed (circumscribed) in the unit circle. Thus, for each model of EG and any line of the model the surd field F_s

---

[3] Taken from his commentary on Archimedes in *Archimedes Opera Omnia cum commentariis Eutociis*, vol. 3, p. 266. Quoted in: Davide Crippa (Sphere, UMR 7219, Université Paris Diderot), 'Reflexive knowledge in mathematics: The case of impossibility'.

is embeddable in the field definable on that line. So we can interpret the Greek theory of proportionality by way of cuts in the ordered surd field F_s. Each pair of proportional pairs of magnitudes determines a cut (e.g. [Coolidge 1963], 33–34).

Definition 10.1.1 (Axioms for π)

(1) *Add to the vocabulary a new constant symbol π. Let i_n (c_n) be the perimeter of a regular $3 * 2^n$-gon inscribed[4] (circumscribed) in a circle of radius 1. Let $\Sigma(\pi)$ be the collection of sentences (i.e. type)*

$$i_n < 2\pi < c_n$$

for $n < \omega$.

(2) *EG_π denotes the deductive closure in the vocabulary $\tau \cup \{0, 1, \pi\}$ of the axioms EG of a Euclidean plane and $\Sigma(\pi)$.*

Dicta 10.1.2 (Constants 2: Naming π) The situation here is very different from that in Dicta 9.4.2. In the previous cases any two distinct elements could be named. Here we fix a cut in the surd field and the named point must be in that cut. Since we have forced the realization of a non-principal type, not all τ-structures that satisfy EG expand to a model of EG_π.

We formulated these axioms as properties of the point π rather than of the segment 0π because it is slightly more compact notation and more congenial to a modern reader. But shortly, we will describe the polygons approximating a circle in terms of segments in the more convenient geometrical rather than the field language.

Theorem 10.1.3 *EG_π is a consistent but incomplete theory. It is not finitely axiomatizable.*

Proof. A model of EG_π is given by closing $F_s \cup \{\pi\} \subseteq \Re$ under Euclidean constructions. To see EG_π is not finitely axiomatizable, for any finite subset Σ_0 of Σ choose a real algebraic number p satisfying Σ_0 when p is substituted for π; the field closure of $F_s \cup \{p\} \subseteq \Re$ is a model of $EG \cup \Sigma_0$ that is not a model of EG_π.

Dicta 10.1.4 (Definitions or Postulates) We now extend the ordering on segments by adding the lengths of 'bent lines' and arcs of circles to the domain. Two approaches[5] to this step are (a) our approach to introduce

---

[4] I thank Craig Smorynski for pointing out that it is not so obvious that the perimeter of an inscribed n-gon is monotonic in n and reminding me that Archimedes started with a hexagon and doubled the number of sides at each step, thus avoiding this issue.

[5] We could define $<$ on the extended domain or, in style (b), we could add $<^*$ to the vocabulary and postulate that $<^*$ extends $<$ and satisfies the properties of the definition.

an explicit but inductive definition or (b) add a new predicate to the vocabulary and new axioms specifying its behavior. The second alternative reflects in a way the trope that Hilbert's axioms are *implicit definitions*. We can use choice (a) in the following definitions only because we have already established a certain amount of geometric vocabulary. Crucially the definition of bent lines (and thus the perimeter of certain polygons) is not a single formal definition but a schema of formulas $\langle \phi_n : n < \omega \rangle$ defining an approximation for each n.

Definition 10.1.5 *By a* bent line[6] $b = X_1 \ldots X_n$ *we mean a sequence of straight line segments $X_i X_{i+1}$, for $1 \leq i \leq n-1$, such that the right hand end point of one is the initial point of the next.*

(1) *Each bent line $b = X_1 \ldots X_n$ has a length $[b]$ given by the straight line segment composed of the sum of the segments of b.*

(2) *An approximant to the arc $X_1 \ldots X_n$ of a circle with center P is a bent line satisfying:*

    (a) *$X_1, \ldots X_n, Y_1 \ldots Y_n$ are points such that the X_i are on the circle and each Y_i is in the exterior of the circle.*

    (b) *Each of $Y_i Y_{i+1}$ $(1 \leq i < n)$, $Y_n Y_1$ is a straight line segment.*

    (c) *For $1 \leq i < n$, $Y_i Y_{i+1}$ is tangent to the circle at X_i; $Y_n Y_1$ is tangent to the circle at X_1.*

Definition 10.1.6 *Let $\mathcal{S}$ be the set (of equivalence classes under congruence) of straight line segments. Let $\mathcal{C}_r$ be the set (of equivalence classes under congru-ence) of arcs on circles of a given radius[7] r. Extend the linear order on $\mathcal{S}$ to a linear order $<_r$ on $\mathcal{S} \cup \mathcal{C}_r$ as follows. For $s \in \mathcal{S}$ and $c \in \mathcal{C}_r$*

(1) *The segment $s <_r c$ if and only if there is a chord XY of a circular arc $AB \in c$ such that $XY \in s$.*

(2) *The segment $s >_r c$ if and only if there is an approximant $b = X_1 \ldots X_n$ to c with length $[b] = s$ and with $[X_1 \ldots X_n] >_r c$.*

It is easy to see that this order is well-defined as each chord of an arc is shorter than the arc and any approximant to the arc is longer than the arc (by repeated use of the triangle inequality, Euclid I.20).

---

[6] This is less general than Archimedes ([Archimedes 1897], 2) who allows segments of arbitrary curves 'that are concave in the same direction.'

[7] It at least requires some work to compare the length of arcs on circles of different radius and with chords of different lengths. We work around the issue now; our assignment of angle measure in Definition 10.3.6 solves the problem in some models.

Now we want to argue that π, as implicitly defined by the theory EG_π, serves its geometric purpose. For this, we add a new unary function symbol C mapping our fixed line to itself and satisfying the following scheme.

Definition 10.1.7 *A unary function $C(r)$ mapping $\mathcal{S}$, the set of equivalence classes (under congruence) of straight line segments, into itself that satisfies the conditions below is called a circumference function. Let C_r denote a circle of radius r.*

ι_n: *$C(r)$ is greater than the perimeter, $i_n(r)$, of a regular 3×2^n-gon inscribed in C_r.*

γ_n: *$C(r)$ is less than the perimeter, $c_n(r)$, of a regular 3×2^n-gon circumscribing C_r.*

Definition 10.1.8 *The theory $EG_{\pi,C}$ is the extension of the $\tau \cup \{0, 1, \pi\}$-theory EG_π obtained by the explicit definition: $C(r) = 2\pi r$.*

By similarity of the polygons, $i_n(r) = ri_n$ and $c_n(r) = rc_n$; the ordering specified in Definition 10.1.7 will be satisfied if $C(r)$ is replaced by 'the circumference of a circle of radius r'. Note that while the approximations are given by standard 3×2^n-gons, defined by a schema, the translation to circles of different radius is done by multiplication within the geometry. So the approximations can be calculated for circles of any radius (including infinite or infinitesimal radius if the field is non-Archimedean.)

Thus we have shown that for each r there is an $s \in \mathcal{S}$ whose length, $2\pi r$, is less than the perimeters of all circumscribing polygons and greater than those of the inscribed polygons. We can verify that by choosing n large enough we can make $i_n(r)$ and $c_n(r)$ as close together as we like (more precisely, for given m, make them differ by $< 1/m$). Our definition of EG_π then makes the following metatheorem immediate.

Theorem 10.1.9 *In $EG^2_{\pi,C}$, $C(r) = 2\pi r$ is a circumference function (i.e. satisfies all the conditions ι_n and γ_n).*

We have computed the circumference; we compute the length of an arbitrary arc in Theorem 10.3.7. In an Archimedean field there is a unique interpretation of π and thus a unique choice for a circumference function with respect to the vocabulary without the constant π. Since we added the constant π to the vocabulary we get a formula which satisfies the conditions in every model. But in a non-Archimedean model, any point in the monad[8] of $2\pi r$ would equally well fit our condition for being the circumference.

---

[8] The *monad* of a point is the set of points infinitesimally close to it.

There may be automorphisms of a model of EG_π, but they must fix F_s pointwise and just move π in its cut in F_s.

We reported in Fact 9.5.2 Hilbert's use of segment multiplication [Hilbert 1971] to establish the usual formulas for the area of polygons. We now extend that method to compute the area of circles.

Lemma 10.1.10 (Encoding a second approximation of π) *Let I_n and C_n denote the area of the regular 3×2^n-gon inscribed or circumscribing the unit circle and for $n < \omega$ let σ_n denote the sentence: $I_n < \pi < C_n$. Then EG_π proves each σ_n.*

Proof. The intervals $(2I_n, 2C_n)$ define the cut for 2π in the surd field F_s as do the intervals (i_n, c_n). (That is, for every t, there exists an N_t such that if $k, \ell, m, n \geq N_t$ the distances between any pair of i_k, c_ℓ, I_m, C_n is less than $1/t$.)

Now, as in the circumference case, by formalizing a notion of equal area, including a schema for approximation by finite polygons (which for conciseness we omit), we can define a formal area function $A(r)$ which gives the area of a circle just if its value is squeezed between the areas of a family of inscribing and circumscribing polygons.

Definition 10.1.11 *The theory $EG_{\pi,C,A}$ is the extension of the $\tau \cup \{0, 1, \pi\}$-theory $E_{\pi,c}$ obtained by the explicit definition $A(r) = \pi r^2$.*

Having named this function, since the $I_n(C_n)$ converge to one half of the limit of the $i_n(c_n)$, we have:

Theorem 10.1.12 *In $EG_{\pi,C,A}$, $A(r) = \pi r^2$ is an area function.*

10.2   From Descartes to Tarski

We step back from π in this section and study the extensions of Euclidean geometry implicit in Descartes. It is not our intent to give a detailed account of Descartes' impact on geometry. We want to first bring out the changes from the Euclidean to the Cartesian data set and later those between Descartes and Hilbert/Tarski. As noted after Notation 9.3.2 the most significant are, first, he explicitly ([Descartes 1637], 1) defines the multiplication of line segments to give a line segment which breaks with Greek tradition.[9] And later on the same page he announces constructions for the extraction of *nth* roots for all *n*. The second of these cannot be

---

[9] His proof is still based on Eudoxus. See page 216.

done in EG, since EG is satisfied in the geometry over the field which has solutions for all quadratic equations but not all of odd degree.[10]

Marco Panza formulates a key observation about the ontological importance of these innovations.

The first point concerns what I mean by 'Euclid's geometry'. This is the theory expounded in the first six books of the *Elements* and in the *Data*. To be more precise, I call it 'Euclid's plane geometry', or EPG, for short. It is not a formal theory in the modern sense, and, a fortiori, it is not, then, a deductive closure of a set of axioms. Hence, it is not a closed system, in the modern logical sense of this term. Still, it is no more a simple collection of results, nor a mere general insight. It is rather a well-framed system, endowed with a codified language, some basic assumptions, and relatively precise deductive rules. And this system is also closed, in another sense ([Julien 1964], 311–312), since it has sharp-cut limits fixed by its language, its basic assumptions, and its deductive rules. In what follows, especially in section 1, I shall better account for some of these limits, namely for those relative to its ontology. More specifically, I shall describe this ontology as being composed of objects available within this system, rather than objects which are required or purported to exist by force of the assumptions that this system is based on and of the results proved within it. This makes EPG radically different from modern mathematical theories (both formal and informal). One of my claims is that Descartes' geometry partially reflects this feature of EPG. ([Panza 2011], 43)

We take Panza's 'open' system[11] to refer to the 'linked constructions'[12] considered by Descartes, which greatly extend the ruler and compass licensed in EG. Descartes endorses such 'mechanical' constructions as those used in the duplication of the cubic as geometric. According to Molland ([Molland 1976], 38) 'Descartes held the possibility of representing a curve by an equation (specification by property)' to be equivalent to its 'being constructible in terms of the determinate motion criterion (specification by genesis).' But as Crippa points out ([Crippa 2014a], 153) Descartes did not prove this equivalence; there is some controversy as to whether the 1876 work of Kempe solves the precise problem. Descartes rejects as non-geometric any method for quadrature of the circle. See ([Descartes 1637], 48) for his classification of problems by degree. Unlike Euclid, Descartes does not develop his theory axiomatically. But an advantage of the 'descriptive axiomatization' rubric is that we can take as data the theorems of Descartes' geometry.

---

[10] See section 12 of [Hartshorne 2000].
[11] See [Rodin 2014] for the open/closed system distinction.
[12] The types of constructions allowed are analyzed in detail in Section 1.2 of [Panza 2011] and the distinctions with the Cartesian view in Section 3. See also [Bos 2001] and [Rodin 2014].

For our purpose we take a common identification of Cartesian geometry with 'real' algebraic geometry, the study of polynomial equalities and inequalities in the theory of real closed fields. To ground this geometry we adapt Tarski's 'elementary geometry'. This move makes a significant conceptual step[13] away from Descartes, whose constructions were on segments and who did not regard a line as a set of points; Tarski's axioms are given entirely formally in a one-sorted language of relations on points. Writing in 1832, Bolyai ([Gray 2004], appendix) wrote in his 'explanation of signs,' 'The straight AB means the aggregate of all points situated in the same straight line with A and B.' This is one of the earliest indications I know of the transition to an extensional version of incidence.[14] Tarski's [Tarski 1959] axiom system for *elementary geometry*, denoted $\mathcal{E}^2$, formalizes this conception.

Theorem 10.2.1 *Tarski's $\mathcal{E}^2$ [Tarski 1959] can be axiomatized by the following system of axioms for plane geometry.[15] It is first order complete.*

(1) *Euclidean geometry (EG)*
(2) *Either of these sets of axioms which are equivalent over (1).*

    (a) *The field is formally real and an infinite set of axioms declaring that every polynomial of odd degree has a root.*
    (b) *The axiom schema of continuity described below.*

Tarski's (b) is a first order analog of Dedekind's axiom: require that for any two sets A and B, if beyond some point a all elements of A are below all elements of B, then there is a point b which is above all of A and below all of B. Tarski [Givant & Tarski 1999] restricts this idea to definable sets and postulates the following formal *Axiom Schema of Continuity*:

$$(\exists a)(\forall x)(\forall y)[\alpha(x) \wedge \beta(y) \rightarrow B(axy)]$$

$$\rightarrow (\exists b)(\forall x)(\forall y)[\alpha(x) \wedge \beta(y) \rightarrow B(xby)],$$

where α, β are first order formulas, the first of which does not contain any free occurrences of a, b, y nor the second any free occurrences of a, b, x. Recalling that $B(x, z, y)$ represents 'z is between[16] x and y', the hypothesis asserts the solutions of the formulas α and β behave as the A, B above.

---

[13] It is also very different from Hilbert, as the first order theory does not require the existence of transcendentals.
[14] William Howard showed me this passage. See also Bolzano quote ([Rusnock 2000], 54)).
[15] The translation between the representations is routine. Thus, we are careless below whether our vocabulary is that of Notation 9.3.1 or the bi-interpretable one of Tarski.
[16] More precisely in terms of the linear order $B(xyz)$ means $x \leq y \leq z$.

This schema enforces the solution of odd degree polynomials. By the completeness of real closed fields, this theory is also complete[17] so (a) and (b) are equivalent over (1).

Remark 10.2.2 (Gödel-completeness) In Detlefsen's terminology Tarski constructed a Gödel-complete first order axiomatization of (in our terminology) Cartesian plane geometry. This guarantees that if we keep the vocabulary and continue to accept the same data set no axiomatization can account for more of the data. There are certainly open problems in plane geometry [Klee & Wagon 1991]. But however they are solved, the proof will be formalizable in $\mathcal{E}^2$. Of course, more perspicuous axiomatizations may be found. Or, one may discover the entire subject is better viewed as an example in a more general context. Thus, in contrast to Detlefsen's assertion in Section 5 of [Detlefsen 2014], Gödel-completeness of a theory that implies the elements of a given first order data set does imply this theory is descriptively complete.

Our axioms (1) are in the spirit of Descartes – asserting the solutions of certain equations. They provide a complete descriptive axiomatization of the Cartesian data set. Of course this makes sense only if we allow the translation from segments to points. (2) If the real algebraic numbers are the natural model for Cartesian geometry then we have axiomatized exactly the first order sentences true in that geometry. So we claim Tarski has a *modest* complete descriptive axiomatization of Cartesian geometry. In the case at hand, however, there are more specific reasons for accepting the geometry over real closed fields as 'the best' descriptive axiomatization. It is the only one which is decidable and 'constructively justifiable'.

Remark 10.2.3 (Undecidability and Consistency) Ziegler has shown that every nontrivial finitely axiomatized subtheory of RCF[18] is not decidable [Ziegler 1982]. This result extends easily to the theories EG_π and $EG_{\pi,C,A}$ in the next section.

Thus, both to more closely approximate the Dedekind continuum and to obtain decidability, we restrict to planes over RCF and thus to Tarski's $\mathcal{E}^2$ [Givant & Tarski 1999]. The bi-interpretability between RCF and the theory

---

[17] Tarski [Tarski 1959] proves that planes over real closed fields are exactly the models of his elementary geometry, $\mathcal{E}^2$.

[18] RCF abbreviates 'real closed field'; these are the ordered fields such that every positive element has a square root and every odd degree polynomial has at least one root. The theory is complete and recursively axiomatized so decidable. A *nontrivial* subtheory is one satisfied in $\mathfrak{R}$. For the context of the Ziegler result and Tarski's quantifier elimination in computer science see [Makowsky 2013].

of all planes over real closed fields yields the decidability of $\mathcal{E}^2$. Decidability is possible only because the natural numbers are *not first order definable* in the real field. As Tarski noticed and Friedman [Friedman 1999] proved, RCF is provably consistent in exponential function arithmetic (EFA).

Of course, the vital contribution of Descartes is coordinate geometry. Tarski (like Hilbert) provides a converse; his interpretation of the plane into the coordinatizing line [Tarski 1951] underlies our smudging of the study of the 'geometry continuum' with axiomatizations of 'geometry'.

Three post-Descartes innovations are largely missing from our account: (a) higher-dimensional geometry, (b) projective geometry, (c) definability by analytic functions. The first is a largely nineteenth century innovation which significantly impacts Descartes' analytic geometry by introducing equations in more than three variables. The second is essentially bi-interpretable with affine geometry. So both of these threads are more or less orthogonal to our development here which concerns the actual structure of the line (and moves more or less automatically to higher-dimensional or projective geometry). Although Dieudonné ([Dieudonné 1970], 140) proclaims the third to be essential to the (at least his) understanding of the meaning of 'analytic geometry' in the late twentieth century, it is hard to differentiate such a nineteenth century development from real and complex analysis. So, it seems again that the (mathematical as opposed to metamathematical) motivation for Hilbert to include the completeness axiom comes from analysis not geometry. Nevertheless, the work described in Chapter 6.3 shows a first order approach to analysis is useful.

10.3 π in Geometries over Real Closed Fields

We now combine the theories from Chapter 10.1 and Chapter 10.2, to find a modest descriptive axiomatization of a variant of Tarski's elementary geometry plus the elements of angle measure. As in Chapter 10.2, we will obtain a complete first order theory. This will illuminate the delicate dependence of descriptive axiomatization on the vocabulary in which the facts are expressed. Dedekind ([Dedekind 1963], 37–38) observes that what we would now call the real closed field with domain the field of real algebraic numbers is 'discontinuous everywhere' but 'all constructions that occur in Euclid's elements can …be just as accurately effected as in a perfectly continuous space.' Strictly speaking, for *constructions* this is correct. But the proportionality constant, 2π, between radius and circumference is absent. So if a segment is the diameter there can be no segment in the model with the length of the circumference.[19] We want to find a theory which proves the circumference and area formulas for circles and countable models of the geometry over RCF, where 'arc length behaves properly'.

Descartes eschews the idea that there can be a ratio between a straight line segment and a curve. As [Crippa 2014b] writes, 'Descartes[20] excludes the exact knowability of the ratio between straight and curvilinear segments':

la proportion, qui est entre les droites et les courbes, n'est pas connue, et mesme ie croy ne le pouvant pas estre par les hommes, on ne pourroit rien conclure de là qui fust exact et assurè.

Hilbert[21] asserts that there are many geometries satisfying axioms I–IV and V1 but only one, 'namely the Cartesian geometry' that also satisfies V2. Evidently, the conception of 'Cartesian geometry'[22] changed radically from Descartes to Hilbert; even the symbol π was not introduced until 1706 (by Jones). Nevertheless, we now define a theory $\mathcal{E}_\pi^2$ analogous to EG_π that does not depend on the Dedekind axiom but has only first order axioms.

Given Descartes' proscription of π, the new system will be immodest with respect to the Cartesian data set. But we will argue at the end of this section that both of our additions of π are closer to Greek conceptions than the Dedekind axiom. At this point we need some modern model theory to guarantee the *completeness* of the theory we are defining.

A first order theory T for a vocabulary with a binary relation $<$ is *o-minimal* (Chapter 6.3) if every 1-ary formula is equivalent in T to a Boolean combination of equalities and inequalities [Dries 1999]. Anachronistically, the o-minimality of the reals is a main conclusion of Tarski in [Tarski 1931].

Theorem 10.3.1 *Form $\mathcal{E}_\pi^2$ by adjoining $\Sigma(\pi)$ (Definition 10.1.1) to $\mathcal{E}^2$. $\mathcal{E}_\pi^2$ is first order complete for the vocabulary τ along with the constant symbols $0, 1, \pi$.*

Proof. By Hilbert, there are well-defined field operations on the line through 01. By Tarski, the theory of this real closed field is complete.

---

[19] Thus, the protractor postulate (page 246) from [Birkhoff 1932] is violated.

[20] Descartes, *Oeuvres*, Vol. 6, p. 412. Crippa suggests a translation of the early French as, 'the proportion between the line and the curve is not known, and I even believe that it cannot be known by men, and one can conclude nothing about that is exact and assured.' Crippa also quotes Averroes as emphatically denying the possibility of such a ratio and notes that Vieta held similar views.

[21] See ([Hallett & Majer 2004], 429–430).

[22] One wonders whether it had actually changed when Hilbert wrote. Had readers at the turn of the twentieth century already internalized a notion of Cartesian geometry which entailed Dedekind completeness that had been formulated in the late nineteenth century (Bolzano–Cantor–Weierstrass–Dedekind)?

The field is bi-interpretable with the plane [Tarski 1951] so the theory of the geometry $\mathcal{E}^2$ is complete as well. Further by Tarski, the field is o-minimal. So the type over the empty set of any point on the line is determined by its position in the linear ordering of the surd field F_s. Each i_n, c_n is an element of the field F_s. This position of 2π in the linear order on the line through 01 is given by Σ. Thus, by o-minimality, $\mathcal{E}^2_\pi$ is complete.

We now rely on the definitions of bent line, circumference function, etc. from Chapter 10.1. Using them, we extend the theory $\mathcal{E}^2_\pi$.

Definition 10.3.2 *We define two new theories expanding* $\mathcal{E}^2_\pi$.

(1) *The theory* $\mathcal{E}^2_{\pi,C}$ *is the extension of the* $\tau \cup \{0, 1, \pi\}$*-theory* $\mathcal{E}^2_\pi$ *obtained by the explicit definition* $C(r) = 2\pi r$.
(2) *The theory* $\mathcal{E}^2_{\pi,C,A}$ *is the extension of the* $\tau \cup \{0, 1, \pi, C\}$*-theory* $\mathcal{E}^2_{\pi,C}$ *obtained by the explicit definition* $A(r) = \pi r^2$.

As an extension by explicit definition, $\mathcal{E}^2_{\pi,C,A} = \mathcal{E}^2_{\pi,A,C}$ is complete and o-minimal. As in Theorem 10.1.9, our definition of $\mathcal{E}^2_\pi$ then gives:

Theorem 10.3.3 *The theory* $\mathcal{E}^2_{\pi,C,A}$, *is a complete, decidable, and o-minimal extension of* $EG_{\pi,C}$ *and* $\mathcal{E}^2_\pi$.

(1) *In* $\mathcal{E}^2_{\pi,C}$, $C(r) = 2\pi r$ *is a circumference function.*
(2) *In* $\mathcal{E}^2_{\pi,C,A}$, $A(r) = \pi r^2$ *is an area function.*

Proof. We are adding definable functions to $\mathcal{E}^2_\pi$ so o-minimality and completeness are preserved. The theory is recursively axiomatized and complete so decidable. The formulas continue to compute area and circumference correctly as they extend $EG_{\pi,C,A}$.

'π is transcendental' is a theorem of $\mathcal{E}^2_\pi$. Lindemann proved that π does not satisfy a polynomial of degree n for any n. Thus for any polynomial over the rationals $p(\pi) \neq 0$ is a consequence of the complete type generated by $\Sigma(\pi)$ and so is a theorem of $\mathcal{E}^2_\pi$.

We now extend the known fact that the theory of real closed fields is 'finitistically justified' ([Simpson 2009], 378) to $\mathcal{E}^2_{\pi,A,C}$. For convenience, we lay out the proof with reference to results[23] recorded in [Simpson 2009].

---

[23] We use RCOF here for what we have called RCF before as the argument here is quite sensitive to adding the order relation to the language. Note that Friedman [Friedman 1999] strengthens the results for PRA to exponential function arithmetic (EFA). Weak König's lemma is denoted WKL_0; the basic system of reverse mathematics, Robinson arithmetic with induction for σ^0_1-formulas and comprehension for π^0_1-formulas, is denoted RCA_0. The theories discussed here, in increasing proof strength, are EFA, PRA, RCA_0, and WKL_0.

Fact 10.3.4 The theory $\mathcal{E}^2$ is bi-interpretable with the theory of real closed fields. And thus it (as well as $\mathcal{E}^2_{\pi,C,A}$) is finitistically consistent (i.e. provably consistent in primitive recursive arithmetic (PRA)).

Proof. By Theorem II.4.2 of [Simpson 2009], RCA_0 proves the system $(Q, +, \times, <)$ is an ordered field and by II.9.7 of [Simpson 2009], it has a unique real closure. Thus the existence of a real closed ordered field and so $Con(RCOF)$ is provable in RCA_0. (Note that the construction will embed the surd field F_s.)

Lemma IV.3.3 of [Friedman et al. 1983] asserts the provability of the completeness theorem (and hence compactness) for countable first order theories from WKL_0. Since every finite subset of $\Sigma(\pi)$ is easily seen to be satisfiable in any RCOF, it follows that the existence of a model of $\mathcal{E}^2_\pi$ is provable in WKL_0. Since WKL_0 is π^0_2-conservative over PRA, we conclude PRA proves the consistency $\mathcal{E}^2_\pi$. As $\mathcal{E}^2_{\pi,C,A}$ is an extension by explicit definitions, its consistency is also provable in PRA.

We stressed in Chapter 9 that modern mathematics replaces the phrase 'area is proportional to' in Euclid by formulas which specify the proportionality constant. We have so far found the proportionality constant only for the entire circle. Crippa describes Leibniz's distinguishing two types of quadrature,

universal quadrature of the circle, namely the problem of finding a general formula, or a rule in order to determine an arbitrary sector of the circle or an arbitrary arc; and on the other [hand] he defines the problem of the particular quadrature, ..., namely the problem of finding the length of a given arc or the area of a sector, or the whole circle... ([Crippa 2014a], 424)

George Birkhoff's 'metric' geometry[24] proposes a uniform answer to all such problems, the *protractor postulate*:

POSTULATE III. The half-lines ℓ, m, through any point O can be put into $(1, 1)$ correspondence with the real numbers $a(\mathrm{mod}\, 2\pi)$, so that, if $A \neq O$ and $B \neq O$ are points of ℓ and m respectively, the difference $a_m - a_\ell(\mathrm{mod}\, 2\pi)$ is $\angle AOB$. [Birkhoff 1932]

This is a parallel to Birkhoff's *ruler postulate* which assigns each segment a real number length. Thus, Birkhoff takes the real numbers as an unexamined background object. At one swoop he has introduced multiplication, and assumed the Archimedean and completeness axioms. So even 'neutral' geometries studied on this basis are actually greatly restricted.[25] He argues

---

[24] This is the axiom system used in virtually all US high schools since the late 1960s.
[25] That is, they must be metric geometries.

that his axioms define a categorial system isomorphic to $\mathfrak{R}^2$. So his system (including an axiomatization of the real field that he hasn't specified) is bi-interpretable with Hilbert's.

Euclid's Postulate III, 'describe a circle with given center and radius,' implies that a circle is uniquely determined by its radius and center. In contrast Hilbert simply defines the notion of circle and proves the uniqueness (see Lemma 11.1 of [Hartshorne 2000]). In either case we have the basic correspondence between angles and arcs: two segments of a circle are congruent if they cut the same central angle. So establishing an angle measure is the same as assigning a straight line segment as the length of each arc.

Birkhoff's postulate conflates three distinct ideas: (i) the rectifiability of arcs, the assertion that each arc of a circle has the same length as a straight line segment; (ii) the claim there is an algorithm for finding the segment; and (iii) the measurement of angles, identifying the measure of an angle with the arc length of the arc it determines.

The next task is to find a more modest version of Birkhoff's postulate: a first order theory with countable models where every angle determines an arc that corresponds to the length of a straight line segment. That is, which assigns to each angle a measure between 0 and 2π. Recall that we have a field structure on the line through 01 and the number π on that line.

Definition 10.3.5 *A measurement of angles function is a map μ from congruence classes of angles into $[0, 2\pi)$ such that if $\angle ABC$ and $\angle CBD$ are disjoint angles sharing the side BC,*

$$\mu(\angle ABD) = \mu(\angle ABC) + \mu(\angle CBD).$$

If we omitted the additivity property this would be trivial: given an angle $\angle ABC$ less than a straight angle, let C' be the intersection of a perpendicular to BC through A with BC and let $\mu(\angle ABC) = 2\pi \cdot \sin(\angle ABC) = \frac{2\pi \cdot BC'}{AB}$. (It is easy to extend to the other angles.)

For the actual definition note that we can code each angle with vertex at the origin O by two points A, B on the unit circle so that the angle is AOB.

Definition 10.3.6

(1) *Define the function μ on the upper half circle by coding points on the unit circle with positive y-value by their x-coordinates and setting, for $x_1 > x_2$,*

$$\mu((x_1, y_1), (x_2, y_2)) = \cos^{-1}(x_1 - x_2).$$

Extend the domain to the rest of the circle by, with $y_1, y_2 > 0$, letting

$$\mu((x_1, -y_1), (x_2, -y_2)) = \mu((x_1, y_1), (x_2, y_2)).$$

(2) The theory $\mathcal{E}^2_{\pi,A,C,\mu}$ is obtained by adding to $\mathcal{E}^2_{\pi,A,C,\mu}$, the assertion that μ is a continuous[26] additive map from congruence classes of angles to $(0, 2\pi]$.

Theorem 10.3.7 *The theory $\mathcal{E}^2_{\pi,A,C,\mu}$ is consistent and complete.*

Proof. Showing consistency is easy. Since, as we noted in justifying Theorem 9.4.6, the sine and cosine of angles in $[0, \pi]$ are definable on any Euclidean plane we can define the μ-function in $\mathcal{E}^2_{\pi,A,C}$, μ is satisfied in $\mathfrak{R}$, so the axioms are consistent. Since the theory of the real field expanded by the sine function restricted to $(0, 2\pi]$ is o-minimal (Chapter 6.3), $\mathcal{E}^2_{\pi,A,C,\mu}$ is complete.

While Definition 10.3.6 solves (i), the rectifiability problem, merely assuming the existence of a μ does not solve (ii). With the addition of the cosine restricted to $(0, 2\pi]$, we can calculate arc length as in calculus, but we don't have a geometric algorithm for this. But a nice axiom system remains a dream.[27]

Blanchette [Blanchette 2014] distinguishes two approaches to logic, deductivist and model-centric. In her scheme, Hilbert represents the deductivist school and Dedekind the model-centric. Essentially, the second creates theories trying to describe an intuition of a particular structure. We briefly consider the opposite question: are there 'canonical' models of the various theories we have been considering?

By modern tradition, the continuum is the real numbers and geometry is the plane over it. Is there a smaller model which reflects the geometric intuitions discussed here? For Euclid II, there is a natural candidate, the Euclidean plane over the surd field F_s. Remarkably, this does not conflict with Euclid XII.2 (circles are to one another as the squares of their diameters); the model is Archimedean. π is not in the model; but Euclid only requires a proportionality which defines a type ($\Sigma(x)$), not a realization π of $\Sigma(x)$. Plane geometry over the real algebraic numbers plays the same role for $\mathcal{E}^2$. Both are categorical in $L_{\omega_1,\omega}$ (Theorem 3.2.1). In the second case, add the Archimedean axiom and say every field element is algebraic.

Now we argue that the methods of this section better reflect the Greek view than does Dedekind. Mueller makes an important point distinguishing the Euclid/Eudoxus use of cuts from Dedekind's.

---

[26] With a little effort we can express continuity of μ in $\mathcal{E}^2_{\pi,A,C,\mu}$ and it could fail in a non-Archimedean model so we have to require it to have chance at a complete theory.

[27] However, there is no known axiomatization of the real field with restricted cosine and Marker tells me the theory is unlikely to be decidable without assuming the Schanuel conjecture (Definition 7.1.3). See [Dries 1999].

One might say that in applications of the method of exhaustion the limit is given and the problem is to determine a certain kind of sequence converging to it, ... Since in the *Elements* the limit always has a simple description, the construction of the sequence can be done within the bounds of elementary geometry; and the question of constructing a sequence for any given arbitrary limit never arises. ([Mueller 2006], 236)

But what if we want to demand the realization of various transcendentals? Mueller's description suggests the principle that we should only realize cuts in the field order that are *recursive*[28] over a finite subset. Call these the *Eudoxian transcendentals*. So a candidate would be a recursively saturated model of $\mathcal{E}^2$. Remarkably, almost magically,[29] this model would also satisfy $\mathcal{E}^2_{\pi,A,C,\mu}$. A recursively saturated model is necessarily non-Archimedean. There are however many different countable recursively saturated models depending on which transcendental cuts are realized.

Here is a more canonical candidate for a natural model which admits the 'Eudoxian transcendentals'; take the smallest elementary submodel of $\mathfrak{R}$ closed[30] under A, C, μ containing the real algebraic numbers and all realizations of recursive cuts in F_s. The Scott sentence[31] of this structure is a categorical sentence in $L_{\omega_1,\omega}$. The models in this paragraph are countable; we cannot give an axiomatization of the Hilbert model, the plane over the reals, by a *sentence* of $L_{\omega_1,\omega}$; it has no countable $L_{\omega_1,\omega}$-elementary submodel.

We return to the question of modesty. Mueller's distinction can be expressed in another way. Eudoxus provides a technique to solve certain specified problems. In contrast, Dedekind's postulate solves $2^{\aleph_0}$ problems at one swoop. Each of the theories $\mathcal{E}^2_\pi, \mathcal{E}_{\pi,A,C}, \mathcal{E}_{\pi,A,C,\mu}$, and the later search for canonical models reflect this difference. They solve at most a countable number of recursively stated problems.

In summary, we regard the replacement of 'congruence class of segment' by 'length represented by an element of the field' as a *modest* reinterpretation of Greek geometry. But this treatment of length becomes *immodest* relative even to Descartes when this length is a transcendental. And *most immodest* of all is to demand lengths for arbitrary transcendentals.

---

[28] In more modern parlance, computable.

[29] The magic is called resplendency. M is recursively saturated if every recursive type over a finite set is realized. M is resplendent if any formula $\exists A \phi(A, \mathbf{c})$ that is satisfied in an elementary extension of M is satisfied by some A' on M. Examples are the formulas defining C, A, μ. Every countable recursively saturated model is resplendent [Barwise 1975].

[30] Interpret A, C, μ on $\mathfrak{R}$ in the standard way.

[31] For any countable structure M there is a 'Scott' sentence ϕ_M such that all countable models of ϕ_M are isomorphic to M (page 173).

11 | Complete: The Word for All Seasons

In Chapters 3.3 and 4.2, we discussed the gradual development of notions of completeness in the first third of the twentieth century. We recounted confusions amongst deductive and semantic completeness (for theories and/or logics) and categoricity. In this chapter we look from another angle on the distinction between descriptive and Gödel completeness and another sense of completeness identified by Kreisel. Today the difference between the *topological* and the logical notions of completeness is clear. But, since the topological notions are deeply entangled with categoricity concerns about the real numbers, the ideas intermingled at the beginning of the twentieth century. We discuss mathematical and logical attempts to clarify the notion from Dedekind to contemporary logicians.

We consider first Kreisel's[1] 'Logical foundations, a lingering malaise.' A quick reading of the excerpt below might suggest Kreisel refers to the necessity of the Dedekind axiom for semantic completeness of Hilbert's geometry. But we argue he is actually referring to a deeper uneasiness on the part of Poincaré. Kreisel remarks of Hilbert's rules for geometry:

True, [they] provided a genuine surprise at the time, and it does not seem to be well-known that Poincaré still doubted the completeness of Hilbert's rules. Presumably everybody else did too who had taken Kant's view on the role of visualization (*Anschauung*) in mathematical reasoning literally (that is, some kind of logical need) and not in the sense of effective use – a distinction already stressed earlier. [Kreisel 1984]

In the first paragraph of the piece, Kreisel had identified one element of the malaise: 'a preoccupation with a universal framework (a universal language, for example) and thus with logical possibilities. This preoccupation is at the heart of the malaise; it concerns a potential conflict between pursuing these logical ideas and effective knowledge.'

In his review of the 1899 edition of the *Grundlagen*, Poincaré wrote,

But this geometry, strange to say, is not quite the same as ours, his space is not our space, or at least is only a part of it. In the space of Professor Hilbert we do not have

---

[1] See Chapter 2.4 footnote 30.

all the points which there are in our space, but only those which we can construct by ruler and compass, starting from two given points. In this space, for example, there would not exist, in general, an angle which would be the third part of a given angle.

I have no doubt that this conception would have been regarded by Euclid as more rational than ours. At any rate it is not ours. To come back to our geometry it would be necessary to add an axiom. ([Poincaré 1903], 8)

He then suggests adding the completeness axiom in the form: a decreasing sequence of closed intervals has a non-empty intersection. As noted in Chapter 9, in the later editions Hilbert added a form of the completeness axiom. Thus, it is unlikely that Kreisel in 1984 took Poincaré to be 'still' concerned with that lacuna. Rather, Poincaré's concern was deeper, with the formal nature of Hilbert's approach:

Thus Professor Hilbert has, so to speak, sought to put the axioms into such a form that they might be applied by a person who would not understand their meaning because he had never seen either point or straight line or plane. It should be possible, according to him, to reduce reasoning to purely mechanical rules, and it should suffice, in order to create geometry, to apply these rules slavishly to the axioms without knowing what the axioms mean. We shall thus be able to construct all geometry, I will not say precisely without understanding it at all, since we shall grasp the logical connection of the propositions, but at any rate without seeing it at all. We might put the axioms into a reasoning apparatus like the logical machine[2] of Stanley Jevons, and see all geometry come out of it. ([Poincaré 1903],4–5)

Poincaré clearly understands and respects Hilbert's goals. A paragraph later, we read,

This notion may seem artificial and puerile; and it is needless to point out how disastrous it would be in teaching and how hurtful to mental development; how deadening it would be for investigators, whose originality it would nip in the bud. But, as used by Professor Hilbert, it explains and justifies itself, if one remembers the end pursued. Is the list of axioms complete, or have we overlooked some which we apply unconsciously? This is what we want to know.

Poincaré now appears to discuss descriptive completeness. His objection, as Kreisel notes, is not to Hilbert's formal methods as a logical tool. It is an unease[3] with these methods as a way of advancing mathematics that worries Poincaré. He thinks Hilbert is overlooking the role of concept and meaning in the continuing development of mathematical ideas. In fact,

[2] The translator references *Lond. Phil. Trans.* vol. 160, (187) pp. 497–518.
[3] I find this word more fitting than malaise.

I think Poincaré, while correctly identifying the uses Hilbert makes of formal methods, overlooks important words of the introduction to the *Grundlagen*.

These fundamental principles are called the axioms of geometry. The choice of the axioms and the investigation of their relations to one another is a problem which, since the time of Euclid, has been discussed in numerous excellent memoirs to be found in the mathematical literature. This problem is tantamount to the logical analysis of our intuition of space. [Hilbert 1971]

Indeed, the coordinatization of the field in the *Grundlagen* and thus the development of the theories of proportion, similarity, and area without reliance on the Archimedean axiom illustrate not logic-chopping but mathematical creativity unleashed by analysis of fundamental concepts. The basic issue is the *choice of axioms*; this choice is not arbitrary but grounded in a 'logical analysis of our notion of space.' The construction and comparison of possible formal axiom systems allows one to develop this intuition. Further this process can and will be revised as the area develops; is there a better example than the continuing investigation of that most formal of all subjects, set theory, whose axioms (Chapter 14) continue to interact with informal mathematics?

Our argument in this book has been that a hundred years of work (and especially in the decades since Kreisel wrote) on these formal methods provides tools in modern mathematics, as described more specifically in Chapter 6, that might assuage Kreisel's doubts about obtaining 'effective knowledge'.

11.1   Hilbert's Continuity Axioms

We discuss first the role of the Archimedean postulate in the *Grundlagen* and then discuss various formulations of the 'completeness of the continuum'.

The Role of the Axiom of Archimedes in the *Grundlagen*. The discussions of the axiom of Archimedes in the *Grundlagen* fall into several categories. (i) Those, in Sections 9–12 (from [Hilbert 1971]) are metamathematical – concerning the consistency and independence of the axioms. (ii) In his Section 17, the axiom of Archimedes is used to justify the coordinatization of the plane by pairs of real numbers. We have already remarked that Hilbert's discussion of the compatibility and independence of axioms prefigures the syntax–semantics distinction that is central to model theory. Coordinatization is certainly a central geometrical notion. But it does not require the axiom of Archimedes to coordinatize the plane by the line in the plane. This second use is to establish a correspondence between an object defined in the geometry and an extrinsic notion of real number. Thus, it is not a proof in the system advanced by Hilbert.

(iii) In Sections 19 and 21, it is shown that the Archimedean axiom is necessary to show equicomplementable (equal content in Definition 9.5.3) is the same as equidecomposable (in two dimensions). This use of the Archimedean axiom is certainly a proof in the system. But an unnecessary one. As we argued in Chapter 9.5, Hilbert could just as easily have defined 'same area' as 'equicomplementable' (as this is a natural reading of Euclid). The first two uses are metatheoretical results and the third is an unnecessary (for the exposition of Euclid) refinement of terminology

Thus, we find no theorems in the *Grundlagen* proved from its axiom system that essentially depend on the axiom of Archimedes. Rather Hilbert uses the axiom to create examples to show independence and, in conjunction with the Dedekind axiom, identify the field defined in the geometry with the independently existing real numbers as conceived by Dedekind.

Formulating 'Order' Completeness: We now compare various formulations of the completeness axiom. Hilbert wrote [Hilbert 1971]:

V.2 Axiom of Completeness (*Vollständigkeitsaxiom*): To a system of points, straight lines, and planes, it is impossible to add other elements in such a manner that the system thus generalized shall form a new geometry obeying all of the five groups of axioms. In other words, the elements of geometry form a system which is not susceptible of extension, if we regard the five groups of axioms as valid. [Hilbert 1971]

We have used in this book the following version of Dedekind's postulate for geometry (DG):

DG: The linear ordering imposed on any line by the betweenness relation is Dedekind complete.[4]

While this formulation is convenient for our purposes, it misses an essential aspect of Hilbert's system. Note that the Archimedean axiom is not a property of a linear order; it requires the ability to make copies of segments. Today it is generally taken as an axiom about ordered groups. In the ordered group context, DG implies the Archimedean axiom while Hilbert was aiming for an independent set of axioms. Hilbert's axiom does

---

[4] See footnote 12 of Chapter 9.

not imply Archimedes. A variant VER (see [Cantú 1999]) on Dedekind's postulate that does not imply the Archimedean axiom was proposed by Veronese in [Veronese 1889].[5] If we substituted VER for DG, our axioms would also satisfy the independence criterion.

Hilbert's completeness axiom in [Hilbert 1971] that asserts any model of the rest of the theory is maximal, is inherently model theoretic. The later line-completeness [Hilbert 1962] is a technical variant.[6] Building on commentary in [Hallett & Majer 2004], Giovannini [Giovannini 2013] includes a number of points already made here and at least three more. First, Hilbert's completeness axiom is not about deductive completeness, but about maximality of every *model* (page 145). Secondly, ([Hilbert 1971], last line of page 153) Hilbert expressly rejects Cantor's intersection of closed intervals axiom because it relies on a sequence of intervals and 'sequence is not a geometrical notion.' A third intriguing note is an argument due to Baldus in 1928 that the parallel axiom is an essential ingredient in the categoricity of Hilbert's axioms.[7]

Here are two reasons for choosing Dedekind's (or Veronese's) version. First, Dedekind's formulation, since it is about the mathematical structure, not about its axiomatization, directly gives the kind of information about the existence of transcendental numbers that we discussed in Chapter 10. Even more basic, one *cannot* formulate Hilbert's version as a sentence Φ_H in second order logic[8] with the intended interpretation $(\Re^2, \mathbf{G}) \models \Phi_H$.

---

[5] The axiom VER asserts that for a partition of a linearly ordered field into two intervals L, U (with no maximum in the lower L or minimum in the upper U) and a third set in between with at most one point, there is a point between L and U just if for every $e > 0$, there are $a \in L, b \in U$ such that $b - a < e$. Veronese derives Dedekind's postulate from his along with Archimedes' axiom in [Veronese 1889] and the independence in [Veronese 1891]. In [Levi-Civita 1892] Levi-Civita shows there is a non-Archimedean ordered field that is Cauchy complete. I thank Philip Ehrlich for the references and recommend section 12 of the comprehensive [Ehrlich 2006]. See also the insightful reviews [Pambuccian 2014a] and [Pambuccian 2014b] where it is observed that Vahlen [Vahlen 1907] also proved this axiom does not imply Archimedes.

[6] Since any point is in the definable closure of any line and any one point not on the line, one can't extend any line without extending the model. Since adding the Dedekind postulate and/or Hilbert completeness gives a categorical theory satisfied by a geometry whose line is order isomorphic to $\Re$ the two axioms are equivalent (over HP5 + AA).

[7] Hartshorne ([Hartshorne 2000], Sections 40–43) gives a modern account of Hilbert's argument that, replacing the parallel postulate by the axiom L of limiting parallels, gives a hyperbolic geometry that is bi-interpretable with the underlying (definable) field. Thus there is a 1–1 correspondence between fields and the Poincaré geometry over them. With V.2 this gives a natural categorical axiomatization for the Poincaré plane over $\Re$. Hartshorne outlines a more complicated situation for other hyperbolic planes.

[8] Of course, this analysis is anachronistic; the clear distinction between first and second order logic did not exist in 1900. By $\mathbf{G}$, we mean the natural interpretation in $\Re^2$ of the predicates of geometry introduced in Chapter 9.3.

The axiom requires quantification over subsets of an extension of the model which putatively satisfies it. Here is a second order statement[9] Θ, where ψ denotes the conjunction of Hilbert's first four axiom groups and the axiom of Archimedes'.

$$(\forall X)(\forall Y)\forall \mathbf{R})[[X \subseteq Y \wedge (X, \mathbf{R} \restriction X) \models \psi \wedge (Y, \mathbf{R}) \models \psi] \rightarrow X = Y]$$

whose validity expresses Hilbert's V.2 but which is a sentence in pure second order logic rather than in the vocabulary for geometry. Väänänen investigates this anomaly by distinguishing (on page 94 of [Väänänen 2012]) between $(\mathfrak{R}^2, \mathbf{G}) \models \Phi$, for some Φ and the validity of Θ. He expounds in [Väänänen 2014] a new notion, 'Sort Logic', which provides a logic with a sentence Φ'_H which, by allowing a sort for an extension, formalizes Hilbert's V.2 with a more normal notion of truth in a structure.

Philip Ehrlich has made several important discoveries concerning the connections between the two 'continuity axioms' in Hilbert and develops the role of maximality. First, he observes ([Ehrlich 1995], 72) that Hilbert had already pointed out that his completeness axiom would be inconsistent if the maximality were only with respect to the first order axioms. Secondly, he [Ehrlich 1995, Ehrlich 1997] systematizes and investigates the philosophical significance of Hahn's notion of Archimedean completeness. Here the structure (ordered group or field) is not required to be Archimedean; the maximality condition requires that there is extension which fails to extend an Archimedean equivalence class.[10]

11.2 Against the Dedekind Postulate for Geometry

Our fundamental claim is that (slight variants on) Hilbert's first order axioms provide a modest descriptively complete axiomatization of most of Greek geometry. We spelt out in Chapter 9.3 a careful collection of different data sets and showed in Chapters 9.4, 9.5, 9.6 that appropriate sets of first order axioms are modest descriptively complete axioms for each of them. We then showed a slight extension of Tarski's first order axiomatization accounts not only for the Cartesian data set but the basic properties of π.

---

[9] I am leaving out many details, $\mathbf{R}$ is a sequence of relations giving the vocabulary of geometry and the sentence 'says' they are relations on Y; the coding of the satisfaction predicate is suppressed.

[10] In an ordered group, a and b are *Archimedes-equivalent* if there are natural numbers m, n such that $m|a| > |b|$ and $n|b| > |a|$.

In Chapter 11.1, we laid out the reasons that the discussion here concerns Dedekind's rather than Hilbert's formulation of the continuity postulate.

As we pointed out in Theorem 3.1.1 various authors have proved under $V = L$, any countable or Borel structure can be given a categorical axiomatization in second order logic. We argued there that this fact undermines the notion of categoricity as an independent desideratum for an axiom system. In Chapter 3 we gave a special role to axiomatizing canonical systems. Here we go further, and suggest that even for a canonical structure there are advantages to a first order axiomatization that trump the loss of categoricity.

We argue then that the Dedekind postulate is inappropriate (in particular immodest) as an attempt to axiomatize the Euclidean or Cartesian or Archimedean data sets for several reasons:

(1) The requirement that there be a straight-line segment measuring any circular arc is clearly contrary to the intent of Descartes.

(2) Categoricity is not in the data set but rather an external limitative principle. The notion that there was 'one' geometry (i.e. categoricity) was implicit in Euclid. But it is not a geometrical statement. Indeed, Hilbert described his completeness axiom ([Hilbert 1962], 23), 'not of a purely geometrical nature.' This is most clearly seen in Hilbert's initial meta-mathematical formulation: that the model of Axiom groups I–IV and Archimedes' axiom must be maximal.

Bernays [Bernays 1967], citing the Frege–Hilbert correspondence, presents another argument for categoricity, that Hilbert's full axiom system provides an *explicit* definition of the model of geometry. While true enough for the categorical full system, it doesn't seem to me that this contradicts the notion that various other axiom systems provide *implicit definitions* – define a class of structures rather than a specific isomorphism type.

(3) It is not needed to establish the properly geometrical propositions in the data set. We showed in Theorem 9.3.4 that the first two data sets in Chapter 9.3 are provable from the axioms we labeled in Notation 9.3.3 as EG (HP5 + CCI). In Chapter 10 we extended to Tarski's $\mathcal{E}^2$ and our $\mathcal{E}^2_{\pi,C,A}$ to give first order axioms accounting for the Cartesian and Archimedean data sets, respectively.

(4) Proofs from Dedekind's postulate obscure the true geometric reason for certain theorems. Hartshorne writes:

there are two reasons to avoid using Dedekind's axiom. First, it belongs to the modern development of the real number systems and notions of continuity, which is not in the spirit of Euclid's geometry. Second, it is too strong.

By essentially introducing the real numbers into our geometry, it masks many
of the more subtle distinctions and obscures questions such as constructibility
that we will discuss in Chapter 6. So we include the axiom only to acknowledge
that it is there, but with no intention of using it. ([Hartshorne 2000], 177)

(5) The use of second order axioms undermines a key proof method –
informal (semantic) proof – the ability to use higher order methods
to demonstrate that there is a first order proof. A crucial advantage of
a first order axiomatization[11] is that it licenses the kind of argument[12]
described in Hilbert and Ackermann:[13]

Derivation of Consequences from Given Premises; Relation to Universally Valid
Formulas

So far we have used the predicate calculus only for deducing valid formulas.
The premises of our deductions, viz Axioms a) through f), were themselves of
a purely logical nature. Now we shall illustrate by a few examples the general
methods of formal derivation in the predicate calculus. ... It is now a question
of deriving the consequences from any premises whatsoever, no longer of a
purely logical nature.

The method explained in this section of formal derivation from premises
which are not universally valid logical formulas has its main application in
the setting up of the primitive sentences or axioms for any particular field
of knowledge and the derivation of the remaining theorems from them as
consequences. ... We will examine, at the end of this section, the question of
whether every statement which would intuitively be regarded as a consequence
of the axioms can be obtained from them by means of the formal method of
derivation.

We exploited this technique in Chapter 10.1 to provide axioms for the calculation of the circumference and area of a circle. This method is fully
exploited in [Väänänen 2012, Väänänen & Wong 2015].

When the Φ in the definition of internal categoricity (page 72) is ZFC^2,
the conjunction of the axioms of second order set theory,[14] models with
the same ordinals are provably isomorphic, so internal categoricity can be
formulated as $\vdash \Psi$ where Ψ is:[15]

---

[11] See Chapter 11.1 for more detail on this argument.

[12] We noted that Hilbert proved that a Desarguesian plane into 3-space by this sort of argument
in Section 2.4 of [Baldwin 2013a].

[13] Chapter 3, §11 of [Hilbert & Ackermann 1938]. Translation taken from [Blanchette 2014].

[14] For simplicity we absorb the comprehension schema and the axiom of infinity into the axioms
of second order logic. See [Väänänen & Wong 2015, Shapiro 1991].

[15] $ISO(F, X, Y)$ means F is an isomorphism between X and Y.

$$
\begin{aligned}
&\big((\exists V, E) ZFC^2(V, E) \wedge \\
&(\forall V, E, V', E')(ZFC^2(V, E) \wedge ZFC^2(V', E') \wedge \\
&(\exists \pi)\, ISO(\pi, ord(V, E), ord(V', E')) \\
&\rightarrow\ (\exists F)\, ISO(F, (V, E), (V', E'))\big)\big)
\end{aligned}
$$

[Väänänen & Wong 2015] proceed by showing that for each general (page 114) model $(M, \mathcal{G})$ for pure second order logic, Zermelo's argument shows $(M, \mathcal{G}) \models \Psi$; whence, by Henkin completeness, $\vdash \Psi$. Complicating the argument slightly formalizes Zermelo's result that ZFC^2 is quasi-categorical (for any two models, one is an initial segment of the other). Thus internal categoricity determines that there is only one reference *without* specifying it. As such it is a counter to Feferman's position

I have long held that CH in its ordinary reading is essentially indefinite (or 'inherently vague') because the concepts of arbitrary set and function needed for its formulation can't be sharpened without violating what those concepts are supposed to be about. [Feferman 2012]

that the continuum hypothesis is 'indefinite' and not 'a definite mathematical problem'. Since the continuum is an initial segment of any model of ZFC^2, internal categoricity arguably implies that CH is determinate; although we don't know what the value is. Precisely because it is in pure logic, this argument (as was observed in [Button & Walsh 2016]) is relevant to determinateness (but not determinacy, page 72) of propositions (i.e. addresses issues in [Martin 2001, McGee 1997, Parsons 1990b]), but does not identify a specific reference for the model of the theory ZFC^2 nor determine the truth value of CH.

Section 5 of [Väänänen & Wong 2015] again uses Henkin completeness to show consistency of ZFC^2 (from a stronger metatheory than the ZFC used so far). But, in this case, they deduce that a second order proposition can be derived from the formal system of second order logic by employing third order concepts to provide semantic proofs.

We summarize our geometry discussion; distinct intuitions have evolved over the years that are modestly axiomatized by different axiom sets.

For the geometry Euclid I (basic polygonal geometry), Hilbert's first order axioms meet this goal. $\mathcal{E}_\pi^2$ provides a modest complete descriptive axiomatization even including the basic properties of π. The Archimedes and Dedekind postulates secure a different goal: the nineteenth century conception of $\mathfrak{R}^2$ as *Euclidean geometry* and further to ground mathematical analysis.

....

Methodology

But attempts to find compelling extralogical principles that can provide us with perfect knowledge of abstract mathematical objects have largely stalled. ... It seems likely that if one is aiming to justify one's axioms and basic concepts on broader grounds, one will ultimately have to attend to the roles they play in organizing our mathematical knowledge, and the role that knowledge plays in organizing and structuring our scientific experiences. [Avigad 2010]

We renounced on page 11 Tarski's goal of creating a methodology of the deductive sciences. More modestly, we have argued directly for the significance in *mathematics* of a particular methodology – formalization. In this last part, we heed Avigad's admonition and 'attend to the role our concepts' play in organizing mathematical knowledge. Having adopted a stance of local foundations, the choice of those axioms will be a part of the organization.

In Chapter 12 we explore the role of formalization in evaluating the purity of proofs, focusing on Hilbert's analysis of Desargues' theorem, and conclude such a method is not sufficient. In Chapter 13, we consider Shelah's method of finding dividing lines as a general technique in developing an area of mathematics. We distinguish definition, classification, and taxonomy as different modes of organizing information and identify Shelah's dividing as creating a taxonomy. While Shelah originally applied his strategy only to the classification of first order formal theories, the technique readily extends to the formalism-free notion of abstract elementary class. Chapter 14 acts as a foil to the rest of the book. As opposed to our emphasis on the fruitfulness of formalizing mathematics, in this chapter we elaborate Kennedy's notion [Kennedy 2013] of formalism-freeness. It might seem that formalism-freeness is incompatible with the syntax–semantics distinction fundamental to model theory. We argue otherwise by considering several formalism-free approaches to specific areas of model theory that are formulated entirely semantically. Finally, we return to the relation between set theory and model theory by outlining connections of work in abstract elementary classes with large cardinal hypotheses. Chapter 15 rehearses the main points of the book to show how they support the argument.

In this chapter, we draw on our earlier analysis of the epistemological role of vocabulary (Chapter 1.2) as a tool for considering the question of whether/how formalization can be used to capture the notion of 'purity of method'. Our attention was drawn to this topic by the papers of [Hallett 2008] and [Arana & Mancosu 2012]. We first use our perspective of the role of vocabulary to set out the mathematical issues involved in the study of Desargues' theorem in a somewhat different way than theirs and to use the notion of explicit definition to investigate purity concerns. However, careful analysis (Chapter 12.5) of Hilbert's embedding theorem shows that even first order explicit definition can introduce impurities. This emphasizes the impotence of purely formal methods (such as logical and strong logical purity discussed in Chapter 12.3) to characterize purity. We see that to diagnose the impurity of Hilbert's embedding theorem one must consider the content of what is in fact a 'formalism-free' (Chapter 14) proof. The appendix to [Baldwin 2013a] provides what I argue is a pure proof.

We discuss the relation between the choice of vocabulary and basic axioms and the 'content' of a subject in Chapter 12.1. In Chapter 12.2 we lay out some of the mathematical background in terms of formal axiomatic systems to clarify the relation between the Desargues' proposition in affine and projective geometry. This analysis is based rather directly on a first order axiomatization of geometry. Consideration in terms of Manders' [Manders 2008] 'diagram proofs' would raise other pertinent issues. Chapter 12.3 analyzes several attempts to define the notion of purity and concludes that formalization can highlight impure methods but does not give a criteria for 'purity'. In Chapter 12.5, we come to grips with the essence of Hilbert's work on Desargues' theorem and assert the impurity of his argument that the three-dimensional proof is impure.

The discussion of purity below will make clear that the context of a proposition is crucial for the issues we are discussing. But, as discussed in Chapter 9, the notions of what geometry actually is have changed radically over the centuries. The distinction between the coordinate geometry of Descartes and synthetic geometry is crucial for our purposes. As [Arana & Mancosu 2012] makes clear the study of geometry in the late nineteenth century involved multiple dimensions, surfaces of various curvatures, and coordinate and coordinate-free approaches. These distinctions remain today although perhaps fields of mathematics are laid out differently: algebraic geometry, real algebraic geometry, differential geometry, topology, etc., etc. Hilbert's work on Desargues' theorem led directly to one subfield of geometry; the coordinatization theorems which are a main focus of this paper developed into the study of projective planes over (usually) finite fields as a distinct area of mathematics. In such major sources for this area as [Artin 1957, Hughes & Piper 1973, Dembowski 1977] the proof of the Desargues theorem in three-dimensional geometry is not central, if it appears at all. Rather, the importance of the *Desargues property* is an indicator of the algebraic properties of the coordinatizing field. The latter two books connect the projective geometries with combinatorics. There are important links with the statistical field of experimental design.

12.1   Content and Vocabulary

Following our template in Chapter 1.1 for formalizing an area of mathematics, we laid out in Chapter 9.3 the choice of primitive notions (vocabulary), the choice of logic,[1] and the choice of basic assumptions (axioms) for plane geometry. In this chapter, we restrict to first order logic.

Several authors [e.g. Arana & Mancosu 2012, Hallett 2008] draw on the following remark of Hilbert in his *Lectures on Geometry* of 1898/99 in their discussions of purity in geometry.

This theorem gives us an opportunity now to discuss an important issue. *The content of Desargues' theorem belongs completely to planar geometry; for its proof we needed to use space.* Therefore we are for the first time in a position to put into practice *a critique of means of proof.* In modern mathematics such criticism is raised very often, where the aim is to preserve *the purity of method,* i.e. to prove the theorem using means that are suggested by the content of the theorem.[2]

We stress that the task is to 'critique a means of proof'. We are given both a conclusion and some premises. Both the premises and the intermediary steps are at issue. And the issue is 'method'. It is not the mere existence of a valid implication; there can be pure and impure proofs of the same implication.

A key issue is whether concepts are introduced in the proof that are not inherent 'in the statement of the problem'. But 'in the statement of the problem' is incomplete without some specification of the context. In

---

[1] The choice of logic includes the components discussed in Chapter 1.1: formulas, truth, proof.

[2] ([Hallett & Majer 2004], 316); translation from [Arana & Mancosu 2012].

Hilbert's approach (Chapter 9.1) such a specification is made by asserting axioms for the geometry which 'implicitly define' the primitive concepts. We follow that line here.

We somewhat refine below the 'standard account of definability' from [Suppes 1956] which argues that a proper definition must satisfy the eliminability and non-creative criteria.

As pointed out in [Shoenfield 1967], an extension by explicit definitions, is conservative over the base theory, i.e. non-creative. We will justify below the need to allow adding to the vocabulary of a given formalization (even in pure proofs) in order to prove a result. The next examples show that axioms for additional relations can change content; we provide directly relevant examples in Chapter 12.3. The crux is the word 'content'. What is the content of a proposition? Surely, the content changes if the models in the base vocabulary in which the proposition holds change when additional axioms are added.

Example 12.1.1

(1) A vocabulary to study an equivalence relation contains only $=$ and a binary relation symbol E_1. There we can assert by a theory T_1 that E_1 is an equivalence relation. But suppose the vocabulary is expanded by a binary relation symbol E_2 and T_1 to a theory T_2 asserting that E_2 is also an equivalence relation and each E_2 class intersects each E_1 class in a single element. Now all the reducts of models of the full theory are equivalence relations in which all equivalence classes have the same size.

This example makes the point very clearly. Surely the first theory tells us exactly what an equivalence relation is. And the concept of equivalence relation entails nothing about the relative size of the equivalence classes. Specifically, the sentence:

$$[(\exists x)(\forall y)(xE_1y \rightarrow x = y)] \rightarrow [(\forall x)(\forall y)(xE_1y \rightarrow x = y)]$$

is a consequence of the added information about E_2.

(2) Consider the proposition $x \cdot x \cdot x = x$ about the class of abelian groups formulated in a vocabulary with a binary operation $\cdot$ and a constant symbol 1. Now expand the vocabulary to a vocabulary for fields by adding new symbols $+$ and 0; add the axioms for fields and the axiom $x + x + x = 0$.

The meaning of the expression $x \cdot x \cdot x = x$ is very different in an arbitrary abelian group (where there may be elements of arbitrary order) than it is in a field of characteristic 3 (where it is a law of the multiplicative group).

(3) Consider the class of linear orders in a vocabulary with a single binary relation symbol $<$. Add an operation symbol $+$ and constant 0 and assert the structure is an ordered abelian group. The original class contained $2^{\aleph_0}$ countable models; but only $\aleph_1$ can be expanded to a group. See ([Hodges 1987], 207).

Of course, our assertion that content changes if the class of models changes is based on the Hilbertian notion that the primitive concepts are implicitly defined by the axioms. For Hilbert, the axioms are not arbitrary; they are developed to clarify intuition. But the geometries we are able to study formally are whatever happens to satisfy the axioms. One might think of content in a more Fregean way: the axioms are describing geometry. If we are speaking of a fixed geometry, the various properties are just true; one cannot be meaningfully said to strictly imply the other without access to counter-models. I don't see how one can back off from describing a geometry to describing a family of geometries without embracing Hilbert.[3]

The sense in which the Arana–Mancosu attempt to found 'purity' on the 'intuitive content' of geometry falls victim to Pierce's paradox (page 38). They argue in Section 4.5 of [Arana & Mancosu 2012] 'against construing purity in terms of formal content'. We agree that 'purity' concerns cannot be fully resolved in terms of formal content. But we object to portions of their justification. On their page 237, they write 'Formal content entails, however, a radical contextualism regarding the content of statements like Desargues' theorem.' We argue at several points in this chapter for the necessity of describing a context rather than a particular statement that has radically different meanings in different contexts.[4] But our complaint here is: Arana–Mancosu reject the Hilbertian view that represents the first horn of the Pierce paradox. But now compare the theorem of Pappus with the theorem of Desargues. They are both true in 'intuitive geometry'. And Pappus implies Desargues by strictly geometric arguments.[5] The two propositions are not equivalent but we can only see this by adapting Hilbert's view of axioms admitting multiple interpretations.

---

[3] Smith [Smith 2010] ascribes priority for this viewpoint to Peano and especially Pieri.

[4] Most glaringly, the Desargues proposition is contingent in 2-space and true in 3-space.

[5] This proof is a common exercise in undergraduate projective geometry courses. E.g. in [Vis 2009] students are asked to draw and make connections between the diagrams for Pappus and Desargues. When writing the *Grundlagen*, Hilbert did not know this implication. It was proved by Hessenberg [Hessenberg 1905] in 1905. This raises the question of why Hilbert's concern is with the purity of Desargues rather than Pappus. Of course Hilbert's argument equally well shows that any Pappian plane is embeddable in 3-space. But the converse fails; not every subplane of 3-space is Pappian.

12.2 Projective and Affine Geometry

Before turning to our specific analysis of the purity of proof of the Desargues proposition we set some notation and clear away some extraneous matters. In Chapter 9.3 we discussed some first order axiomatizations of geometry. As in [Hilbert 1971], we are trying to formalize the 'field of geometry (in the traditional sense).' But here we use a very incomplete axiom system that enables us to focus on the distinction and connections between affine and projective geometry to clarify that while the two situations are distinct they behave the same with respect to purity of the Desargues propostion.

We now lay out the definitions and relations between affine and projective planes. We have slightly modified the statement of the axioms for projective planes from [Heyting 1963] and the analog for affine geometry from [Hilbert 1971]. Crucially the axioms given here for a projective plane actually imply the structure is planar; any two lines intersect, while the affine axioms hold in 3-space.

Definition 12.2.1 (PP) *A projective plane is a structure for a vocabulary with one binary relation R. We interpret the first coordinate to range over points and the second to range over lines. The axioms for a projective* plane *assert:*

(1) *Any two lines intersect in a unique point.*
(2) *Dually, there is a unique line through two given points.*
(3) *There are four points with no three lying on a line.*

These axioms are far from complete; analogous axioms for an affine plane, which are agnostic about the parallel postulate, assert:

Definition 12.2.2

(1) *There is a unique line through two given points.*
(2) *There are four points with no three lying on a line.*

We described in Chapter 9.3 how to pass from the informal axiomatization of the *Grundlagen* to fully formalized axiomatizations based on the conventions for first order logic with a precise formal vocabulary. For the simpler notion of projective plane here the translation is more direct. Note that the axioms for spatial geometry in [Heyting 1963] and [Hilbert 1971] explicitly introduce a new primitive term: plane.

Following [Henderson & Taimina 2005] by the 'high school parallel postulate' we mean the assertion: for any line ℓ and any point p not on the line, there *is a unique line ℓ'* through p and parallel to ℓ.[6]

Remark 12.2.3 There is an easy translation between projective and affine geometry.

Given a projective plane $\mathcal{P} = (\Pi, R)$, eliminate one line, ℓ, and all points that lie on it. Now two lines ℓ_1, ℓ_2 whose intersection point was on ℓ are parallel. It is easy to see that this affine plane satisfies the parallel postulate.

Similarly, suppose $\mathcal{A} = (\Pi, R)$ is an affine plane satisfying the high school parallel postulate. Add a new line ℓ_∞ and let all members of an equivalence class of parallel lines in Π intersect at a point on ℓ_∞; let these be the only points on ℓ_∞.

Theorem 12.2.4

(1) *The affine Desargues proposition asserts: if ABC and A'B'C' are triangles with AC $\parallel$ A'C', AB $\parallel$ A'B', and BC $\parallel$ B'C' then AA', BB', and CC' are parallel or the three intersect in a single point.*

(2) *The projective Desargues proposition asserts: if ABC and A'B'C' are triangles such that the points of intersection of AC with A'C', AB with A'B', and BC with B'C' are collinear then AA', BB', and CC' intersect in a point p.*

It is easy to check:

Claim 12.2.5 *Under the translation in Remark 12.2.3, the affine plane satisfies affine Desargues if and only if the projective plane satisfies projective Desargues.*

Hilbert used a ternary betweenness relation to fill what he regarded as gaps in Euclid; this extension is essentially irrelevant to our discussion here. While betweenness is appropriate for affine geometry, to consider both affine and projective geometry requires the more general quaternary separation predicate introduced by Pasch (see e.g., [Arana & Mancosu 2012]) representing that the four points are cyclically ordered. Unlike betweenness this relation is projectively invariant. For the axioms of this relation see e.g. [Heyting 1963]. Notably, neither betweenness nor cyclic order appears in [Heyting 1963] until after the discussion of Desargues and coordinatization.

---

[6] Euclid proves the existence of a parallel line on the basis of his first four axioms; in this context, the fifth postulate asserts uniqueness. In the context of projective plane geometry existence fails. See [Henderson & Taimina 2005] for an amusing and informative account of professional confusion over the difference between existence and uniqueness of parallel lines and the actual statement of Euclid's fifth postulate.

12.3 General Schemes for Characterizing Purity

In this section, we discuss several suggestions for more clearly specifying the notion of a 'pure' proof, consider how they evaluate the purity of certain arguments, and then draw some conclusions about these specifications.

Detlefsen and Arana distinguish a notion of *topical purity*. Rephrasing their discussion in Section 3.5 of [Detlefsen & Arana 2011], there are certain resources which *determine* a problem (for a given investigator). In mathematics, the determinants include definitions, axioms concerning primitive terms, and inferences. These are referred to as the 'commitments of the problem' and specify what we call the context of the problem. The topic of a problem is a set of commitments. 'A purity constraint restricts the resources available to solve a problem to those which determine it.' They then analyze the topical purity of a solution in terms of its stability under changes in the commitments. We want here to connect topical purity with several related notions considered by Arana [Arana 2008a, Arana 2008b]. Our general conclusion (as Arana's) is that it is not possible to translate the problem of purity into a proposition about formal systems in the most traditional sense; it is essential to retain a notion of 'meaning' in the discussion. Our argument can be seen as a model theoretic analog of the proof theoretic discussion in [Arana 2008b]. Arana introduces a notion of logical purity in [Arana 2008a].

Definition 12.3.1 (Logical Purity A)

(1) *The axiom set S is logically minimal for P if $S \vdash P$ but there is no proper subset of S proves P.*

(2) *The proof of P is pure if it is a proof from an S which is logically minimal for P.*

He points out that there are some obvious difficulties with this definition, since we could conjoin a set of axioms and get something that is logically minimal. Here is a more robust formulation.

Definition 12.3.2 (Logical Purity B)

(1) *The axiom set S is fully logically minimal for P if $S \vdash P$ and there is no S' such that $S \vdash S'$, $S' \vdash P$ and $S' \nvdash S$.*

(2) *The proof of P is logically pure if it is a proof from an S which is fully logically minimal for P.*

The difficulty with the second formulation is that it turns out to be an even stronger version of the following notion of Arana. [Arana 2008a].

Definition 12.3.3 (Strong Logical Purity) *The proof of P from S over T is strongly logically pure over some basis theory T if also $T \vdash P \to S$.*

S can only be minimal in the sense of Definition 12.3.2 if S is logically equivalent to P; otherwise choose P as the S' to show S is not minimal. Thus the existence of logically pure proof of P from S in the sense of Definition 12.3.2 requires that S and P are logically equivalent.

Strong logical purity has a long history including Sierpinski's equivalents of the continuum hypothesis in the 1920s, Rubin's 101 equivalents of the axiom of choice, and Friedman's reverse mathematics. Pambuccian [Pambuccian 2005] pursues a similar 'reverse geometry', finding a minimal weak axiom system for four results in Euclidean geometry. These are searches for the weakest hypotheses in terms of proof theoretic strength. They are *not* what Hilbert or Hallett claims for the Desargues property. And as Arana rightly points out 'reverse mathematics' is not an issue of purity. Each of these notions of logical purity, which I have just described, is about equivalence of statements not about proofs; none of them address the key issue: '*Which, if any, proof of a theorem is pure?*' Nevertheless, these *formal* notions are sometimes capable of detecting the non-existence of pure proofs.

Proposition 12.3.4 *No proof of the Desargues proposition from the assumption of three dimensions is strongly logically pure.*

If there were such a proof, every geometry satisfying the Desargues proposition would actually be three-dimensional. This is clearly false; we investigate the subtly different consequence (embeddability) of the Desargues proposition in Chapter 12.5.

The notion of topical purity builds on an earlier formulation of Arana, 'a proof ... which draws only on what must be understood or accepted in order to understand that theorem' ([Arana 2008a], 38). Two issues arise: What does it mean to draw on? How can one determine 'must be understood or accepted to understand'? We follow Arana in leaving the second question to individual cases. But there is a more uniform way to understand 'draw on'. We say that a 'concept' is drawn on in a proof when it is named in the proof. We are going to discuss arguments below which could be formalized as derivations in a basic language. We first observe that introducing relations that are *not* definable in the base language is either a definite sign of impurity or an admission that the formalization of the hypothesis omitted a necessary concept. In the latter case there might be pure argument although the one originally given was not. On the one hand, we argued in Chapter 12.1 that adding both additional relations and axioms about them often distorts meaning. More strongly if non-definable relations are added, even with no additional axioms, any claim that this is a pure move, is simply a claim that the original formalization left out crucial concepts. An example shows how this issue focuses on the importance of specifying what is being formalized. I take Presburger arithmetic as an attempt to formalize *addition* on the natural numbers. If one moves from the schema of induction to introducing multiplication by a recursive definition, the topic has vastly changed. $(N, +)$ is decidable; $(N, +, \times)$ is not.

But we argue more strongly that an explicit first order definition may violate purity concerns. We will discuss what we claim are 'pure' and 'impure' proofs, both invoking explicit definition, of the same fundamental result, explaining the reasons for this diagnosis. And then we argue for the value of each of these proofs.

We seek now a more mathematical formulation of the topical purity introduced in [Detlefsen & Arana 2011].[7] Although earlier in the book we avoided the word language because of its multiple meanings in similar situations, we now introduce a specific meaning clarifying one of the three possibilities on page 44 that is useful in the current context (see also Chapter 14).

Definition 12.3.5 *The* language *for a mathematical topic A is a vocabulary for A with symbols for each of the primitive notions identified by the investigator and axioms for the relations which are sufficient to delimit these concepts.*

Arana and Detlefsen introduced topical purity as a concept about the solution of a problem. In accordance with the last couple of pages, we require the context in which this problem arose must be formalized. There is an issue of whether the following definition should be construed in the Euclid–Hilbert or Hilbert–Gödel–Tarski frameworks laid out in Chapter 1.1. For most of the argument it makes no difference. However, Lemma 12.5.4 and its application in Theorem 12.5.2 are mathematical theorems invoking the completeness theorem and so require full formalization. Thus officially we see the definition from the second standpoint. I think Arana and Detlefsen saw it more from the Euclid–Hilbert perspective.

Topical Purity. Choose a first order formalization for the resources which determine the problem in the Detlefsen–Arana sense used at the start of this section. That is, specify a language including a set of primitive concepts and axioms needed to describe the particular problem and its context A.

More formally, fix a vocabulary τ and a theory T_0 that implicitly defines the concepts named by the symbols τ. Now a *topically pure* proof of ϕ from ψ where ϕ and ψ are τ-sentences is a proof of ψ from ϕ in T_0 that *invokes* only concepts from the context A. We will interpret 'invoke' in our notion of topical purity as 'introduce by explicit definition'.

---

[7] Arana suggested the specification was close enough to topicality to not deserve a new name.

The particular first order formalization is a real choice. Pambuccian provides examples in [Pambuccian 2001, Pambuccian 2009] of distinct interpretations of the basic notions of a proposition which lead to distinct, indeed incompatible, systems; each system can be thought to provide a pure proof for the understanding of the concepts that has been formalized. [Pambuccian 2009] studies the Sylvester–Gallai theorem: If the points of a finite set are not all on one line, then there is a line through exactly two of the points. One might conceive of a line in terms of betweenness or as 'the shortest distance between two points'. These provide different contexts; Pambuccian explains three distinct proofs, one using the first concept and two the second. These are based on incompatible axiom systems. He remarks that still another proof holds for planes satisfying a certain Artin–Schreier condition.

To show this notion differs from strong logical purity, we exhibit in Corollary 12.5.6 a proposition which has both topically pure and impure proofs from the same base theory.

It is tempting to insist that the analysis of the context and conclusion should elicit all relevant concepts and thus the set of concepts used in the proof should be fixed in the choice of language. Sobociński [Sobociński 1955] discusses this criterion for a formal system, primarily in the context of propositional logics. Givant and Tarski [Givant & Tarski 1999] argue that including defined concepts in an axiom system leads to a misleading appearance of simplicity of the axioms. They discuss simplicity in terms of both the 'length' of the axiom system and its complexity in terms of the number of quantifier alternations. While the second has important structural consequences, as we discussed in Chapter 1.3, both measures of simplicity relate to technicalities of the formalization and appear irrelevant to the notions of purity considered here.

For purity purposes, the 'resources which determine the problem' in the definition of topical purity must be seen as a limitation – 'which determine the problem and no more'. But this condition is too strong. If it were accepted, the formalizations of Heyting and Hilbert [Hilbert 1971, Heyting 1963] would be impure for studying Desargues' theorem because they take 'plane' as a primitive concept. This however is by no means necessary. While Hilbert takes *plane* as a primitive, [Robinson 1959b] makes the following intrinsic definition.

Definition 12.3.6 *The plane generated by a, b, c is the collection of all points on lines that contain a and intersect bc.*

And Desargues' theorem is about lines, points, and incidence. But according to the just cited definition of plane, these points and lines all lie in the same plane. So 'plane' should be viewed as part of the context and it is not important for purity considerations whether it is taken as a primitive or introduced by explicit definition. This is a judgement about this specific case.[8] Such examples are another reason for demanding that base language should reflect both the particular problem and the context.

We reject the criterion of demanding all relevant concepts be fixed among the primitive terms for a number of reasons. It is contrary to the goal of finding 'basic' vocabulary. It is not true to mathematical practice. Mathematical proofs are not carried out as derivations in a fixed formal language. In particular, new concepts are introduced by definition for the purpose of particular proofs. We explore examples of this type of extension in detail in Chapter 12.5. If these new definitions are mere abbreviations it seems they should be harmless. Certainly if axioms are added about the new relations, this is no longer harmless (see Example 12.1.1.). In fact, we will argue that, even without additional axioms, explicit definitions can violate purity. That is why we add the requirement that the new definitions remain within context of the original topic. We further illustrate the meaning of this phrase in Chapter 12.5.

The first use of our characterization of topical purity is to determine cases where there is no topically pure proof of a proposition. Here is an example. We generalize Proposition 12.3.4 from 'strongly logically pure' to the much broader notion of topical purity.

Claim 12.3.7 *There is no topically pure proof of the Desargues proposition in the plane (from PP).*

A proof of this proposition requires three steps. Choose a formalization; then argue that this formalization is appropriate for the context and that there is no topically pure proof in that formalization.

Clearly the Desargues proposition is stated in terms of points, lines, and incidence. And the basic properties of points, lines, and incidence in a projective plane are given by PP (Definition 12.2.1), so this is an appropriate formalization. Thus for the projective plane the existence of a topically pure proof would entail that every model of the axioms PP is a Desarguesian projective plane. Many counterexamples to this assertion have been exhibited in the last century. The result extends to the affine plane by Claim 12.2.5.

This claim is controversial. It is well-known that there are many distinct completions of PP, so it does not exhaust our intuitions about the relations

---

[8] As another justification of considering 'plane' to be part of the context, we could simply ask from the standpoint of three-dimensional geometry whether the Desarguesian proposition can be proved using only resources from the plane.

of lines in the projective plane. But it does set the context in the sense of purity issues. We want to argue that certain sufficient conditions clearly add information that is not in the context. Arana and Mancosu[9] make the following counterclaim: affine Desargues can be proved from planar axioms so there is a pure proof in the plane.

To evaluate this assertion, we first clarify the mathematical situation. In [Hilbert 1971], Hilbert proves two mathematical results:

Fact 12.3.8

(1) *In three-dimensional (affine or projective) geometry, Desargues' theorem holds. This depends only on the incidence and order[10] axioms.*

(2) *In two dimensions, the affine Desarguesian theorem can be proved from the incidence axioms, the parallel axiom, and the congruence axiom.*

We showed in Example 12.1.1 that adding additional relations and structure can change the interpretations of the basic structure. Here, we note that the problems arise in the specific geometric context. It is true that affine Desargues can be proved from the parallel postulate and congruence axioms (basically side-angle-side) and these are surely planar concepts. But while parallelism is necessary to understand affine Desargues, congruence is not. The proof of Fact 12.3.8(2) requires extensions of the basic geometric axioms in two distinct ways. First there are additional axioms in the same vocabulary, the parallel postulate. But secondly a new concept of congruence must be introduced; this requires both adding several relation symbols to the vocabulary (Notation 9.3.1) and positing a congruence axiom such as SAS. The fact that additional axioms are introduced is immediate evidence of impurity. In fact congruence is definitely foreign to the situation as the theorem of Desargues holds in an affine plane over any algebraically closed field. There is no notion of congruence definable in the geometry over such fields. To define congruence one must introduce further relations; e.g., regard the complexes as a two-dimensional real vector space. Thus, in the spirit of [Detlefsen & Arana 2011], there can be no topically pure proof of the Desargues theorem in the plane, even for affine geometry.

The basic point here is that two distinct geometrical contexts are being considered. Geometry with parallels and congruence is a different subject than projective geometry which encodes only the properties of lines and incidence. But in fact there is no pure proof in the context of affine metric

---

[9] See sections 4.1 and 4.7 of [Arana & Mancosu 2012].

[10] In fact, the order axioms are a red herring. They are used only to guarantee that the coordinatizing field is ordered. See Bernays Supplement IV in [Hilbert 1971].

geometry, because the congruence axioms require 'flatness'; the Desargues theorem fails in various non-Euclidean geometries. This illustrates an important attribute of the search for pure proofs. It forces the clarification of hypotheses.

12.4 Modesty, Purity, and Generalization

We introduced in Chapter 9.1 *modest descriptive axiomatization*, an axiomatization that captures a data set without proving too much more, and argued that Hilbert's axioms are immodest for several systems, specifically Euclid II. A more modest axiomatization usually proves the old results by introducing new terms by explicit definition. Since modesty concerns only provability, not the particular proof, the constraints imposed by topical purity do not apply.

When there is no question of reinterpreting a data set the notions of finding a more modest axiomatization and generalizing a topic are very close together. A theorem or a topic is introduced with more special hypotheses than are necessary and this is remedied. Thus the theory of convergence was first studied from the standpoint of the real numbers, then in context of metric spaces, and finally topological axioms grasp the concept. Here, the area described is widened but also a clearer understanding of the original topic is given.

Generalization connotes increasing the number of examples. While a more modest axiomatization allows more models, the emphasis is that a more basic explanation had been found for an existing situation.

The referee of [Baldwin 2017b]) asked whether the key results to show Euclid I admitted a first order axiomatization imply the similar triangles theorem and justification of area could be given pure proofs. The purity of the similar triangle theorem depends on intricate but well-studied questions about the acceptability to the Greeks of various properties of proportions. For area, the range of an area function must be numbers with units rather than magnitudes, a thoroughly modern idea. This problem bears further investigation.

12.5 Purity and the Desargues Proposition

Although the definition of topical purity allows the introduction of new terms by definition, this introduction is restrained by the given informal context A. We will now see that without this restriction proofs using manifestly impure notions would meet the requirement for topical purity.

As reported by Hallett ([Hallett 2008], 227), Hilbert argues that although the content of the Desargues proposition is manifestly two-dimensional, three-dimensional methods are necessary for its proof. We explore the role of vocabulary versus axioms in understanding this claim. We first formalize Hilbert's results on the strength of planar Desargues. Thus the following notation is introduced in the parlance of contemporary model theory. We will further explain the motivation for the introduction of Σ after stating Theorem 12.5.2.

Definition 12.5.1 *Fix a vocabulary of points, lines, and the incidence relation.*

(1) *PG is the theory of projective geometry*[11] *(asserting the existence of at least 3 dimensions) and PP is the theory of projective planes as in Definition 12.2.1.*[12]

(2) *Let Σ be the collection of sentences σ that are satisfied in some projective plane such that $PG \vdash \sigma$. I.e., $\sigma \in \Sigma$ just if σ is true in at least one projective plane and in every projective geometry of dimension at least 3.*

Hilbert conjectured and later proved two results which establish the pivotal role of the Desargues theorem from a deductive standpoint.

Is Desargues' Theorem also a sufficient condition for this? i.e. can a system of things (planes) be added in such a way that all Axioms I, II are satisfied, and the system before can be interpreted as a sub-system of the whole system? Then the Desargues Theorem would be the very condition which guarantees that the plane is distinguished in space, and we could say that everything which is provable in space is already provable in the plane from Desargues. ([Hallett 2008], 227) or ([Hallett & Majer 2004], 240)

Using Definition 12.5.1, we formulate the two assertions of this quote in modern terms; the first is 'formalism-free', the second is formal.

Theorem 12.5.2 (Hilbert)

(1) *If Π is a Desarguesian projective plane, Π can be embedded in 3-space.*
(2) *If $\sigma \in \Sigma$ then (PP + Desargues) $\vdash \sigma$.*

---

[11] To be precise, we take the formalization in [Robinson 1959b] since it has the same vocabulary: points, lines, and incidence for two- and multi-dimensional geometry. The notion of plane is introduced by explicit definition as in Definition 12.3.6.

[12] Note that the union of the theories PP and PG is inconsistent.

We place the situation in a more general framework. For a formula $\theta(\mathbf{y}, x)$ a tuple $\mathbf{a}$ and a sentence ψ, $\psi^{\theta(\mathbf{a},x)}$ denotes the relativization[13] of ψ to $\theta(\mathbf{a}, x)$. The Hilbert quote above is somewhat ambiguous. What does 'everything' in 'everything which is provable in space' mean? It must mean 'everything about planes'; there is no intent to assert that an arbitrary statement about 3-space (e.g. properties of spherical geometry) is provable from the planar axioms and Desargues. We introduced the set of sentences Σ in Definition 12.5.1 to formalize this observation. Letting $\theta(\mathbf{y}, x)$ define the plane[14] generated by $\mathbf{y}$, we rephrase the key sentence in formal terms as, 'If $T_2 \models (\forall \mathbf{y})\sigma^{\theta(\mathbf{y},x)}$ then $T_1 \models \sigma$.'

Definition 12.5.3 *[Interpretation] Let T_1 and T_2 be two theories in the same vocabulary. We say T_2 is interpretable in T_1 if there is a formula $\theta(\mathbf{y}, x)$ such that*

(1) *for every $\psi \in T_2$, $T_1 \models \forall \mathbf{y} \psi^{\theta(\mathbf{y},x)}$;*
(2) *if $M \models T_2$ and $M' \models T_1$ and $M \subseteq M'$, there is an $\mathbf{a} \in M'$ such that $M = \theta(M', \mathbf{a})$.*

In our current situation, take T_2 as $PP + Desargues$, T_1 as PG, $\theta(\mathbf{y}, x)$ as the formula: x is on the plane generated by $\mathbf{y} = \langle y_1, y_2, y_3 \rangle$. Note the projective plane M is merely a substructure not an elementary submodel of M'; the structure M satisfies that every two lines intersect. When $\theta(\mathbf{a}, x)$ defines M, $M \models \sigma$ if and only if $M' \models \sigma^{\theta(\mathbf{a},x)}$.

Lemma 12.5.4 *Suppose T_2 is interpretable in T_1. Suppose further that if $M \models T_2$ there is an $M' \models T_1$ with $M \subseteq M'$. If $T_2 \models (\forall \mathbf{y})\sigma^{\theta(\mathbf{y},x)}$ then $T_1 \models \sigma$.*

Proof. Fix $\sigma \in \Sigma$. Let $M \models T_2$, then M extends to a model M' of T_1. By Condition (2) of Definition 12.5.3, there is an $\mathbf{a} \in M'$ such that $M = \theta(M', \mathbf{a})$. By Condition (1), for each such $\mathbf{a}$, $M' \models \sigma^{\theta(\mathbf{a},x)}$; in particular, $M \models \sigma$. Since we have shown every model of T_1 is a model of σ, by the extended completeness theorem $T_1 \vdash \sigma$.

Proof of 12.5.2: We discuss Hilbert's proof of part (1) at length in the next few pages. For part (2), let $\theta(\mathbf{y}, x)$ assert that x is on the plane[15] generated

---

[13] The relativization ψ^ϕ to ψ of $\exists x \phi$ is $\exists x(\phi \wedge \psi)$ and of $\forall x \phi$ is $\forall x(\phi \to \psi)$.

[14] Hilbert's informal use of plane in [Hilbert 1971] reinforces our concern about vocabulary. Axiom II,5 refers to a line a lying in the plane ABC. Theorem 5 refers to a line a lying in a plane α (without parameters). Axiom III refers to both points and lines lying in a plane α. This can all be naturally formalized by a 4-ary predicate $P(\mathbf{y}, x)$ which holds if the point x lies on the plane generated by $\mathbf{y} = \langle y_1, y_2, y_3 \rangle$, where for Hilbert this would be an implicit definition of 'generated'. We give an explicit definition in [Baldwin 2013a].

[15] Thus θ formalizes Definition 12.3.6.

by y_1, y_2, y_3. To apply Lemma 12.5.4, take T_2 as $PP + Desargues$ and T_1 as PG. Condition (1) of Definition 12.5.3 holds since every subplane of a three-dimensional space is Desarguesian. Clearly T_2 is interpretable in T_1. Condition (2) holds since planes are definable. Theorem 12.5.2.1 gives us the second hypothesis of Lemma 12.5.4. Note that $\sigma \in \Sigma$ (Definition 12.5.1) implies $PG \vdash \sigma$ and thus $PG \vdash (\forall \mathbf{y})\sigma^{\theta(\mathbf{y},x)}$. By Lemma 12.5.4, $(PP + Desargues) \vdash \Sigma$.

Thus if θ is a sentence about projective planes that we show (perhaps in a formalism-free way) to be true in every plane that can be embedded in 3-space, then θ can be formally derived from PP plus the Desarguesian property. The Pappus theorem is an example of a statement concerning projective planes, which is *false* in some planes that can be embedded in 3-space.

Hilbert's analysis of the quality of a proof extends beyond topical purity. He wrote,

Nevertheless, drawing on differently constituted means has frequently a *deeper and justified* ground, and this has uncovered beautiful and *fruitful relations*; e.g. the prime number problem and the $\zeta(x)$ function, potential theory and analytic functions, etc. In any case one should never leave such an occurrence of the mutual interaction of different domains unattended.[16]

The role of 'spatial assumptions' is better seen by a more careful examination of Hilbert's proof of Fact 12.3.8 and Theorem 12.5.2. He begins [Hilbert 1971] by noting that the three-dimensional proof of Desargues' theorem (Fact 12.3.8(1)) from the axioms of connection, order, and parallels is well-known. Hallett clearly outlines the structure of Hilbert's proof of embeddability from Desargues in ([Hallett 2008], 228). We place Hilbert's argument for Fact 12.3.8(2) and Theorem 12.5.2 and their formalization in a more modern context. A ternary field is a structure with a single ternary operation; roughly $t(a, x, b)$ corresponds to $ax + b$, which satisfies a set of axioms as specified in [Dembowski 1977, Hughes & Piper 1973]. But for this correspondence to be literally true the plane coordinatized by the ternary field must satisfy the Desargues property.[17]

---

[16] This is from unpublished notes of Hilbert that are quoted in [Hallett 2008].

[17] In [Baldwin 1994] I constructed a non-Desarguesian projective plane which is $\aleph_1$-categorical. In [Baldwin 1995], I proved that despite its well-behaved nature from a model theoretic standpoint, this plane admits little 'algebraic' structure; in particular the ternary operation can not be decomposed into two well-behaved binary operations and no group is interpretable in the structure. I also proved this projective plane is in the definable closure of any line (with no parameters). That is, the plane admits no perspectivities. The task of giving a geometric proof of this last result remains open.

Hilbert's argument involves three steps.

(1) *Any* geometry can be coordinatized by a ternary field.
(2) If the geometry satisfies
  (a) the Desargues proposition or
  (b) the parallel postulate and SAS (the congruence axiom in Hilbert's parlance)

  then the coordinatizing ring is associative (and in fact a skew field[18]).
(3) An n-dimensional affine (projective) geometry can be constructed as a set of n $(n+1)$-tuples from a skew field and the plane can be embedded in the 3-space.

Hilbert's embedding theorem is an example[19] of what we call a *fully semantic proof*: the entire proof consists of operations on mathematical structures. Geometric statements like Desargues' theorem or that two points determine a line or algebraic properties such as the associative law are treated as properties of mathematical objects. These objects might be geometric or algebraic. The proof proceeds from a plane to a 3-space by an explicit definition of binary functions that satisfy the axioms for a skew field. Now the original geometric properties are ignored and a 3-space structure is defined on the product of three copies of the skew field.

This proof from [Hilbert 1971] introduces a different set of purity concerns. In Hilbert's argument, a field is defined whose elements are equivalence classes of segments.[20] These are not geometric notions and the objects are not in the model but are 'imaginary elements' in the sense of Definition 4.6.3. This objection is somewhat reduced by Heyting's proof. Heyting still defines a field, but its elements are points of the given plane. Even if we have the fields as the points on a line, the construction of the three-dimensional model goes far afield from geometry. These new objects do not have 'geometric interpretations'. On the one hand, the objects of the field are viewed as numbers. In Hilbert's formulation, they are equivalence classes of segments (hardly a geometric notion). Even accepting the domain of the field (as points on a single line), the operation of multiplication while explained geometrically is hardly geometrical in a Greek sense. The modern geometry of homogeneous quadruples is employed in the construction.

---

[18] A skew field or division ring is a structure for the vocabulary $(+, \times, 0, 1)$ which satisfies all the axioms for a field except commutativity of multiplication.

[19] Most mathematical proofs are of this sort. Note in particular Theorem 14.0.1.

[20] This is a deliberate decision of Hilbert so as to study the geometry of segments. Already in his 1893–94 lectures he had established a correspondence between the points on a line and numbers. See ([Hallett & Majer 2004], 68–69).

This is essentially a metamathematical argument constructing a 3-space out of whole cloth and embedding the original plane in it. This seems to be a really new method introduced by Hilbert.[21] It is very different from Hilbert's geometric construction of counterexamples to Desargues or of the geometric arguments for Theorem 12.5.2(1) as given by Levi [Levi 1939] or in the appendix to [Baldwin 2013a]. At the least it is a precursor of the modern notion of the interpretation of one theory in another. Moreover, Hilbert's proof of Desargues in an affine plane (Fact 12.3.8(2)) with congruence also goes through this metamathematical trick of embedding in 3-space and deducing the result from the known proof of Desargues in 3-space. Desargues gave a geometric proof of his theorem (see [Arana & Mancosu 2012]) in the three-dimensional affine case using the theorem of Menelaus.

Our use of 'metamathematical' in the last paragraph has two senses. Metaphorically, Hilbert is constructing a model and so this is a precursor of model theory. But he has also given a 'formalism-free' proof of Theorem 12.5.2(2). (That is with the conclusion expressed as in the quotation before Theorem 12.5.2.) In the weakest sense this proof is formalism-free since there is no formal proof system. But even more, the 'interpretation' makes no reference to the formal vocabulary or any notion of a formal language. But as we noted in proving Lemma 12.5.4, this formalism-free proof translates to the existence of a formal proof by the extended completeness theorem. (Of course, this translation was not available to Hilbert in 1900.) That is, in the proof of the formal Theorem 12.5.2(2), we invoked Hilbert's semantic proof of Theorem 12.5.2(1). Such a translation is a standard consequence of the extended completeness theorem and a routine model theoretic tool. This does not mean that the semantical proof is tacitly formal; this translation just expresses the content of the completeness theorem (page 257).

Consideration of some of the standard texts in projective geometry of the last half century [Artin 1957, Dembowski 1977, Hughes & Piper 1973] reveals an interesting phenomenon. The proof of the Desargues proposition is at best barely mentioned.[22] The crux is the understanding of the Desargues proposition in terms of the properties of the group of collineations and in terms of the properties of the coordinatizing ternary ring.

As we noted in outlining the argument and in Proposition 12.3.4, the Desargues proposition does not imply there are non-coplanar points.

---

[21]   This was remarked by Hallett [Hallett 2008].

[22]   Hartshorne [Hartshorne 1967] is an exception.

Thus, it is not true that the Desargues proposition implies there is a third dimension. Rather, Hilbert showed, by a *fundamentally non-geometric construction*, one can embed the given plane in 3-space. But we presented a 'geometric' construction (joint with Howard) of this embedding in the appendix to [Baldwin 2013a]. Thus we have an example where an impure proof provides very significant information. Indeed the very impurity of Hilbert's argument is crucial for the twentieth century development of the theory of plane projective geometry. An important mathematical impact of a proof of impurity is to focus attention on the proposition in question as an axiom for selecting a new field of study. For example, the fact that there is no pure proof of the Desargues proposition in the plane calls attention to the importance of studying Desarguesian planes. The crucial property, as Hilbert saw, is not the geometric configuration itself but the associated algebraic structure; it was later codified in terms of transitivity properties of the automorphism group (Lenz–Barlotti classification). The new 'algebraic' concepts introduced by Hilbert are all introduced by explicit definition but they stray far from the geometric topic under consideration.

In contrast, I claim that the geometric argument for Theorem 12.5.2 in the appendix to [Baldwin 2013a] is topically pure. The crucial point is that Hilbert's argument introduces the notions of coordinatization and field which are foreign to synthetic geometry. In that appendix, we reinterpreted the words point, line, and plane in terms of certain planar configurations to interpret a 3-space containing π in a Desarguesian plane π but didn't introduce significantly new concepts. As Hilbert's proof is impure, we conclude:

Fact 12.5.5 *The assertion that every Desarguesian plane is embedded in 3-space has both topically pure and topically impure proofs from the axioms PP of projective planes.*

Corollary 12.5.6 *The notions of* strong logical purity *and* topical logical purity *differ.*

This difference is evident. Topical purity is a property of a particular argument. While strong logical purity was defined in terms of the existence of a derivation (Definition 12.3.3) and does not depend on the particular argument advanced for the proposition.

The existence of a good theory of area provides a similar example. Eduardo Giovannini pointed out in correspondence that establishing a theory of area only on the basis of the theory of equal content for plane polygons appears to require de Zolt's axiom as a new geometrical axiom.

As in the embeddability problem, this shows that there is a topically impure proof. But here the question of a topically pure proof remains open. Hartshorne notes ([Hartshorne 2000], 210) that he knows no 'purely geometric' (without segment arithmetic and similar triangles[23]) proof for justifying the omission of de Zolt's axiom.

The Desarguesian proposition is a dividing line in the sense of Chapter 13. Its truth implies strong coordinatization properties; its failure implies planarity (in an axiom system that is agnostic on dimension). Specifically, the associativity of the coordinatizing field is used to prove that the relation of tuples from the field $\mathbf{x} \sim \mathbf{y}$ if there is a 'number' c with $\mathbf{y} = c\mathbf{x}$ by coordinate-wise multiplication (used to introduce homogeneous coordinates) is transitive and thus an equivalence relation. Thus, the following are equivalent:

A projective plane

(1) is coordinatized by an *associative* skew field;
(2) satisfies the Desargues property;
(3) can be embedded in 3-space.

There is another connection between spatial axioms and associativity. In three-dimensional Euclidean geometry the volume of a cube can be computed. Interpreting XI.32 of [Euclid 1956] in modern language yields the formula $V = \ell wh$. (Euclid proves that the volume of a parallelepiped is determined by the area of the base and the height.) The fact that the geometric notion is independent of which side is chosen as the base of the parallepiped implies the associative law for the coordinatizing field.[24]

Finding the associative field is, in modern terms, an interpretation (Definition 12.5.3) of the 'field' into the geometry. It proceeds by a sequence of explicit definitions. The proof of the algebraic axioms follows from the geometry. And then the plane is interpreted back into the 3-space over the field. Thus, if there is any distinction between algebra and geometry the deduction of (3) from (2) via (1) fails to be a topically pure proof. But this conclusion cannot be established by a characterization of purity such as strong logical purity which concerns the mere existence of proof. In fact, since we know (3) implies (2), the proof that (2) implies (3) is strongly logically pure.[25] The failure of topical purity is seen by consideration of the

---

[23] Giovannini emphasized in correspondence that some measure of area function is involved in the non-geometric proofs.

[24] Serendipitously, this argument was given by Ken Gross in a professional development program for elementary school teachers while I was working on this material.

[25] We can take the base theory T in Definition 12.3.3 as PP. The formalization of the embedding is clearer from the argument of the appendix to [Baldwin 2013a] than from Hilbert.

meaning of concepts introduced in the proof: the introduction of the notion of a skew field which is not a geometric notion is decisive.

This illustrates Tait's maxim: the notion of formal proof was invented to study the *existence* of proofs, not methods of proof. Or as Burgess [Burgess 2010] puts it, 'For formal provability to be a good model of informal provability it is not necessary that formal proof should be a good model of informal proof.'

We should not ignore the virtues of a demonstration that there is no pure proof. It shows that additional resources are needed for a particular claim. The Desargues proposition is particulary instructive in showing the value of the searching for the content of those additional resources. Hilbert isolated the ability to coordinatize in terms of the Desargues configuration and its connections with the interpretability of division rings. A significant part of twentieth-century mathematics, the further development of projective planes, particularly finite projective planes, relied both on the algebraization and on the discovery of the underlying properties of the group of perspectivities of the plane.

Hilbert succeeded in showing a deep connection between algebraic and geometric conceptions by identifying both the algebraic (associativity) and geometric (Desargues proposition) conditions necessary and sufficient for Descartes coordinatization to succeed.

12.6 Distinguishing Algebraic and Geometric Proof

We comment here on the appendix (with William Howard) to [Baldwin 2013a], arguing that it provides a pure geometric proof of the embeddability theorem while Hilbert's proof is manifestly not pure. For this we need some distinction between algebra (generalized arithmetic) and geometry. Algebra deals with abstract objects (often called numbers) under well-defined rules; geometry deals with magnitudes. Geometric arguments admit and (as the writing of the appendix demonstrated) often demands pictures. The distinction is clearly made in the quotation from Newton in [Detlefsen & Arana 2011]; there should be no arithmetical computations of 'Quantities truly geometrical' except the checking for equality and inequality. The essence of coordinatization, fundamental to Hilbert's proof of the embedding theorem, is to reject this notion.

Thus in the appendix to [Baldwin 2013a] the crucial vocabulary remains points, lines, and planes. There is no introduction of multiplication and addition and no reliance on the development of coordinate geometry. Crucially, however, new 'points' are introduced as certain triples of points and new 'lines' are introduced as sets of these 'points'; explicitly defined equivalence relations play a significant role. This is a more complex argument than the construction of models of non-Desarguesian planes that interpret (pieces of) curves as lines. This level of complexity is implicit in Hilbert.

The number of special cases that appear in the proof are characteristic of geometric arguments. Algebraic methods (as is clear in twentieth century algebraic geometry) can clarify the notion of a 'generic configuration'. Thus, the coordinatization of a Desarguesian plane requires a not-quite-arbitrary choice of coordinate points. In contrast Claim 4.3.2 of [Baldwin 2013a] requires a delicate argument to replace 'arbitrary' points by ones that are in general position. A similar situation concerns the relation between Pappian and Desarguesian fields. The first is coordinatized by a commutative field, the second by a division ring (which will not be commutative if the plane is not Pappian). Hessenberg [Hessenberg 1905] proved this in 1905. But he missed some possible intersections in the Desargues configuration and the proof was completed by Cronheim [Cronheim 1953] in 1953.

While [Levi 1939] gave a pure geometric proof of the Hilbert embedding theorem, Baldwin and Howard make clear the geometric picture that motivates the coding of points in 3-space by triples in the plane. Levi does not provide such a clear picture. He gave the proof in the affine case and then extended to projective planes on general grounds as in Remark 12.2.3.

13 | On the Nature of Definition: Model Theory

As promised in the introduction, we now consider the following goal of Maddy, in the light of classification theory.

> In sum, then, the Second Philosopher sees fit to adjudicate the methodological questions of mathematics – what makes for a good definition, an acceptable axiom, a dependable proof technique – by assessing the effectiveness of the method at issue as means towards the goal of the particular stretch of mathematics involved. ([Maddy 2007], 359)

We focus on how to make useful definitions. That is, we study, not the role of formal definition,[1] but the strategy for making informal definitions in mathematics and particulary in model theory. We analyze the motivation for the dividing line strategy.[2]

For context, we consider some earlier work on the justification of definitions. In [Tappendon 2008a, Tappendon 2008b], Tappenden argues for the necessity of studying the notion of definition in context, asserting 'the math matters.' In explaining how the actual mathematics matters, he proposes that 'fruitfulness' and 'naturality' ('possesses a certain kind of appropriateness or correctness[3]') are markers of 'good' definitions. We expand below on the meaning of 'appropriate and correct' in our context. But first we emphasize the point of one of Tappenden's examples: the role of the Legendre symbol.[4] With this symbol, the law of quadratic reciprocity, which is complicated to state because of many cases, becomes a single equation. But, Tappenden argues, this ease of statement is not the reason for its significance; rather the symbol simplified proofs and was a tool for the eventual solution of Hilbert's 9th problem.[5] The crucial point for us

---

[1] See Chapters 4.6 and 12.5.

[2] Recall this strategy was the key to proving the main gap theorem, solving Morley's conjecture (Chapter 5.5), and the freeing of first order model theory from axiomatic set theory (Chapter 8).

[3] He opposes this notion to 'related to the natural world.' See ([Tappendon 2008b], 281–282).

[4] If $n \not\equiv 0 \bmod p$ the term $(\frac{n}{p})$ is evaluated as 1 if $x^2 \equiv n$ is solvable mod p and 1 if not; if $n \equiv 0 \bmod p$, $(\frac{n}{p}) = 0$.

[5] Hilbert's 9th problem asked for the most general formulation of quadratic reciprocity.

is that while a specific technical definition (Legendre symbol) might be a focus at one stage, the significance of the definition is investigated in the context of its role in a wider theory (algebraic number theory).

Similarly, Lakatos emphasizes variations of *proof-centered* definition.[6] His guiding illustration is the development of the concept of a polyhedron. Lakatos expounds, in his Socratic exploration, the attempt to verify Euler's assertion that for any 'polyhedron' with E edges, V vertices, and F faces, $E = V + F - 2$. He shows how the interpretation of these variables changed[7] in the 2,000-year passage from the Greek picture of a polyhedron to chains of simplices and the notion of Euler characteristic in topology. The test of a definition is how it performs in proving the result. One may avoid 'unintended counterexamples' with such adjustments to the proof/definition as *exception-barring*, or *barring or adjusting to monsters*.[8] The role of the conjecture in the foundation of algebraic topology supports Lakatos' characterization of the process as 'concept-formation'.

In her case studies of the development of the notion of randomness in ergodic theory, Werndl [Werndl 2009] introduces three notions of definition justification to clarify Lakatos' 'proof-centered': natural-world justification ('capturing a pre-formal idea for describing the natural world'), condition-justification (proved 'equivalent in an allegedly natural way to a previously specified condition which is regarded as mathematically valuable'), and redundancy-justification[9] ('eliminates at least one redundant condition in an already accepted definition'). In all these cases, she evaluates the role of a definition in the context of a more general developing mathematical theory. Rather than continue constructing a list of examples, we want to focus on a particular strategy advocated by Shelah for making definitions: choosing dividing lines. We will see that some of Werndl's notions aptly describe aspects of Shelah's project.

To put our investigation of the role of classification techniques in model theory in context, we first discuss the classification of structures in other areas of mathematics. We exhibit the versatility of the dividing line strategy in Chapter 13.2 and probe its advantages and limitations in Chapter 13.3.

---

[6] Lakatos introduces the term in a footnote ([Lakatos 1976], 122). Werndl [Werndl 2009] gives a more detailed analysis.

[7] We provide a similar analysis for the notion of number in Chapter 9.

[8] The Hrushovski–Zilber study (Chapter 13.1) of Zariski geometries to regain the trichotomy conjecture (page 307) is a textbook example of monster-barring.

[9] A nice example in model theory is the elimination of the adjective 'excellent' from the notion of quasiminimal excellent class by [Bays et al. 2014]; see Chapter 7.1.

13.1 Methodology of Classification

We began Chapter 5.5 with Gowers' account of the role of classification in mathematics and with some basic examples of the classification of mathematical structures. Here, we consider several kinds of classification *programs* in mathematics:

(1) classification of structures: groups, rings, fields;
(2) classification of structures with further (usually) second order properties: topological groups, Lie groups, manifolds;
(3) classification of definable sets: real or complex algebraic varieties.

Consider for example the classification[10] of finite groups. We described on page 150 the reduction of the classification problem for finite groups to understanding the extension problem[11] and to the classification of finite simple groups – one of the most lengthy proofs in mathematics. That classification proof employs a dividing line strategy. The first dividing line in the classification of finite simple groups is the Feit–Thompson odd order theorem.[12] Does the group have odd order? If yes, it is solvable; thus it is not simple and we finish. If not, the analysis begins.[13]

One should note the solution of this problem splits the simple groups into four classes of finite groups, which are parameterized by (finite) cardinal invariants (analogous to the structure theorems in model theory) plus 26 exceptional finite simple groups.[14] Making this analogy precise is a far distant dream (page 289). Because of the higher order nature of such structures,[15] the classification of differentiable, topological groups, etc., seems at first glance to be a completely different subject from the model

---

[10] See also Chapter 6.1.

[11] The extension problem asks: given two groups H and Q, find all groups G such that $G/H \approx Q$.

[12] The Feit–Thompson theorem (published in a 150 page paper in 1963) launched the program for classifying finite simple groups; it asserts every finite group of odd order (cardinality) is solvable.

[13] Here, Cherlin has pointed out by email, 'I think just one level to it [the analysis]: semisimple vs. unipotent.' This comment reveals the essence of a mathematical classification problem. The investigators knew that they were aiming at a collection of groups somewhat larger than a family of matrix groups. 'Semisimple' and 'unipotent' are standard adjectives about matrices over a field. But at this stage in the classification, the notions can only be defined using group theoretic vocabulary. However, since there are cases (e.g. the exceptional groups) that are not matrix groups, the group theoretic notion must be a generalization of the matrix notion which is strong enough to enable classification but still cover the outlying cases.

[14] The first two classes are (i) simple cyclic abelian groups (parameterized by p) and (ii) the alternating groups A_n. The third-class Lie-type groups, which include Chevalley groups (page 48), fit better in a model theoretic context. Each is a matrix group over a finite field fitting one of a finite number of patterns. The fourth class contains the 26 exceptional finite simple groups.

[15] There is quantification over the set of open *subsets* of a space X.

theoretic program. But sometimes the topology is definable (for example, many o-minimal structures). See *definable analysis* in Chapter 6.3. In studying Zariski geometries Hrushovki and Zilber (page 307) introduce a 'topological model theory'. This is not a new logic for studying topologies; it is using topology to get more information about a structure. Definability remains a central tool for this project.

Finally, there is the classification of definable sets (varieties of various sorts) in real and complex algebraic geometry. Here the notion of 'isomorphism' is 'birational correspondence'. The exact meaning of this term is not important here; the key point is that the 'birational correspondences' the geometer studies are *first order definable* in the ambient structure. This kind of situation underlies much of 'algebraic' model theory.

One of the key points of the stability approach is an entirely new notion of decomposition. Rather than build up the larger models via such operations as direct or subdirect product or group extension, each model is decomposed into a tree of very similar (i.e. elementarily equivalent) but smaller models. The goal is not the analysis of models, perhaps of the same size, with 'simpler', i.e. indecomposable, structures. Rather, models of arbitrary cardinality are realized as a direct limit of a tree of countable models, where each extension is governed by a clear notion of generated models ('prime over', page 82) and the structure of the countable models is taken as given.

An amusing example of proof-centered definition is the evolution of the name for the concept of dependent extension in stability theory. Shelah successively defined: splitting, strong splitting, dividing. All of these words describe a 'bad behavior' of an extension of a type; more information has been introduced. That is, it is 'good' when p is a non-splitting extension of q. When he finally came on the right notion he asked Chang for another English[16] word in this family. Chang suggested forking and that stuck. However, when Lascar and Poizat approached the concepts, they looked first at the positive side and thinking of an extension as a 'fils' (son), the non-forking extension became the 'fils aîné'. And Harnik translated 'fils aîné' as heir. I mention this string of names to emphasize that finding a name for a concept is a crucial part of making the definition. Finding an evocative short phrase to sum up a concept forces the author to inquire in a non-technical way into the 'main idea' that is being expressed and to consider the relationship with similar already known concepts. Co-authors often have long and sometimes contentious discussions of this issue.

---

[16] He had one in Hebrew.

13.2 The Fecundity of the Stability Hierarchy

We now exhibit applications of the stability hierarchy to problems far from the original test problem, computing the spectrum function (Chapter 5.5).

Finite axiomatizability: The techniques of stability theory also address a basic philosophical problem. How can one describe infinity? More precisely, how can a first order sentence prescribe that its models are infinite? At first glance stability theory should have little to do with characterizing such sentences. Much of stability theory seems to deal specifically with *un*countable models. Can those methods shed light on how a sentence requires all models to be infinite?

W. W. Tait pointed out to me three basic intuitions about how infinite sets arise: linear order (dense or unbounded), successor, and the observation that a finite number is smaller than its square.[17] Are there other such intuitions?

Here are three theories that have only infinite models.

(1) $<$ is a dense (or an unbounded) linear order.
(2) $t(x, y)$ is a pairing function from $M \times M$ onto M.
(3) $f(x)$ is a bijective function; exactly one element does not have a predecessor; there are no n-cycles.

The first two of these are expressed by a *single* first order sentence; i.e. they represent a *finitely axiomatized* theory. The third is not; can we understand the difference? To sharpen the problem and provide more tools for analysis require that the theory be complete. Dense linear order is nearly a complete theory; it becomes a complete finitely axiomatized theory by deciding whether there are end points. Every extension of the theory of (infinite) linear order is unstable; so none is $\aleph_1$-categorical; but dense linear order (with the number of end points fixed) is $\aleph_0$-categorical. There is an extension of a sentence describing a pairing function to a complete theory[18] that is strictly stable but none that is superstable. The third theory is a complete $\aleph_1$-categorical theory as given but requires infinitely many axioms. If the completeness hypothesis is strengthened to $\aleph_1$-categoricity, the answer to the finite axiomatizability problem (posed in [Morley 1965a]) is known but represents two major projects distinguished by $\aleph_0$-categoricity.

---

[17] See footnote 40 of Chapter 2.4.
[18] Lachlan's example is presented in [Baldwin & McKenzie 1982].

Theorem 13.2.1 (Zilber, Cherlin–Harrington–Lachlan) *No first order* sentence *with only infinite models is categorical*[19] *in all infinite cardinalities.*

Nevertheless, such theories are *quasi-finitely axiomatizable*; that is, axiomatized by a single sentence plus an 'infinity scheme' ($\exists^{\geq n}x$ for each n) and there is detailed structure theory for *both* finite and infinite models [Cherlin & Hrushovski 2003]. But, for finite axiomatizability, the total categoricity is essential.

Theorem 13.2.2 (Peretyatkin) *There is an $\aleph_1$-categorical first order* sentence *with no finite models.*

Peretyatkin [Peretyatkin 1997] was motivated by trying to capture a tiling problem[20] but his example really seems to capture 'pairing'. The following remains open: Is there a finitely axiomatizable strongly minimal set?

Connecting the Finite and the Infinite: The proof of Theorem 13.2.1 ([Zilber 1984b, Zilber 1991, Cherlin et al. 1985] engendered an analysis of countable structures that can be approximated by finite structures. A crucial point for Zilber's axiomatizability result is that if two strongly minimal sets are not orthogonal and each is modular (Zilber's trichotomy, 136) then they are not weakly orthogonal (Definition 5.6.3). That is, the dependence of the types is recognized without adding parameters. Zilber's lemma that in a *totally* categorical theory, all strongly minimal sets are modular, makes the previous result available for showing each sentence with parameters from a finite subset A that is true in an infinite totally categorical structure is true in a finite substructure (the *envelope* of A).

This result spurred Lachlan's program to classify finite homogeneous[21] structures [Cherlin 2000]. He obtained a system of invariants for the finite structures (essentially the envelopes) patterned on that for infinite structures in Chapter 5.5 and wrote the infinite structure as a direct limit of them. This pursuit was generalized by [Kantor et al. 1989] to the notion of a *smoothly approximable structure*. Rather than defining it, we locate this notion among other relevant concepts. Any smoothly approximable structure M has the finite submodel property.[22] Further any ω-stable

---

[19]   The impetus for this theorem is due to Zilber [Zilber 1984a]; the three authors
     [Cherlin et al. 1985] filled a gap in his first proof by appealing to the classification of finite
     simple groups; Zilber (and Evans) [Evans 1986, Zilber 1991] later filled the gap by deducing
     the necessary finite group theory without passing through the classification (Chapter 5.4).
[20]   That is, he wanted to finitely axiomatize an $\aleph_1$-categorical theory of a finite set of tiles which
     give an aperiodic tiling of the plane.
[21]   A finite structure is homogeneous if any two isomorphic finite substructures are automorphic.
[22]   If $M \models \phi$ then for some finite submodel $N \subset M$, $N \models \phi$.

ω-categorical structure is smoothly approximable and M is smoothly approximable if and only if M is *Lie-coordinatizable* [Cherlin & Hrushovski 2003]. The last technical condition is the culmination of a profound connection among model theory, the classification of finite simple groups and finite geometries. Moreover, the finite submodel property is characterized by properties of the geometry induced on the finite substructures in [Koponen 2006]. These structures along with the *random graph* are ω-categorical and simple. [Macpherson & Steinhorn 2011] define an *asymptotic class* as a class of finite models in which the number of solutions of a formula $\phi(x; a)$ in a finite model M can be uniformly approximated as $\mu M^d = N$ where N is a parameter of the class and μ, d are uniformly defined depending on a. This generalizes results of [Chatzidakis et al. 1992] on finding the number of solutions of Diophantine equations in finite fields. Defining a *measurable structure* as one having the properties of an ultraproduct from an asymptotic class, later work by various authors shows a measurable structure is supersimple of finite rank and a stable measurable structure is 1-based. This provides a framework to try to explain the families of finite simple groups in terms of their definability. A precise conjecture is that every measurable simple group (i.e. limit of a family of finite simple groups) is a (twisted) Chevalley group over a pseudofinite field.

In this case, the solution of a natural philosophical problem led to deep investigations of a related problem, how does one pass from the finite to the infinite?

Quantifier Reduction and Computability Theory: Model theorists usually regard stability theory methods as orthogonal to questions of decidability (page 132). However, there is a large literature in *computable model theory*. Until about 2000, this subject primarily[23] used methods of 'Vaughtian model theory'. The deep connection between the two areas in [Goncharov et al. 2003] illustrates the wide applicability of classification theory. The theorem in the title, 'Trivial strongly minimal sets are model complete after naming constants', shows an unexpected connection of the geometry of strongly minimal sets with the quantifier hierarchy: triviality of the geometry (page 136) implies every formula is equivalent to an existential formula (Theorem 4.4.2). The conclusion from the eponymous theorem[24] is even more unexpected: 'We conclude that all countable models of a trivial, strongly minimal theory with at least one computable

---

[23] Andrews' [Andrews 2011] use of the Hrushovski construction is a prominent counterexample.
[24] See the website of Steffen Lempp: www.math.wisc.edu/~lempp/papers/list.html.

model are $0''$-decidable, and that the spectrum of computable models of any trivial, strongly minimal theory is Σ_5^0.' Now, the geometric consideration of triviality limits computational complexity of the class of *computable models*.

Another unexpected application is to the notions of resplendency and recursive saturation [Barwise & Schlipf 1976, Kossak 2011]. Recursive saturation (page 249) requires that all recursive types over the empty set are realized. The 'recursive' restriction on the kind of type is totally foreign to the usual model theoretic standpoint. And resplendency arose as a kind of generalized recursive saturation. Nevertheless Poizat points out in his review of [Baldwin 1990] (my italics) that the recursive aspect disappears under appropriate stability hypotheses:

In the first section the author shows that an $\aleph_1$-homogeneous resplendent model of an unstable, or stable nonsuperstable, or small superstable non-omega-stable theory is weakly saturated. This is a generalization of a result of Julia Knight for the unstable case, which indicates that, *in these contexts, the recursive flavor which is attached to the notion of resplendency is neutralized.* [Poizat 1991]

Other Notions of Complexity: Applications of the first order stability classification have extensions well beyond first order logic. In the introduction to [Shelah 2009a], Shelah provides several examples (Laskowski, Shelah, Hyttinnen, Tuuri, Väänänen, etc.) that concern the difficulty of finding invariants defined in various ways (different logics, games, etc.) that characterize models. Slightly different is the Baldwin–Shelah classification of first order theories by monadic logic and the applications by Baldwin–Benedict of the stability classification to a very abstract theory of databases. Shelah makes further attempts to apply this program in the study of the categoricity problem for abstract elementary classes (e.g. by introducing the further dividing line 'smoothness' [Jarden & Shelah 2013]). Chapter 7.3 shows the totally unexpected connection between categoricity in infinitary second order logic and first order stability theory.

Malliaris and Shelah have recently made remarkable advances in the study of the Keisler order[25] which have had very surprising consequences. Keisler's order preceded Shelah's classification theory. But, partly because of its clear syntactic content, Shelah's stability classification became the central

---

[25] For complete countable first order theories T_1, T_2, we write $T_1 <_\kappa T_2$ if for any $A_1 \models T_1, A_2 \models T_2$, and regular ultrafilter D on κ, if A_2^κ/D is κ^+-saturated if A_1^κ/D is κ^+-saturated. Moreover $T_1 < T_2$ if for every infinite cardinal κ, $T_1 <_\kappa T_2$. Keisler showed this notion is well-defined. The striking results surveyed here are described in more depth in [Keisler 2017].

model theoretic tool. Shelah [Shelah 1978] showed that all countable stable theories fell into two classes under the Keisler order and found three additional classes. For over 30 years, no more were found.[26]

Malliaris [Malliaris 2009] showed that, like the stability hierarchy, the Keisler order reduces to syntactic properties of single formulas. She assigned to each first order theory T a 'characteristic sequence' of hypergraphs and so translates classification of theories into classification of the hypergraphs. In particular, each of the stability theoretic notions: order property, independence property, tree property, SOP_2 has such a translation. Tools from graph theory around Szemerédi's study of regular graphs can be applied to the hypergraph side of this correspondence. This insight provides a powerful link between model theory and sophisticated finite combinatorics. Thus, there is a new dividing line between simple theories and those with the strict order property in terms of a 0–1 law-like property of the characteristic sequence. An important methodological principle arises: the Keisler order really establishes a correspondence between syntactic properties of theories and the fine structure of ultrafilters. Malliaris and Shelah [Malliaris & Shelah 2013, Malliaris & Shelah 2016] have used this correspondence in both directions. By detailed combinatorial work they established the first dividing line for the Keisler order within simple theories, lowness. In the other direction, they give a model theoretic proof of the striking result that SOP_2 (the 2-strict order property) is a sufficient condition for a theory to be maximal in the Keisler order. Then in an astonishing turn of events, this model theoretic analysis solves a 1948 problem, showing the cardinal invariants of the continuum[27] $\mathfrak{p}$ and $\mathfrak{t}$ are equal. This is particularly impressive as the other conjectured relations among cardinal invariants that remained open for any length of time were invariably shown independent of ZFC. Malliaris and Shelah [Malliaris & Shelah 2015] have shown, contrary to expectation, that there are countably many different classes in the Keisler order.

Thus, the first order stability example of finding dividing lines, apparently centered on a narrow question of counting the number of models, is much more fruitful. As we argue throughout the book, the counting is just a test question for whether structure theorems are adequate. The examples

---

[26] In [Shelah 1990], Shelah says, 'It would be very desirable to prove (by a particular specified strategy) that there are 5 classes. This will complete the model theoretic share of investigating Keisler's order for countable theories.'

[27] This subfield of set theory investigates the relations among cardinals that are between $\aleph_0$ and $2^{\aleph_0}$ and defined as invariants of various structures on the reals. A typical such invariant is the least cardinality of a non-measurable set.

in this section show the effect of the first order stability hierarchy on finite axiomatizability, applications to quantifier complexity and complexity of axioms, counting models in more expressive logics, other tools for classifying first order theories, and links with combinatorial graph theory. This wide applicability of the stability hierarchy fulfills one of the four criteria for successful dividing lines discussed in the next section.

13.3   Dividing Lines

Chapter 5.5 looked at the scheme of definitions comprising the stability hierarchy and how they resolved the main gap problem. Now we are considering Shelah's *strategy* in drawing up the scheme. Shelah's apologia in the section 'Why to be interested in dividing lines?' ([Shelah 2009a] 2–9) propounds four theses:

(a) *It is very interesting to find dividing lines and it is a fruitful approach in investigating quite general classes of models.* The initial claim is somewhat restricted. The investigation is of classes of models. The 'dividing lines' that Shelah exhibits are the syntactic ones we have discussed in Chapter 5.5 that apply when the classes are collections of first order theories. The use of syntactic properties of *formal* theories distinguishes these dividing lines from, say, the classification of simple groups. But the overall strategy is similar to the group theory program. Later in the same book, when the classes are defined in purely mathematical terms, Shelah[28] introduces semantic versions of the order property (i.e. ordering a sequence of Galois types – orbits of the automorphism group of the monster model).

We quoted Shelah's explanation of the method of dividing lines for first order theories (page 61). A *dividing line* is a property such that both it and its negation have strong mathematical consequences. It is here that the seemingly technical theorems showing many equivalences of the notion of stable show their power; one form of the negation is used to find many models; one positive form gives prime models, another a notion of rank, a third leads to the exchange property for the notion of independence ...

A canonical example of a *non-dividing line* is linear ordering. If a theory admits a linear order then it is certainly unstable and so its models cannot be characterized by invariants as in Chapter 5.5. But, failing to be linearly ordered gives little information. Models of such ill-behaved theories as set theory are not linearly ordered. It is the more subtle notion, 'not the order property', with its consequences bounding the size of Stone spaces, that is

---

[28] See Chapter 14.

a useful dividing line. The balance point both proves many models in one direction (Chapter 8.3) and gets a nice dependence relation in the other. Such choices are motivated by his second thesis.

(b) *It is desirable to have an exterior,*[29] *a priori, test problem.* Exterior problems, for example, the Morley conjecture that the number of models of a theory increases with cardinality, gave a specific goal to focus the work.

So the direct aim was to solve the test question (e.g. main gap) but the motivation has always been the belief that solving it will be rewarded with worthwhile dividing lines and developing a theory for both sides of each. ([Shelah 2009a], 5)

Already, the main gap is a more structural criterion than just proving the spectrum function is increasing. The main gap specifies the line between having and not having invariants (Chapter 5.5). Here his criterion echoes Lakatos' concern with 'proof-centered' definition.

Thus, the real problem is not 'counting models'; it is to determine how syntactical properties of a complete theory influence the structure of their models. Being able to count the models is an external question that can be used to determine whether the assignment of invariants is adequate.

(c) *Suggested dividing lines will throw light on problems not considered when suggesting them.* We discussed this criterion in detail in Chapter 13.2, exhibiting many applications of the stability hierarchy dividing lines to unexpected topics.

(d) *Non-structure is not so negative.* At some steps in the solution of the Morley conjecture, the following phenomenon occurs. In order to make the next dichotomy within the structure side of a dividing line, further refinement of the structure theory may be necessary to find a property whose negation implies that classes have the maximal number of models. One of the most complicated such arguments is Shelah's solution of the Vaught conjecture for ω-stable theories and refinements by Buechler and Newelski for superstable theories of finite U-rank.

Moreover, it may be possible to transfer non-structure in one situation to structure in a second. Thus Göbel and Trlifaj [Göbel & Trlifaj 2006] give representations of certain rings as rings of automorphisms, applying the black-box for non-structure arising from Chapter VIII of [Shelah 1978].

Although having the maximal number of models prevents classification by the kinds of invariants considered in Chapter 5.5, there may be more refined invariants. Thus, even after resolving Vaught's conjecture for a particular theory one might investigate the Borel complexity of isomorphisms of models of T (Chapter 7.2).

---

[29] Compare with the discussion of Vaught's conjecture as a test problem in Chapter 7.2.

Shelah's four theses illustrate the analysis of Lakatos, Tappenden, and Werndl. The word 'fruitful' in thesis (a) reminds us of Tappenden and fruitfulness is measured by mathematical consequences; thesis (b) gives a general prescription for generating 'proof-centered' definitions; thesis (c) emphasizes the fruitfullness further by valuing unexpected consequences. The value ascribed to test problems shows that the notion of proof-generated definition is conscious in the design of the definition. The somewhat defensive condition (d) still draws on the fruitfulness of the definition. There is no explicit mention of naturality, but since we took natural as appropriate, the demand that dividing lines have both positive and negative consequences is a kind of naturality – the definition must be appropriate to enable the proofs.

It is not pretended that the 'dividing lines' are always an important criterion for choosing definitions, nor that 'dividing lines' only make sense for classifying models.

Probably, the most important innovation of the entire stability analysis is the choice of level; the goal is to produce properties of *classes of theories* which enable (on the positive side) the development of a structure theory. The study of o-minimality and ω-stability, especially groups of finite Morley rank (Chapter 6.1), build on the lesson of the last sentence. Isolate a class of theories that have good properties; but seek structure not only in terms of few models or a structure theorem of the models but in terms of the structure of definable sets. We quoted in Chapter 6.3 Wilkie's careful analysis of the mathematical reasons for adapting the notion of o-minimality. He emphasizes the fruitfulness of a notion having consequences on spaces of all finite dimensions; this is accomplished for o-minimality by the cell-decomposition theorem. The prescience in choosing this definition is witnessed by the great fruitfulness of the concept. But not every useful line is a dividing line. As often (perhaps usual) in mathematics, the failure of o-minimality has no useful consequence and a number of attempts to generalize the notion are ad hoc. Similarly, the only direct consequence of failure of ω-stability is many models in $\aleph_1$.

13.4   Definition, Classification, and Taxonomy

We now place Shelah's dividing line methodology in both a mathematical and a philosophical context. Gregory Cherlin[30] pointed out an interesting

---

[30] I thank him and David Pierce for conversations on this point.

analogy between Shelah's method of dividing lines and Plato's analysis of classification in *The Statesman*. Jowett[31] writes, 'the dialog might have been designated by two equally descriptive titles – either the "Statesman", or "Concerning Method". Dialectic, which in the earlier writings of Plato is a revival of the Socratic question and answer applied to definition, is now occupied with classification.' The most obvious distinction between definition and classification is that a definition aims at one concept while classification concerns the partitioning of a collection.[32] In this dialog the main role is taken by the Visitor and Young Socrates (the child of Socrates) is his foil.[33]

Stranger. We must not take a single small part, and set it off against many large ones, nor disregard species in making our division. [262b] On the contrary, the part must be also a species. It is a very fine thing to separate the object of our search at once from everything else, if the separation can be made correctly, and so, just now, you thought you had the right division and you hurried our discussion along, because you saw that it was leading towards man. But, my friend, it is not safe to whittle off shavings; it is safer to proceed by *cutting through the middle*, and in that way one is more likely to find classes. [262c] This makes all the difference in the conduct of research.

To explain the phrase 'cutting through the middle,' Plato makes a subtle distinction between a 'part' and a 'class'. Essentially a 'class' is a useful cut, while a 'part' just picks out some of the objects. He gives a suggestive example of a part which is not a class: the 'Hellene–Barbarian' distinction. Nothing is really known of the Barbarian except that he is not Hellene. To be a class,[34] something significant must be said about both the members and the nonmembers. Coming from a mathematical perspective, Shelah sees the same point, his analog to Plato's *class* being a *dividing line*.

Plato proposes[35] to 'seek for the statesman' by successive approximations that list properties of 'the statesman', with the approximating properties chosen by 'cutting through the middle.' Shelah proposes finding a

---

[31] See II of the introduction to [Plato 2016].

[32] Evidence that this dialog is concerned with classification rather than definition is the joke told by Socrates at the beginning of the dialog comparing the relative value of statesman, philosopher, and sophist, 'I mean that you rate them all at the same value, whereas they are really separated by an interval, which no geometrical ratio can express' (translation in [Plato 2016]).

[33] Lines 262b and 262c of Fowler's translation are displayed along with the Greek at [Plato 1999].

[34] Quine makes essentially the same point [Quine 1977] in his discussion linking Hempel's puzzle of the non-black non-ravens with Goodman's grue emerald; in our terminology non-black, non-raven, and grue are not dividing lines.

[35] Line 258b of Fowler's translation [Plato 1999].

classification by successive approximation by listing properties of 'a *classifiable theory*'. In the mathematical context, the metaphorical 'cut through the middle' is replaced by the requirement that each separating criterion *and its complement* must, in the language of Chapter 2.3, be a *virtuous property*.[36]

Considering this analogy, we now elucidate the distinctions among definition, classification, and taxonomy; we will see that the interaction of reliability (proof) and clarification (definition) arises naturally in seeking to understand these concepts in the mathematical context. Coffa places the relationship between 'reliability and clarity', discussed in the Introduction and in Chapter 5.2, in historical perspective:

[We consider] the sense and purpose of foundationalist or reductionist projects such as the reduction of mathematics to arithmetic or arithmetic to logic. It is widely thought that the principle inspiring such reconstructive efforts was epistemological, that they were basically a search for certainty. This is a serious error. It is true, of course, that most of those engaging in these projects believed in the possibility of achieving something in the neighborhood of Cartesian certainty for principles of logic or arithmetic on which a priori knowledge was to be based. But it would be a gross misunderstanding to see in this belief the basic aim of the enterprise. A no less important purpose was the clarification of what was being said. ...

The search for rigor might be, and often was, a search for certainty, for an unshakable 'Grund'. But it was also a search for a clear account of the basic notions of a discipline. ([Coffa 1991], 26)

We begin with classification. The classical sense of 'classification' is given by Merriam-Webster[37] as 'systematic arrangement in groups or categories according to established criteria.' This definition of 'classification' immediately raises the question of how such criteria are established. Plato's example approximates by finding successively what are judged important criteria of similarity. But similarity is a notoriously slippery notion. Murphy and Medin[38] observed, 'Suppose that one is to list the attributes that plums and lawnmowers have in common in order to judge their similarity. It is easy to see that the list could be infinite: Both weigh less than 10,000 kg (and

---

[36] The mathematical context evades such problematic issues as critiquing Plato's definition of man as a featherless biped by exhibiting a cock shorn of its feathers or wings.

[37] www.merriam-webster.com/dictionary/classification.

[38] ([Murphy & Medin 1985], 292–294). They also distinguish a probabilistic from a classical view. 'The classical view has it that categories are defined by singly necessary and jointly sufficient features.' While they (properly by my lights) object to that view in general, within mathematics I hold that it is correct; we can give explicit definitions of Shelah's approximating dividing lines.

less than 10,001 kg, . . .), [etc., etc.].' They compare the 'similarity-based' approach to concepts with the 'theory-based' approach and fill out Quine's rather vague reference [Quine 1977] to 'scientific theory' as a supplement to similarity. In elaborating the theory-based approach they emphasize the role of 'explanatory' and 'underlying' principles as adjuncts to 'correlated attributes' in determining classes.

The difference between 'similarity-based' and 'theory-based' classifications is illustrated well by competing justifications for the 'tree of life'. Classical morphology defines species by physical similarity; twentieth century biology relied on 'ability to interbreed'; DNA sequencing[39] is state of the art. Increasing amounts of 'theory' are needed justify the classification.

In the mathematical context the notion of 'natural kind' in an everyday sense is nonsensical; we are examining technical notions. Nevertheless there are explanations for why certain definitions are made. The connection of reliability and clarity arises as we examine an unfortunate double use of the word 'classify' in our description above. The first is the same as Gowers' (Chapter 5.5) description of the classification of structures of a particular type (manifolds, groups, etc.) via attaching invariants. Secondly, Shelah classifies theories via a hierarchy of kinds (of first order theories) that determines whether the models of a theory of a particular kind admit a classification in a precise version of Gowers' sense.

Plato aims at finding the definition of a single concept. He assumes some sort of natural kind. Such an assumption is not necessary in mathematics but it is psychologically valuable. Fields medalist Langlands wrote,[40] 'it [the understanding of mathematics] often comes in the form of intimations, a word that suggests that mathematics, and not only its basic concepts, exists independently of us. This is a notion that is hard to credit, but hard for a professional mathematician to do without.'

Shelah starts with a well-defined concept. That is, he has a working definition of a classifiable theory: a theory T is classifiable if there is an algorithm for attaching invariants to each model of T that determines it up to isomorphism. His goal is to find a more accessible definition that allows the implementation of the ideal that the original version set. However,

---

[39] DNA sequencing relies not just on amount of shared DNA, measured by statistical cluster analysis, but also on molecular phylogeny, which uses such data to build a 'relationship tree' that shows the probable evolution of various organisms. https://en.wikipedia.org/wiki/Molecular_phylogenetics.

[40] 'Is there beauty in mathematical theories?' Robert P. Langlands, University of Notre Dame, January 2010. http://publications.ias.edu/sites/default/files/ND.pdf.

Shelah's methodology has a more general form. Rather than merely aiming at the class of 'classifiable theories', the following template describes the generalization to the task of defining a fixed concept X. Plato's strategy is to construct a sequence of approximations X_n to X, requiring that each X_n is a dividing line. Shelah's strategy refines Plato's by requiring these virtuous properties chosen at each level must further the ultimate goal. We can represent this symbolically by not just a sequence of approximations X_n to X but a series of refined reasons for the failure of X: if X_{n-1} holds but X_n does not then X fails. Thus, the final step is an example of Werndl's 'condition-justification'[41]; we have found an equivalent condition to X. But the intermediate steps are chosen to attain this goal so they exemplify Lakatos' 'proof-centered definition'.

Shelah's procedure is better understood as a strategy for constructing a taxonomy in the following sense:[42] '*Taxonomy* is the process of giving names to things or groups of things according to their positions in a hierarchy.'

Recall the precise statement of the main gap theorem. Every model of a superstable, shallow complete theory with NDOP and NOTOP has a system of cardinal invariants and these properties of the theory show how to find those invariants (Chapter 5.5); if the theory does not satisfy these conditions then it has the maximal number of models in every uncountable cardinality. The proofs for unstable and unsuperstable theories have a similar pattern. Using stationary set technology and Ehrenfeucht–Mostowski models, one is able to construct the maximal number of models in each cardinality. Stable and superstable theories may have many models; but these properties imply conditions (e.g. every stable theory has a good dependence relation) that are tools for classifying models. One might think ω-stability should appear in the definition of classifiable. However, ω-stability is not a dividing line. Many non-ω-stable theories have few models and while ω-stability allows the crucial ability to construct prime models over sets, that property is stronger than needed for the classification. The more refined analysis with the dividing lines of NDOP, NOTOP, and still more technical notions in [Hart & Laskowski 1997, Hart et al. 2000] yield actual dividing lines. The process of proving the main gap requires that Shelah (and the later contributors) refine the phrase 'system of cardinal invariants' by more

---

[41] Page 284.

[42] Lalonde gives this definition and also makes the suggestive comparison with the Merriam-Webster definition of classification at www.ehow.com/info_10074596_difference-between-classification-taxonomy.html.

technical descriptions of how the invariants are organized – most simply by a tree but then specifying more properties of the tree.

The main gap theorem gives an example of a taxonomy, a system of definitions to construct a hierarchy; the essential criteria to use a property in defining this hierarchy are: the property of a theory T contributes to a classification theorem for the models of T *and* its negation, along with the currently assumed conditions, force the maximal number of models. Thus the hierarchy defined by the X_n becomes the partition given by $X_{n-1} \wedge \neg X_n$.

We have touched on psychology and philosophy in delineating the distinctions among definition, classification, and taxonomy. In Chapters 4.6 and 12, we discussed definition primarily as making some intuition more precise and even formal. But here we examined a strategy for making definitions in aid of a proof. In particular, this proof required not merely a classification but a hierarchy of theories. Not only is there a relation between the categories but there are explanatory principles that organize the categories.

The key distinction of the main gap setting is the ability and need to integrate 'definition' and 'proof'. That is, definitions of levels in the hierarchy are tweaked to obtain a taxonomy which answers the question: which theories admit a classification of their models?

14 | Formalism-Freeness (Mathematical Properties)

We discussed 'formalization' at length in Chapter 1. What do we mean by 'formalism-freeness'? Kennedy begins with

That mathematics is practiced in what one might call a formalism-free manner has always been the case – and remains the case. Of course, no one would have thought to put it this way prior to the emergence of the foundational formal systems in the late nineteenth and early twentieth centuries. [Kennedy 2013]

Kennedy goes on to explore what Gödel [Gödel 1946] calls the search for absolute notions of computability, provability, and definability. We try to distinguish such practice from the use of logical methods (either for metamathematical studies or as tools for proving mathematical results in the normal sense (in number theory or geometry or graph theory or ...)). Formalism-freeness does not argue against formal methods per se. Here we point to a number of instances where model theorists have studied informal equivalents of fully formalized frameworks (Tarski) or found such frameworks inadequate for certain goals (Makkai–Reyes, Shelah, Hrushovski–Zilber). Our general theme differs from Kennedy's by focusing specifically on the development of formal theories for parts of mathematics.

Formalism-free is a matter of degree. It involves dropping some of the components of Definition 1.0.1. Clause 1 of Definition 1.0.1 is sacrosanct. We deal only with structures of a fixed vocabulary.[1]

To clarify the situation, we distinguish between 'formalism-free' and 'formalism-independent'. We call a definition *formalism-free* if it is given semantically[2] without any formal distinction between syntax and a semantics. For example 'a group is a set and a binary function satisfying ... ' We call an investigation *formalism-free* if it is a semantic study. Thus there are two aspects: definition and investigation.

---

[1] Note the caveat on page 306.

[2] We accept Burgess' critique in [Burgess 2008] that Tarski's use of 'semantics' with no modifier for the study of interpretation of formal languages leads to confusion in other philosophical investigations. However, after four score years the terminology is fixed for the model theory community.

Here are three illustrative examples of formalism-free definition; we study some such investigations in a few pages. We say a notion, such as 'computability', is *formalism-independent* (formalism-free in Kennedy's terminology) because of the known equivalence among the Turing, Markov, λ-calculus, and Gödel characterizations. Only the Gödel characterization as Δ_1^0-definable functions in number theory is completely formal in the sense defined here; Turing machines are formalism-free.

Secondly, as Kennedy has pointed out,[3] if L is constructed by adding relatively definable sets then the usual L is obtained by several different logics, e.g. by iterating either first order definitions or weak second order logic definitions. This represents *formalism-independence*. But the definition of L as the closure of the ordinals under the Gödel functions provides a fully formalism-free definition.

In a third direction, consider the notion of a polynomial. In a standard high school Algebra I book ([Educational Development Center 2009]) a *polynomial* is defined (with no fanfare of syntax and semantics) as a sum of *monomials* where a monomial is earlier defined as a product of variables raised to non-negative integer powers and a (usually real) number coefficient. In this style of development a *polynomial function* is a map (e.g. from $\mathfrak{R}^2 \mapsto \mathfrak{R}$) defined by a polynomial. The interpretation of the formal language in the structure that is fundamental to Tarski's definition of truth is made matter-of-factly in elementary algebra. Then operations on the ring of polynomials is smoothly defined (not named) in a concrete way by example and previous experience with 'combining terms'. In contrast, the definition of a ring of polynomials in Lang's algebra book [Lang 1964] is much more abstract but avoids even a glimpse of the syntax–semantic distinction. Then polynomial functions are defined by a composition of functions. In our terminology, both of these approaches are formalism-free.

This distinction has always been part of model theory. Tarski begins his fundamental work on definable subsets of the real numbers [Tarski 1931] by saying 'Mathematicians, in general, do not like to deal with the notion of definability; their attitude toward this notion is one of distrust and reserve.' He goes on to focus the cause of this attitude on the failure to fully specify dependence of definability on the choice of primitive terms and particular model.[4] Then he challenges this attitude:

---

[3] This is a sampling from a much more detailed account in [Kennedy 2013, Kennedy et al. 2016].

[4] See also [Feferman 2008b] where Feferman discusses Tarski's attitude toward mathematics and metamathematics. The distinction between formal/logical I discuss here is a separate issue from the controversy Feferman addresses over Tarski's concept of a 'logical notion'.

I believe that I have found a general method to construct a rigorous metamathe-matical definition of this notion. Moreover, by analyzing the definition obtained it proves possible[5] to replace it with a definition solely in mathematical terms. Under this new definition the notion of definability does not differ from other mathematical notions and need not arouse either fears or doubts; it can be discussed entirely within the domain of normal mathematical reasoning. [Tarski 1931]

Vaught summarizes the definition nicely in his historical account:

In the same paper [Tarski 1931], Tarski first made a definition of truth, or rather of the related notion of definability. Moreover, he did so without mentioning a formal language, by a method to which he returned in [Tarski 1950]. Indeed, he defined the class of definable relations over $(\mathfrak{R}, +, *)$, for example, to be the smallest class of finitary relations over $\mathfrak{R}$ containing the ternary relation $(x, y, z) : z = x + y$ and $(x, y, z) : z = x \cdot y$ and closed under the Boolean operations and the 'geometrical' operation of projection. [Vaught 1986]

In his work foreshadowing the notion of *o*-minimality, van den Dries [Dries 1986] took up this idea and called a class of sets, which is closed under the operations Vaught describes, a Tarski System. In the introduction to [Dries 1999] Tarski's name has disappeared in this context but several examples of 'Tarski Systems' are given to motivate the general definition of an o-minimal structure.[6]

In the opening paragraph of what might be viewed as the founding paper[7] of model theory, Tarski writes,

Every set Σ of sentences determines uniquely a class K of mathematical systems. ... Among questions which arise naturally in the study of these notions, the following may be mentioned: Knowing some structural (formal) properties of a set Σ of sentences, what conclusions can we draw concerning the mathematical properties of the correlated set of models? Conversely, ... [Tarski 1954]

Tarski takes his readers to understand the notion of 'mathematical prop-erty'. His sample theorems make his meaning plain – the ordinary stuff of mathematics: subalgebra, homomorphisms, direct products, etc. Tarski gives a number of examples of answers to questions of this sort; two are:

Tarski A class K of structures in a finite relational language is axiomatized by a set of universal sentences if and only if K is closed under

---

[5] He says 'with some reservations'. They are that there is no uniform definition but only one for formulas of quantifier rank n for each n.

[6] In fact the syntactical notions are not introduced. In ([Dries 1999], 3) a set $A \subset M^n$ is said to be *definable* if it is in one of these collections $S_n \subset \mathcal{P}(M^n)$, closed under the natural operations.

[7] Vaught [Vaught 1986] says this is the first time the term *model theory* appeared in print.

isomorphism, substructure and if for every finite substructure B of a structure A, $B \in \mathbf{K}$ then $A \in \mathbf{K}$.

Birkhoff A class $\mathbf{K}$ of algebras is axiomatized by a set of equations if and only it it is closed under homomorphism, subalgebra, and direct product.

Thus formal properties of the axioms for a class of models are shown equivalent to the mathematical property of the class being closed under certain basic mathematical operations. This notion of 'mathematical property' is similar to that which Kennedy [Kennedy 2013] traces as the notion of 'formalism-freeness' in the works of Gödel. She writes, 'one can think of indifferentism[8] or formalism-freeness ... as the simple preference for semantic methods,[9] that is methods *which do not involve or require the specification* of a logic – at least not *prima facie*.' The distinctions in Chapter 1.3[10] clarify this notion. It is tempting to speak of language here. We have avoided the word 'language' because of its several usages including the three we listed on page 44. Kennedy is making this distinction. That is, a formalism-free approach would take language in the first sense given on page 44, what we have termed the vocabulary, not the second or third. An inquiry can be 'formalism-free' while being very careful about the vocabulary but eschewing a choice of logic (in the sense of Definition 1.3.1) and in particular any notion of formal proof. Thus it studies mathematical properties in the sense we quoted from Tarski above.

It is in this sense that certain work of Makkai-Reyes, Zilber, Hrushovski, and Shelah can be seen as developing a formalism-free approach to model theory. Makkai and Reyes [Makkai & Reyes 1977] give a 'formalism-free' presentation of syntax.[11] That is, they associated to each first order theory T in a vocabulary τ a category $\mathbb{T}$, whose objects are (essentially) the τ-formulas such that for each $M \models T$, the function $\phi(\mathbf{x}) \mapsto \phi(M)$ extends to a functor $\mathbb{T} \mapsto \mathbb{S}et$. They define the notion of a *Boolean logical category* so that each such category is equivalent to $\mathbb{T}$ for some T. Thus the syntax is represented in category theory; each theory becomes a category and the

---

[8] I regard Burgess' notion of indifferentism to identity ([Burgess 2010], 9) as a component of formalism-freeness. Indifferentism seems to me to refer to working with structures up to isomorphism rather than caring about the set theoretic construction. Here we take that modus operandi for granted and consider how one is to describe the relations between structures.

[9] Kennedy has later emphasized that is a characteristic of formalism-freeness, not an evaluation of it.

[10] Our articulation of them here was partially motivated by Kennedy's work.

[11] My summary here is largely based on the clear basic exposition in [Harnik 2011].

models of T become a functor category. While motivated by the syntactic–semantic distinction, the presentation is in terms of a pair of categories.

In contrast, Zilber's notion [Zilber 2005b] of a quasiminimal excellent class, the Zilber–Hrushovski [Zilber 2005b] notion of Zariski geometry, and Shelah's concept [Shelah 2009a] of an abstract elementary class each gave axiomatic but mathematical definitions of classes of structures in a vocabulary τ. That is, the axioms are not properties expressed in some formal language based on τ but are mathematical properties of the class of τ-structures and some relations on it.

The last three are examples where difficulties in the normal (from a logician's viewpoint) syntactic approach to a problem led to more semantic methods. The notion of abstract elementary class arose from the difficulties of dealing with the syntax of infinitary logics. The failure of compactness leads to an inability to develop the usual model theoretic tools of stability theory. And underlying similarities among different logics are obscured by technicalities about the meaning of sentence and, more critically, type. In contrast, the examples of Zilber and Hrushovski–Zilber arose from specific mathematical problems: the desire to understand the complex numbers with exponentiation and trying to show all strongly minimal structures are 'canonical'. One difficulty involves the failure of first order model theoretic formalization to catch a key idea of algebraic geometry: positive formulas.[12]

In Shelah's case, the basic relation is a notion of 'strong submodel' relating the members of a class of structures. The axiomatic properties of 'strong submodel' summarize the properties of elementary submodel in first order and various infinitary logics as well as some unexpected contexts. Quasiminimal excellent classes require a combinatorial geometry on each model with specified connections with the basic vocabulary. Zariski geometry adds a topology on the definable sets.

An abstract elementary class (AEC) $(K, \prec_K)$ is a collection of structures[13] for a fixed vocabulary τ and a relation of 'strong substructure',

---

[12] Intuitively, a formula is *positive* if it contains no negations ([Marker 2002], 107).

[13] Here is a precise definition. An abstract elementary class (AEC) $(K, \prec_K)$ is a collection of structures for a fixed vocabulary τ that satisfy the following, where $A \prec_K B$ implies, in particular, that A is a substructure of B:

(1) If $A, B, C \in K$, $A \prec_K C$, $B \prec_K C$, and $A \subseteq B$ then $A \prec_K B$ (coherence);
(2) Closure under direct limits of $\prec_K$-embeddings;
(3) Downward Löwenheim–Skolem. If $A \subset B$ and $B \in K$ there is an A' with $A \subseteq A' \prec_K B$ and $|A'| \leq |A| = \mathrm{LS}(K)$.

Naturally we require that both K and $\prec_K$ are closed under isomorphism.

$\prec_K$ among the members of K obeying certain axioms. Examples of 'strong substructure' might be 'substructure', or elementary substructure, or elementary in $L_{\omega_1,\omega}$. But these various syntactically based properties are replaced by mathematical properties of the relation $\prec_K$. First the class is closed under unions of increasing $\prec_K$-chains and more subtly the union of a $\prec_K$-chain is a strong submodel of any M which is an extension of each element in the chain. From a category theoretic standpoint this just means the class is closed under directed colimits.[14] There is a more technical condition (coherence) which is the semantic version of the Tarski–Vaught condition[15] for elementary submodels.

Finally, countable Löwenheim–Skolem number means that every infinite subset of a model is contained in a strong submodel of the same size. The syntactic notion of type (Chapter 5.1) is replaced by a semantic notion. If the class has a monster model (page 127), $\mathbb{M}$, the Galois type of a over M is the orbit of a under automorphisms of $\mathbb{M}$ fixing M. A slightly more technical definition works for an arbitrary AEC.

In fact, these classes can be defined in another 'formalism-free' way using purely categorical terms. The key difficulty stems from the size requirement built into AEC's by the Löwenheim–Skolem property. Lieberman[16] [Lieberman 2013] proved that every AEC could be seen as a concrete λ-accessible category[17] for any regular $\mu > LS(K)$.

Further elaboration of the category of models approach to AEC arose from the attempts to modify the AEC notion to study functional analysis. Hyttinen and Hirvonen [Hirvonen & Hyttinen 2009] introduced a variant called a metric-AEC. Roughly, this notion arose as a solution to the equation:

$$\frac{AEC}{X} = \frac{\textit{first order logic}}{\textit{continuous logic}}.$$

---

[14] Category theoretic directed colimits are defined by universal mapping conditions; model theoretic direct limits are defined by a quotient of directed unions of structures. In this concrete case the notions are equivalent.

[15] $M \prec N$ if for every formula $\phi(x, \mathbf{m})$ with parameters from M, if $N \models \exists x \phi(x, \mathbf{m})$ then $N \models \phi(a, \mathbf{m})$ for some $a \in M$.

[16] Kirby [Kirby 2008] provided an alternative account. See also [Beke & Rosický 2012].

[17] Quoting [Boney et al. 2016], 'Roughly speaking, an accessible category is one that is closed under certain directed colimits, and whose objects can be built via certain directed colimits of a set of small objects.' More precisely, a category is μ-accessible if it is closed under λ-directed colimits, and every object is a μ-directed colimit of μ-presentable objects, the latter being a purely internal, category theoretic notion of size; a category is accessible if it is μ-accessible for some μ.

As in continuous logic, the models in the AEC are taken to be *complete metric spaces*. The metric AEC formulation has certain advantages over continuous logic in treating unbounded operators.

The difficulty is to observe that while, purely algebraically, a metric AEC **K** is closed under directed colimits, these are guaranteed to be concrete – i.e. the unions are themselves complete – only if they are $\aleph_1$-directed; otherwise the union must be completed to an element of **K**. Lieberman and Rosický ([Lieberman & Rosický 2017]) (see also [Boney et al. 2016]) placed the study in a wider hierarchy of accessible categories, by specializing the notion of an accessible category to pairs $(K; U)$ which form an accessible category with concrete μ-directed colimits.[18] In [Boney et al. 2016] this problem is addressed by a more general notion; a μ-AEC is one which is required to be closed only under μ-directed limits, thus finding a more general notion than AEC which encompasses metric AEC.

By working in abstract category theory this study violates our stricture at the beginning of this chapter that we always work with a fixed vocabulary. However, this concern is somewhat ameliorated by the discovery in ([Lieberman & Rosický 2016], [Rosický 1981]) of a *canonical signature* for a concrete directed accessible category $(K; U)$ with concrete directed colimits whose morphisms are monomorphisms preserved by U.

The connection between AECs and logic is at first only motivational. The AEC notion was developed to simplify the study of infinitary logics by generalizing some of the crucial properties and avoiding syntactical complications. The crucial Löwenheim–Skolem property is derived from thinking of $\prec$ as a kind of elementary submodel. In an AEC there is no explicit syntax[19] and no notion of a definable set.

In contrast, Zilber's notion of a *quasiminimal excellent class* [Zilber 2005b] was developed to provide a smooth framework for proving the categoricity in all uncountable powers of Zilber's pseudo-exponential field. This example itself can be axiomatized in a standard model theoretic framework in $L_{\omega_1,\omega}(Q)$. But this is *not* the way Zilber's proof works. The structure is patterned on (and conjecturally isomorphic to) the complex exponential field $(\mathbb{C}, +, \times, e^x)$. For technical reasons having to do with the simplicity of dealing with relational languages, the vocabulary is taken to include all polynomially definable sets as basic predicates. The fundamental result that a quasiminimal excellent class is categorical in all

---

[18] That is, K is an accessible category with μ-directed colimits and $U : K \to$ *Set* is a faithful functor to the category of sets preserving μ-directed colimits.

[19] Shelah's presentation theorem, discussed below, shows there is an 'implicit' syntax.

uncountable powers can be presented in a formalism-free way. The key point is that there are no axioms in the object language of the general quasiminimal excellence theorem; there are only statements about the combinatorial geometry determined by what are in the application the $(L_{\omega_1,\omega}(Q)$-definable sets. But there is a formal version. Kirby (Theorem 5.5 of [Kirby 2010]) proved that any quasiminimal excellent class with an infinite-dimensional model is the class of models of a set of sentences in $L_{\omega_1,\omega}(Q)$. Analogously to the results of Tarski and Birkhoff that opened this chapter, this result has the form: any class which satisfies specified 'mathematical properties' has a formal definition of a specified sort. In an astonishing development, [Bays et al. 2014] proved every quasiminimal class is quasiminimal excellent. Such a reduction is impossible in the general case of excellence studied by Shelah.

Hrushovski and Zilber [Hrushovski & Zilber 1993] introduced Zariski geometries partly in an attempt to remedy a notorious gap in the model theoretic study of algebraic geometry.[20] Algebraic geometry is concerned with the solution of systems of *equations*. But from a model theoretic standpoint, there is no way to distinguish among definable sets, 'all definable sets are equal'. In particular, the class of definable sets is closed under negation; equations and inequations have the same status. But from the perspective of algebraic geometry, 'some definable sets are more equal than others'. Systems of equations (varieties) are the objects of true interest. Hrushovski and Zilber remedy this situation by introducing a topology. The definition of a Zariski geometry[21] [Zilber 2010] concerns the relations between a family of topologies on the sets D^n for a fixed D. Generalizing the Zariski topology of algebraic geometry the closed sets should be given by conjunction of equations. The main result of [Hrushovski & Zilber 1993] is that every Zariski geometry satisfying sufficiently strong semantic conditions can in fact be realized as a finite cover of an algebraic curve.[22] While we noted in Chapter 13 that this theorem exemplifies what Lakatos called 'monster-barring', the other goal of recognizing 'positive sentences' in a general model theoretic framework provides both coherence and clarity.

---

[20] The immediate impetus was to find a correct version of Zilber's conjecture that all strongly minimal sets were 'set-like', 'group-like' or 'field-like', but we explain a second epistemological motivation.

[21] A collection of sets S_n is defined in each D^n which is closed under Boolean operations and the projection of a set in S_{n+1} to D^n is in S_n. In contrast to footnote 14, the topology is not definable; the topology discriminates between equations and their negations.

[22] Here is the precise theorem from [Hrushovski & Zilber 1993]. **Theorem** [Hrushovski–Zilber] Let X be a very ample Zariski geometry. Then there exists a smooth curve C over an algebraically closed field F, such that X, C are isomorphic as Zariski geometries. F and C are unique, up to a field isomorphism and an isomorphism of curves over F.

To clarify the distinctions between 'formalism-free' and 'logical' treatments, we provide some more examples of results in the study of AEC. Here is an example of a 'purely semantical' theorem. WGCH abbreviates the assertion, for all λ, $2^\lambda < 2^{\lambda^+}$. Shelah's Theorem 14.0.1 is proved (e.g. [Baldwin 2009], Chapter 17) in a 'formalism-free' manner; there is no mention of syntax. This non-structure result combined with seminal positive effects of amalgamation (construction of monster model, usable Galois types, etc.) makes the notion of amalgamation a 'dividing line' for abstract elementary classes while it isn't for complete theories in first order logic; it is just true.

Theorem 14.0.1 (WGCH) *Let K be an abstract elementary class (AEC). Suppose $\lambda \geq \mathrm{LS}(K)$ and K is λ-categorical. If amalgamation fails in λ there are 2^{λ^+} models in K of cardinality $\kappa = \lambda^+$.*

On the positive side, the amalgamation property allows the construction of a monster model and is a key hypothesis in the eventual categoricity line in AEC discussed on the next page.

Shelah's celebrated *presentation theorem* [Shelah 1983a, Baldwin 2009] changes the role of logic from a motivation (AECs are supposed to abstract the properties of classes defined in various infinitary logics) to a tool. The presentation theorem asserts that an AEC with arbitrarily large models can be defined as the reducts of models of a first order theory which omit a family of types. But Morley [Morley 1965b] had calculated the Hanf number[23] (Definition 8.5.2) for such syntactically defined classes. Thus, using the syntactical presentation as a tool, Shelah obtains a purely semantic theorem: if an AEC with Löwenheim number $\aleph_0$ has a model of cardinality $\beth_{\omega_1}$, it has arbitrarily large models.

When spelled out, the syntactic condition in the presentation theorem is a set of sentences in roughly Tarski's sense. We have discussed several modern examples of Tarski's consideration of a duality between description in a formal language and mathematical description. In the case of Kirby's axiomatization of quasiminimal excellent classes and Shelah's presentation theorem we gain a firmer grasp on a certain class of models by seeing a specific logic in which they are definable. In addition, via the presentation theorem, we are able to deduce purely semantical conclusions passing through the syntactic representation. Notably, the vocabulary arising in the

---

[23] The Hanf number for AECs with Löwenheim–Skolem number λ is the least κ such that for every AEC K with $LS(K) < \kappa$, if K has a model of cardinality κ, it has arbitrarily large models.

presentation theorem arises naturally only as a tool to prove that theorem. There is no apparent connection of each symbol of the resulting vocabulary with any basic mathematical properties of the AEC in question.

Recent work on AECs impacts two themes of this book: the role of formalization and the entanglement of model theory and set theory. Further it gives another illustration to clarify the notion of dividing line.

There were two main lines of work in the study of AECs at the beginning of the twenty-first century. In one, Shelah introduces the notion of a 'good λ-frame', a set of conditions on models of a fixed cardinal λ and an independence relation on their Galois types. Frames are an analog of first order superstability in λ, but unlike first order superstability, the transfer of frame existence between cardinalities is a major problem. He then uses hypotheses (e.g. categoricity) on models with cardinality close to λ and some global set theoretic hypotheses to build frames in arbitrarily large cardinalities by induction.

In the other line, Grossberg, Vandieren, and others study 'nice' AECs: those that satisfy amalgamation and joint embedding. Usually these studies also assume the existence of arbitrarily large models. 'Tameness' is a specific further hypothesis[24] for much of this work [Boney & Vasey 2017].

Definition ([Grossberg & VanDieren 2006a]) An AEC is κ-*tame* if for every model M with cardinality $\geq \kappa$ and pair p, q of Galois types over M, if p and q agree on all submodels of cardinality $\leq \kappa$, then they are equal. K is *eventually tame* if it is κ tame for some κ.

In both lines the holy grail is an analog of Morley's theorem: to show that if an AEC K is categorical in unboundedly many cardinals then it is categorical in all sufficiently large powers[25] (Shelah's eventual categoricity conjecture).

Boney [Boney 2014a] unites these two programs by proving that if K is a 'nice' AEC (but no assumption of arbitrarily large models) that is λ-tame and admits a λ-frame then there is a frame on $K_{\geq\lambda}$ (all models above λ). The key to this argument is to realize that tameness is equivalent to the existence of *unique non-forking extensions* and the careful construction by induction of a non-forking relation in larger cardinals.

Boney [Boney 2014b] interprets an arbitrary AEC into a theory of $L_{\kappa,\kappa}$ where κ is sufficiently large. Building on [Makkai & Shelah 1990],

---

[24] This notion is far from our earlier usage as every complete first order theory is tame in the sense here.

[25] Ideally, one would *calculate* (find a cardinal arithmetic formula for [Baldwin & Shelah 2014]) the cardinal at which eventual categoricity is a function of the Löwenheim number of K.

he[26] shows that for any AEC K (with $LS(K) < \kappa$)) if κ is strongly compact then K is κ-tame, and if further K is categorical in a successor cardinal above κ then K satisfies amalgamation for models of cardinality greater than κ. The proof depends heavily on the compactness of the logic $L_{\kappa,\kappa}$. Define the *Hanf number for* $< \kappa$-*tameness* to be the minimal λ such that: if K is an AEC with $LS(K) < \kappa$ that is $(< \kappa, \mu)$-tame for *some* $\mu \geq \lambda$, then it is $(< \kappa, \mu)$-tame for arbitrarily large μ. The results of [Boney 2014b] show that the Hanf number for $< \kappa$-tameness is κ when κ is *strongly compact*. However, this is done by showing a much stronger 'global tameness' result that ignores the hypothesis: *every* AEC K with $LS(K) < \kappa$ is $(< \kappa, \mu)$-tame for all $\mu \geq \kappa$. Boney and Unger [Boney & Unger 2017], building on earlier work of Shelah [Shelah 2013a], have shown that this global tameness result is actually an equivalence (for a variant of strong compactness).

Now applying the earlier ZFC theorem [Grossberg & VanDieren 2006b] transferring categoricity from successors for a tame 'nice' AEC yields the eventual categoricity of any AEC that is categorical in a successor cardinal above a strongly compact cardinal.

These three theorems show that 'tame' is an extremely useful hypothesis. It does not however show it is a dividing line in the sense of Chapter 13; there are no known useful consequences of an AEC failing to be tame. Of course, such may be discovered; Vasey suggested (in conversation) the possibility that one would find that if K satisfies a suitably defined notion of a superstable, then K either is tame or has many models on a tail of cardinals.

Vasey [Vasey 2017b], *working entirely in ZFC*, eliminates three major hypotheses of Boney's categoricity theorem: tameness, amalgamation, and the requirement that the categoricity cardinal is a successor, and shows any universal class (roughly speaking, an AEC closed under substructure, e.g. subclasses of locally finite groups) with arbitrarily large models is either eventually categorical or eventually never categorical.

A further Hanf number calculation involves both syntax and entanglement with set theory. Baldwin and Boney [Baldwin & Boney 2017] show that the Hanf number of amalgamation or JEP in an arbitrary AEC is no more than the first strongly compact cardinal. While the usual proof of this result for first order logic is heavily syntactic – e.g. the joint embedding property for a theory is equivalent to it being complete for universal

---

[26] After his result, Lieberman and Rosický [Lieberman & Rosický 2016] observed that a version for accessible categories with concrete directed colimits could be proven from the Makkai–Paré theorem (Theorem 5.51 of [Makkai & Paré 1989]); this slightly generalizes Boney since the coherence axiom is omitted.

sentences – the AEC proof can be done entirely semantically by taking an appropriate ultraproduct. In fact, the attempt to mimic the first order proof has severe limitations; to remedy this another expansion of the language to restore the adequacy of syntactic expressibility is introduced: the relational presentation theorem.

In the paper [Grossberg & VanDieren 2006a] introducing tameness, Grossberg and VanDieren show that a nice κ-tame AEC K is (Galois)-stable in some λ greater than both κ and the Hanf number of K if and only if it is stable in all λ with $\lambda = \lambda^{<\mu}$. In [Vasey 2016a], Vasey introduces a vastly stronger expansion of a vocabulary than those discussed in Chapter 4.6. For each Galois type over the empty set he adds predicates for the set of realizations of p. With the aid of this notion and embedding in an infinitary logic he succeeds in showing there that the λ in this stability spectrum result can be taken below the Hanf number.[27] Recall that this property of a stable first order theory (Theorem 5.3.5), established by a subtle use of compactness, is the key to the stability hierarchy.

Thus, while Shelah's original notion of abstract elementary class represents a turn away from formalization, recent developments provide more evidence for the importance of formalization.

---

[27] Theorem 3.3 of [Baldwin & Shelah 2008] obtains the same result for syntactic types for a complete sentence of $L_{\omega_1,\omega}$ without assuming tameness.

15 | Summation

We review here the main themes of the book and summarize the argument. Our principal claim is that the process of formalization and the use of fully formalized theories is a useful tool both in the philosophy of mathematical practice and in mathematics. We argued by exhibiting such uses.

Feferman addresses the connection of logic and mathematics as follows:

Logic attempts to provide us with a theoretical analysis of the underlying nature of mathematics as physics provides us with a theoretic analysis of the underlying nature of the physical world. ... In the case of logic, this theoretical analysis is supposed to explain what constitutes the underlying content of mathematics and what is its organizational and verificational structure. [Feferman 1978]

Without engaging ontological issues, we take the underlying content to be the corpus of mathematics: definitions, theorems, programs. But we try to show how logic and the formal method provide not only a verificational structure, namely proof in first order logic, but also through the methods of modern model theory a tool to organize the structure of mathematics. In fact, our account of the verificational structure is minimal. The coherence of the entire project of finding formalizations of specific mathematical topics in *first order theories*, which purport to preserve meaning, depends on the completeness theorem. But formal proof itself, with its permitted redundancy and necessary focus on small points, does not represent a faithful idealization of actual mathematical proofs. Moreover, the attempt to achieve verification by a global foundation leads to uninformative coding. However, by providing the axioms and primitive notions for local areas of mathematics, formalization can focus on the actual ideas of the particular subject. The paradigm shift (page 2) from the study of properties of logics to a systematic search for virtuous properties of theories enables the use of model theoretic principles to choose useful axiomatizations (Chapter 6).

We specified our investigation dealt with the philosophy of mathematical practice (page 5) to emphasize our study of the activities of mathematicians and the corpus of mathematics. Our first two theses read:

(1) Contemporary model theory makes formalization of *specific mathematical areas* a powerful tool to investigate both mathematical problems and issues in the philosophy of mathematics (e.g. methodology, axiomatization, purity, categoricity, and completeness).

(2) Contemporary model theory enables systematic comparison of local formalizations for distinct mathematical areas in order to organize and do mathematics, and to analyze mathematical practice.

The first of these specifies the study of a specific area of mathematics; the second concerns the relationship among areas.

Chapter 6 illustrates the profound impact of formalization as a tool to prove results across mathematics. These applications range from using model completeness to find analogs of Hilbert's Nullstellensatz in various fields (Chapter 4.4) to employing the various tools of modern model theory (geometric stability, ranks, o-minimality, neo-stability, etc.) to algebra, number theory and analysis. From the philosophical side, the geometry chapters, 9 to 11, provide a local focus by examining different formal axiomatizations of Euclidean geometry from the standpoint of descriptive axiomatization. In stressing the importance of Hilbert's *first* order axiomatization of Euclidean geometry and exploring the changing meaning of 'number' and 'magnitude' over the millennia this study illustrates the interplay of technical mathematics and historical perspective in the philosophy of mathematical practice. In discussing modern practice, we highlight two notions of classification: the novel representation of models of certain first order theories by decomposition as trees of countable structures and the taxonomy of theories to determine which theories admit such a structure theory (Chapter 5.5). In Chapter 13, we take a longer view of this methodology, exploring the relation among definition, classification, and taxonomy.

Perhaps most important, contemporary model theory contributes to the philosophy of mathematical practice by giving a more precise description of 'tame mathematics'. The aphorism 'There is a tame mathematics but if you define the ring of integers you are dead'[1] redirects the philosophy of mathematical practice from the traditional foundational priority of natural number arithmetic to the tame subjects of modern mathematics. Note that number theory remains a fundamental topic but much of model theory collaborates with the tools of modern algebraic geometry and number

---

[1] The aphorism, possibly apocryphal, was attributed to Hrushovski by Tibor Beke in conversation. This is of course not merely an aphorism but a theorem under the mild requirement that tame structures should admit a dimension.

theory. Chapter 6 shows how the wild structure of arithmetic is being corralled by the tamer theories of algebraically closed fields ($\aleph_1$-categorical), algebraically closed valued fields (NIP), and interpretation in o-minimal theories.

In making these claims for the mathematical uses of model theory, we contradict the opinion of the editors of the Friedman Festschrift volume who wrote:[2]

> It is generally regarded that much of uncountable model theory displays *substantial amounts* of set theoretic pathology significantly beyond what is usual in mathematics. The countable case is very interesting but does not cover many important mathematical situations, particularly in natural mathematical contexts of the power of the continuum. [Harrington et al. 1985]

Thesis (3) includes a rejoinder to their opinion.

(3) The choice of vocabulary and logic appropriate to the particular topic are central to the success of a formalization. The technical developments of first order logic have been more important in other areas of modern mathematics than such developments for other logics.

We demonstrated in Chapter 8 that these *substantial amounts* were, for first order logic, neither substantial nor pathological. Moreover the stability hierarchy, built to understand the uncountable, profoundly impacts the study of mathematical structures of arbitrary cardinality including countable.[3] Other model theoretic formalizations study mathematical analysis and the continuum more directly. As examples we have sketched the areas of continuous model theory, o-minimality, 'axiomatic' and 'definable' analysis (Chapter 6.3), the ω-stable theory of compact complex manifolds, and metric AECs. Note that this argument is about the practice of mathematics; it is not directed to a 'foundation' for all mathematics, but to the importance of first order model theory in the methodology of mathematics.

Thesis (3) points out the immense importance of the choice of logic by emphasizing the lack of entanglement between *first order logic* with extensions of ZFC while exploring the entanglements of such extensions with other logics (Chapters 1.3, 8, and 14). Here the *paradigm shift* – the partition

---

[2] In fairness, many of these developments and the contributions to other areas came after the quotation. The emphasis is mine.

[3] In a particularly striking example (page 172), the full solution of the Vaught conjecture for differentially closed fields stimulated the discovery of new phenomena in differential algebra.

is proved for first order logic – enables the separation of first order model theory from axiomatic set theory.

It can be objected that these issues arise from our choice of a set theoretic rather than a putative (perhaps HoTT) category theoretic or other metatheory. That is a task for another day; one which seems difficult to me. On the one hand, such proposals often upset the syntactic–semantic distinction that is fundamental to the view of this book. On the other, the combinatorial issues discussed in Chapter 8 seem fundamental to me. Such cardinality issues were replicated in Lawvere's categorical foundations for set theory in the study of accessible categories described in Chapter 14.

We argued in Chapter 3.3 that categoricity in second order logic has no intrinsic mathematical consequences and so lacks virtue. Categoricity in $L_{\omega_1,\omega}$ has some intrinsic mathematical consequences and so some virtue; it provides a natural setting for pursuing traditional philosophical concerns about categoricity. But the truly important notion is categoricity in power for first order theories. Most of the remainder of the book provides justification for that claim. Thesis (4) is an important element of that justification.

(4) The study of geometry is not only the source of the idea of axiomatization and many of the fundamental concepts of model theory, but geometry itself (through the medium of geometric stability theory) plays a fundamental role in analyzing the models of tame theories and solving problems in other areas of mathematics.

We elaborated this thesis in three motifs spread throughout the text. First, we recorded the immense influence of the revolution in geometry in the nineteenth century, solidified in Hilbert's *Grundlagen*, on the development of model theory. Most fundamentally (Chapter 4), exhibiting models to justify non-Euclidean geometry gave rise to the hypothetical approach that is key to the syntax–semantics separation that founds model theory.

Secondly, the notion of (bi)-interpretation arising in the consistency proofs and in Hilbert's coordinatization theorem became a fundamental tool of model theory (Chapter 4.5). Moreover, it provides a foundation of geometry even more rigorous than that attained for analysis in the late nineteenth century. Unlike arithmetic and analysis, first order geometry admits of a finitistic consistency proof (Fact 10.3.4).

Thirdly, the ability to find independence relations (Theorem 5.4.5) and thus combinatorial geometries (Theorem 5.4.6) in stable theories enables the assignment of cardinal invariants to models. The near classification of these geometries by the Zilber trichotomy proved a powerful tool for both pure and applied model theory (Chapters 5.5–6.4).

The ubiquity of geometry is evidenced by both this foundational (in the non-reductionist sense) role for geometry in organizing mathematics and in the analysis of axiomatizations of geometry in Chapters 9–11.

The crucial epistemological issue between foundationalism and the philosophy of mathematical practice is the one/many divide: should there be one foundational scheme for all mathematics or should one undertake the study of various areas of mathematics and develop schema for understanding the relations between areas: relations that depend on their content? We have argued that the, at first sight, purely mathematical paradigm shift in model theory described throughout this book not only reflects this refocusing of epistemological goals but provides tools for attaining them.

The late nineteenth and early twentieth centuries witnessed symbolic logic arise as a powerful tool to clarify the foundations of mathematics. The late twentieth and early twenty-first centuries witnessed symbolic logic, by formalizing particular areas of mathematics, become a powerful tool to study mathematics-at-large.

In 1952, Abraham Robinson signaled the first stage of the paradigm shift:

For example, although it was clearly false to say that every proposition that is true of the algebraic numbers is also true of the complex numbers, on the other hand one could say that all true propositions formulated about the field of algebraic numbers also hold for the field of complex numbers. It was the ability of logic to distinguish certain classes of propositions that made it possible to establish such metamathematical theorems.[4]

Decades later, Shelah stated the aim of the second stage:

The Classification Problem: Classify the T's in a useful way, i.e., such that for suitable questions on the class of models of T the partition to cases according to the classification will be helpful. ([Shelah 1985], 229)

While this general problem was stated in announcing the arguably specialized 'main gap theorem', we now know that 'suitable' stretches from 'exploring the white space' to deep applications across mathematics.

In the Introduction, we stated our hope that this book would demonstrate that the profound technical developments in late twentieth century model theory initiated by Shelah's paradigm shift had philosophical significance. If the reader finds even a portion of this text grist for their philosophical mill, our hopes will be realized.

---

[4] Paraphrased by Dauben ([Dauben 1995], 207) from a 1952 Robinson lecture in Paris.

References

Addison, J., Henkin, L., and Tarski, A., editors (1965). *The Theory of Models*. North-Holland.

Ajtai, M. (1979). Isomorphism and higher order equivalence. *Annals of Mathematical Logic*, 16:181–203.

Aldama, Ricardo de (2013). Definable nilpotent and soluble envelopes in groups without the independence property. *Mathematical Logical Quarterly*, 59: 201–205.

Altinel, T. and Baginski, P. (2014). Definable envelopes of nilpotent subgroups of groups with chain conditions on centralizers. *Proceedings of the American Mathematical Society*, 142:1497–1506.

Altinel, T., Borovik, A., and Cherlin, G. (2008). *Simple Groups of Finite Morley Rank*. American Mathematical Society Monographs Series.

Andrews, U. (2011). A new spectrum of recursive models using an amalgamation construction. *Journal of Symbolic Logic*, 76:883–896.

Angere, S. (2017). Identity and intensionality in univalent foundations and philosophy. *Synthese*, 77:1–41; published online: January 24, 2017.

Apostol, T. (1967). *Calculus*. Blaisdell.

Arana, A. (2008a). Logical and semantic purity. *Protosociology*, 25:36–48.

(2008b). On formally measuring and eliminating extraneous notions in proofs. *Philosophia Mathematica*, III:1–19.

(2014). Purity in arithmetic: Some formal and informal issues. In *Formalism and Beyond: On the Nature of Mathematical Discourse*, pages 315–316. DeGruyter.

Arana, A. and Mancosu, P. (2012). On the relationship between plane and solid geometry. *Review of Symbolic Logic*, 5:294–353.

Archimedes (1897). On the sphere and cylinder I. In *The Works of Archimedes*, pages 1–56. Dover Press. Translation and comments by T. L. Heath (including 1912 supplement).

Artin, E. (1957). *Geometric Algebra*. Interscience.

Aschenbrenner, M., van den Dries, L., and van der Hoeven, J. (2016). The surreal numbers as a universal H-field. *Journal of the European Mathematical Society*. To appear.

(2017). *Asymptotic Differential Algebra and Model Theory of Transseries*. Annals of Mathematics Studies. Princeton University Press.

Avigad, J. (2006). Review of Calixto Badesa, *The Birth of Model Theory: Löwenheim's Theorem in the Frame of the Theory of Relatives*. *The Mathematical Intelligencer*, 28:67–71.

(2007). Philosophy of mathematics. In Boundas, C., editor, *The Edinburgh Companion to the 20th Century Philosophies*, pages 234–251. Edinburgh University Press.

(2010). Understanding, formal verification, and the philosophy of mathematics. *Journal of the Indian Council of Philosophical Research*, 27:161–197 (Special Issue on Logic and Philosophy Today).

Avigad, J., Dean, E., and Mumma, J. (2009). A formal system for Euclid's elements. *Review of Symbolic Logic*, 2:700–768.

Avigad, J. and Morris, R. (2014). The concept of 'character' in Dirichlet's theorem on primes in an arithmetic progression. *Archive for History of Exact Sciences*, 68:265–326.

(2016). Character and object. *The Review of Symbolic Logic*, 9:480–510.

Awodey, S., Pelayo, A., and M., W. (2013). Voevodsky's univalence axiom in homotopy type theory. *Notices of the AMS*, 60:1164–1167.

Awodey, S. and Reck, E. (2002a). Completeness and categoricity, part I: Nineteenth-century axiomatics to twentieth-century metalogic. *History and Philosophy of Logic*, 23:1–30.

(2002b). Completeness and categoricity, part II: Twentieth-century metalogic to Twenty-first-century semantics. *History and Philosophy of Logic*, 23:77–94.

Ax, J. (1968). The elementary theory of finite fields. *Annals of Mathematics*, 88:239–271.

Ax, J. and Kochen, S. (1965). Diophantine problems over local fields II: A complete set of axioms for p-adic number theory. *American Journal of Mathematics*, 87:631–648.

Badesa, C. (2004). *The Birth of Model Theory*. Princeton University Press.

Baginski, P. (2009). Stable $\aleph_0$-categorical alternative rings. PhD thesis, U. California, Berkeley.

Baldwin, J. T. (1979). Stability theory and algebra. *Journal of Symbolic Logic*, 44:599–608.

(1988a). *Fundamentals of Stability Theory*. Springer-Verlag.

(1988b). Classification theory: 1985. In Baldwin, J. T., editor, *Classification Theory: Chicago, 1985. Proceedings of the U.S.-Israel Binational Workshop on Model Theory in Mathematical Logic*, Springer Verlag.

(1989). Diverse classes. *Journal of Symbolic Logic*, 54:875–893.

(1990). The spectrum of resplendency. *The Journal of Symbolic Logic*, 55:626–636.

(1994). An almost strongly minimal non-Desarguesian projective plane. *Transactions of the American Mathematical Society*, 342:695–711.

(1995). Some projective planes of Lenz Barlotti class I. *Proceedings of the American Mathematical Society*, 123:251–256.

(2004). Notes on quasiminimality and excellence. *Bulletin of Symbolic Logic*, 10:334–367.

(2007). Vaught's conjecture: Do uncountable models count? *Notre Dame Journal of Formal Logic*, pages 1–14.

(2009). *Categoricity*. Number 51 in University Lecture Notes. American Mathematical Society.

(2010). Review of *The Birth of Model Theory: Löwenheim's Theory in the Frame of the Theory of Relatives. Bulletin of American Mathematical Society*, 47: 177–185.

(2012). Amalgamation, absoluteness, and categoricity. In Arai, T., Feng, O., Kim, B., Wu, B., and Yang, Y., editors, *Proceedings of the 11th Asian Logic Conference, 2009*, pages 22–50. World Scientific Publishing.

(2013a). Formalization, primitive concepts, and purity. *Review of Symbolic Logic*, 6:87–128.

(2013b). From geometry to algebra. Draft paper: Urbana 2013 http://homepages .math.uic.edu/~jbaldwin/pub/geomnov72013.pdf.

(2014). Completeness and categoricity (in power): Formalization without foundationalism. *Bulletin of Symbolic Logic*, 20:39–79.

(2015). How big should the monster model be? In Hirvonen, A., Kontinen, J., Kossak, R., and Villaveces, A., editors, *Logic without Borders*, pages 31–50. Ontos Mathematical Logic.

(2017a). Axiomatizing changing conceptions of the geometric continuum I: Euclid and Hilbert. *Philosophia Mathematica*, 2017. 32 pages, online doi: 10.1093/philmat/nkx030.

(2017b). Axiomatizing changing conceptions of the geometric continuum II: Archimedes-Descartes-Hilbert-Tarski. *Philosophia Mathematica*, 2017. 30 pages, online doi: 10.1093/philmat/nkx031.

(2018). The explanatory power of a new proof: Henkin's completeness proof. In Piazza, M. and Pulcini, G., editors, *Philosophy of Mathematics: Truth, Existence and Explanation*, Boston Studies in the History and Philosophy of Science, page 14. Springer-Verlag.

Baldwin, J. T. and Boney, W. (2017). Hanf numbers and presentation theorems in AEC. In Iovino, J., editor, *Beyond First Order Model Theory*, pages 81–106. Chapman Hall.

Baldwin, J. T., Friedman, S., Koerwien, M., and Laskowski, C. (2016a). Three red herrings around Vaught's conjecture. *Transactions of the American Mathematical Society*, 368:22. Published electronically: November 6, 2015.

Baldwin, J. T. and Lachlan, A. (1971). On strongly minimal sets. *Journal of Symbolic Logic*, 36:79–96.

Baldwin, J. T. and Larson, P. (2016). Iterated elementary embeddings and the model theory of infinitary logic. *Annals of Pure and Applied Logic*, 167:309–334.

Baldwin, J. T., Larson, P., and Shelah, S. (2015). Almost Galois ω-stable classes. *Journal of Symbolic Logic*, 80:763–784. Shelah index 1003.

Baldwin, J. T., Laskowski, C., and Shelah, S. (2016b). Constructing many atomic models in $\aleph_1$. *Journal of Symbolic Logic*, 81:1142–1162.

Baldwin, J. T. and McKenzie, R. N. (1982). Counting models in universal Horn classes. *Algebra Universalis*, 15:359–384.

Baldwin, J. T. and Rose, B. (1977). $\aleph_0$-categoricity and stability of rings. *Journal of Algebra*, 45:1–17.

Baldwin, J. T. and Shelah, S. (2008). Examples of non-locality. *Journal of Symbolic Logic*, 73:765–783.

(2012). Stability spectrum for classes of atomic models. *Journal of Mathematical Logic*, 12:19 pages.

(2014). A Hanf number for saturation and omission II. *Mathematical Logic Quarterly*, 60:437–443.

Baldwin, J. T. and Kolesnikov, A. (2009). Categoricity, amalgamation, and tameness. *Israel Journal of Mathematics*, 170:411–443.

Baldwin, J. T. and Mueller, A. (2012). A short geometry. http://homepages.math .uic.edu/~jbaldwin/CTTIgeometry/euclidgeonov21.pdf.

Barwise, J., editor (1975). *Admissible Sets and Structures*. Perspectives in Mathematical Logic. Springer-Verlag.

Barwise, J. and Eklof, P. (1969). Lefschetz's principle. *Journal of Algebra*, 13:554–570.

Barwise, J. and Feferman, S., editors (1985). *Model-Theoretic Logics*. Springer-Verlag.

Barwise, J. and Schlipf, J. (1976). An introduction to recursively saturated and resplendent models. *Journal of Symbolic Logic*, 41:531–536.

Bays, M., Hart, B., Hyttinen, T., Kesala, M., and Kirby, J. (2014). Quasiminimal structures and excellence. *Bulletin of the London Mathematical Society*, 46: 155–163.

Bays, T. (1998). Some two-cardinal results for o-minimal theories. *Journal of Symbolic Logic*, 63:543–548.

Beeson, M. (2008). Constructive geometry, proof theory and straight-edge and compass constructions. www.michaelbeeson.com/research/talks/Constructive GeometrySlides.pdf.

Behmann, H. (1922). Beiträge zur Algebra der Logik, insbesondere zum Entscheidungsproblem. *Mathematische Annalen*, 86:163229.

Beke, T. and Rosický (2012). Abstract elementary classes and accessible categories. *Annals of Pure and Applied Logic*, 163:2008–2017.

Ben Yaacov, I., Berenstein, A., Henson, C. W., and Usvyatsov, A. (2008). Model theory for metric structures. In Chatzidakis, Z. et al., editors, *Model Theory with Applications to Algebra and Analysis. Vol. 2*, volume 350 of *London Mathematical Society Lecture Note Series*, pages 315–427. Cambridge University Press.

Ben Yaacov, I. and Pederson, A. (2010). A proof of completeness for continuous first-order logic. *Journal of Symbolic Logic*, 75:168–190.

Berline, C. (1982). Déviation des types dans les corps algébriquement clos. In Poizat, B., editor, *Seminaire Théories Stable: 1980-82*, volume 3, pages 3.01– 3.10. Poizat.

Bernays, P. (1918). *Beitrage zur axiomatischen Behandlung des Logik-Kalküls.* PhD thesis, Habilitationsschrift, Universitat Gottingen. Bernays Nachla, WHS, ETH Zurich Archive, Hs 973.192.

—— (1967). Hilbert, David. In *Routledge Encyclopedia of Philosophy*, pages 496–505. Macmillan. Bernays Project: text 32 www.phil.cmu.edu/projects/bernays/Pdf/bernaysHilbert_2003-05-17.pdf.

—— (1998). Hilbert's significance for the philosophy of mathematics. In Mancosu, P., editor, *From Brouwer to Hilbert*, pages 189–197. Oxford University Press. English transation of German published in 1922.

Beutelspacher, A. and Rosenbaum, U. (1998). *Projective Geometry: From Foundations to Applications.* Cambridge University Press.

Białynicki-Birula, A. and Rosenlicht, M. (1962). Injective morphisms of real algebraic varieties. *Proceedings of the American Mathematical Society*, 13: 200–203.

Birkhoff, Garrett (1935). On the structure of abstract algebras. *Proceedings of the Cambridge Philosophical Society*, 31:433–454.

Birkhoff, Garrett and Kreyszig, E. (1984). The establishment of functional analysis. *Historia Mathematica*, 11:258–321.

Birkhoff, George (1932). A set of postulates for plane geometry. *Annals of Mathematics*, 33:329–343.

Blanchette, P. (2014). The birth of semantic entailment. Lecture notes from ASL European meeting, Vienna.

Blum, L. (1968). Generalized algebraic structures: A model theoretical approach. PhD thesis, MIT.

Bolzano, B. (1810). *Beyträge zu einer begründeteren Darstellung der Mathematik.* Caspar Widtmann. Trans. by S. Russ in *Philosophy of Logic*, Elsevier, edited by Ewald 1996.

Bolzano, B. and Russ, S. (2004). *The Mathematical Works of Bernard Bolzano.* Oxford University Press.

Boney, W. (2014a). Tameness and extending frames. *Journal of Mathematical Logic*, 14.

—— (2014b). Tameness from large cardinal axioms. *Journal of Symbolic Logic*, 79:1092–1119.

Boney, W., Grossberg, G., Lieberman, M., Rosický, J., and Vasey, S. (2016). μ-abstract elementary classes and other generalizations. *Journal of Pure and Applied Algebra*, 220:3048–3066.

Boney, W. and Unger, S. (2017). Large cardinal axioms from tameness in AECs. To appear. PAMS: arXiv:1509.01191.

Boney, W. and Vasey, S. (2017). A survey on tame abstract elementary classes. In Iovino, J., editor, *Beyond First Order Model Theory*. Chapman Hall.

Borovik, A. and Nesin, A. (1994). *Groups of Finite Morley Rank.* Oxford University Press.

Bos, H. (1993). 'The bond with reality is cut' – Freudenthal on the foundations of geometry around 1900. *Educational Studies in Mathematics*, 25:51–58.

(2001). *Redefining Geometric Exactness*. Sources and Studies in the History of Mathematics and the Physical Sciences. Springer Verlag.

Bourbaki, N. (1950). The architecture of mathematics. *American Mathematical Monthly*, 57:221–232.

Bouscaren, E., editor (1999). *Model Theory and Algebraic Geometry: An Introduction to E. Hrushovski's Proof of the Geometric Mordell-Lang Conjecture*. Springer Verlag.

Bouscaren, E. and Hrushovski, E. (2006). Classifiable theories without finitary invariants. *Annals of Pure and Applied Logic*, 142:296–320.

Boyer, C. (1956). *History of Analytic Geometry*. Scripta Mathematica.

Brady, G. (2000). *From Peirce to Skolem: A Neglected Chapter in the History of Logic*. Sources and Studies in the History and Philosophy of Mathematics. North-Holland.

Breuillard, E. (2016). Lectures on approximate groups and Hilbert's 5th problem. In *Recent Trends in Combinatorics*, volume 159 of *The IMA Volumes in Mathematics and its Applications*, pages 369–404. North Holland.

Buechler, S. (1991). *Essential Stability Theory*. Springer Verlag.

    (2008). Vaught's conjecture for superstable theories of finite rank. *Annals of Pure and Applied Logic*, 155:135–172.

Burgess, J. P. (2008). Tarski's tort. In *Mathematics, Models, and Modality*, pages 262–300. Cambridge University Press.

    (2010). Putting structuralism in its place. Preprint.

Burgess, J. P. and Tsementzis, D. (2016). Structuralism and fidelity to mathematical practice. Preprint.

Buss, S., Kechris, A., Pillay, A., and Shore, R. (2001). The prospects for mathematical logic in the twenty-first century. *Bulletin of Symbolic Logic*, 7:169–196.

Button, T. and Walsh, S. (2016). Structure and categoricity: Determinacy of reference and truth-value in the philosophy of mathematics. *Philosophia Mathematica*, 24:283–307.

    (2017). *Philosophy and Model Theory*. Oxford University Press.

Caicedo, X. and Iovino, J. (2014). Omitting uncountable types and the strength of [0,1]-valued logics. *Annals of Pure and Applied Logic*, 165:1169–1200.

Cantú, P. (1999). *Giuseppe Veronese and the Foundations of Geometry. (Giuseppe Veronese e i fondamenti della geometria.) (Italian)*. Biblioteca di Cultura Filosofica 10, Milano.

Casanovas, E. (1999). The number of types in simple theories. *Annals of Pure and Applied Logic*, 98:69–86.

Chang, C. and Keisler, H. (1966). *Continuous Model Theory*. Annals of Mathematical Studies. Princeton University Press.

    (1973). *Model Theory*. North-Holland. 3rd edition 1990.

Chatzidakis, Z. (2000a). Model theory of difference fields and applications to algebraic dynamics. In Jang, S. Y. et al., editors, *Proceedings of the International Congress of Mathematicians, Seoul, 2014, Vol. 2*, pages 1–14. Kyung Moon Sa Co, Ltd, Seoul.

(2000b). A survey on the model theory of difference fields. In Barwise, J., editor, *Model Theory, Algebra, and Geometry*, pages 36–64. MSRI Publications.

(2015). Model theory of fields with operators. In Hirvonen, A., Kontinen, J., Kossak, R., and Villaveces, A., editors, *Logic without Borders*, pages 91–114. Ontos Mathematical Logic.

Chatzidakis, Z., van den Dries, L., and Macintyre, A. (1992). Definable sets over finite fields. *Journal fur die reine und angewandte Mathematik*, 427:107–135.

Cherlin, G. (1976). *Model Theoretic Algebra: Selected Topics*. Lecture Notes in Mathematics. Springer Verlag.

(2000). Sporadic homogeneous structures. In *The Gelfand Mathematical Seminars*, pages 15–48. Birkhuser.

(2004). Algebraicity conjecture. Web page, Rutgers: Algebraicity conjecture.

Cherlin, G., Harrington, L., and Lachlan, A. (1985). $\aleph_0$-categorical, $\aleph_0$-stable structures. *Annals of Pure and Applied Logic*, 28:103–135.

Cherlin, G. and Hrushovski, E. (2003). *Finite Structures with Few Types*. Annals of Mathematics Studies. Princeton University Press.

Chernikov, A. and Starchenko, S. (2016). Regularity lemma for distal structures. *Journal of the European Mathematical Society*.

Church, A. (1956). *Introduction to Mathematical Logic: Volume 1*. Annals of Mathematics Studies. Princeton University Press. 1st edition 1944.

Coffa, A. (1991). *The Semantic Tradition from Kant to Carnap: To the Vienna Station*. Cambridge University Press.

Cohen, P. (1969). Decision procedures for real and p-adic fields. *Communications on Pure and Applied Mathematics*, 22:131–151.

Collins, G. E. (1975). Quantifier elimination for the elementary theory of real closed fields by cylindrical algebraic decomposition. In *Automata Theory and Formal Languages: 2nd GI Conference Kaiserslautern*, volume 33 of *Lecture Notes in Computer Science*, pages 134–183. Springer-Verlag.

Coolidge, J. L. (1963). *A History of Geometrical Methods*. Dover Publications. First published 1940 by Oxford University Press.

Corcoran, J. (1980). Categoricity. *History and Philosophy of Logic*, 1:187–207. http://dx.doi.org/10.1080/01445348008837010.

Corfield, D. (2003). *Towards a Philosophy of Real Mathematics*. Cambridge University Press.

Corry, L. (1992). Nicolas Bourbaki and the concept of mathematical structure. *Synthese*, 92:315–348.

Crippa, D. (2014a). Impossibility results: From geometry to analysis: A study in early modern conceptions of impossibility. PhD thesis, Université Paris Diderot (Paris 7).

(2014b). Reflexive knowledge in mathematics: The case of impossibility results. Slides from lecture at French Philosophy of Mathematics Workshop 6.

Cronheim, A. (1953). A proof of Hessenberg's theorem. *Proceedings of the American Mathematical Society*, 4:219–221.

Dauben, J. (1995). *Abraham Robinson*. Princeton University Press.

Dawson, J. W. (1993). The compactness of first-order logic: From Gödel to Lindstrom. *History and Philosophy of Logic*, 14:15–37.

Dedekind, R. (1888/1963). *Essays on the Theory of Numbers*. Dover. First published by Open Court publications 1901; first German edition 1888.

Deissler, R. (1977). Minimal models. *The Journal of Symbolic Logic*, 42:254–260.

Dembowski, P. (1977). *Finite Geometries*. Springer Verlag.

Demopoulos, W. (1994). Frege, Hilbert, and the conceptual structure of model theory. *History and Philosophy of Logic*, 15:211–225.

Denef, J. and Loeser, F. (2002). Motivic integration and the Grothendieck group of pseudo-finite fields. In *Proceedings of the International Congress of Mathematicians, Beijing 2002*, volume 2, pages 13–23. Higher Education Press.

DePaul, M. and Zagzebski, L., editors (2003). *Intellectual Virtue: Perspectives from Ethics and Epistemology*. Clarendon Press.

Descartes, R. (1637). *The Geometry of René Descartes*. Dover. Translated by David Eugene Smith and Marcia L. Latham: 1924; 1954 edition; French publication 1637.

Detlefsen, M. (2014). Completeness and the ends of axiomatization. In Kennedy, J., editor, *Interpreting Gödel*, pages 59–77. Cambridge University Press.

Detlefsen, M. and Arana, A. (2011). Purity of methods. *Philosophers' Imprint*, 11: 1–20.

Dickmann, M. (1985). Larger infinitary logics. In Barwise, J. and Feferman, S., editors, *Model-Theoretic Logics*, pages 317–363. Springer Verlag.

Dieudonné, J. (1939). Les méthodes axiomatique modernes et les fondements des mathématiques. *Revue Scientifique*, 77:224–232.

(1970). The work of Nicolas Bourbaki. *American Mathematical Monthly*, 77: 134–145.

Dreben, B. and van Heinenoort, J. (1986). Note to 1929, 1930, and 1930a. In Feferman, S. et al., editors, *Kurt Gödel: Collected Works, Volume I*, pages 44–59. Oxford University Press. 1930 article reprinted.

Dries, L. v. d. (1986). A generalization of the Tarksi-Seidenberg theorem, and some non-definability results. *Bulletin (New Series) of the American Mathematical Society*, 15:189–193.

Dries, L. v. d. (1999). *Tame Topology and O-Minimal Structures*. London Mathematical Society Lecture Note Series.

(2005). Stability theory notes. In lecture notes at www.math.uiuc.edu/~vddries/; accessed Feb. 11, 2017.

Dries, L. v. d., Macintyre, A., and Marker, D. (1997). Logarithmic-exponential power series. *Journal of the London Mathematical Society*, 56(3):417–434.

Dries, L. v. d. and Miller, C. (1994). On the real exponential field with restricted analytic functions. *Israel Journal of Mathematics*, 85:19–56.

Ebbinghaus, H.-D. (2007). Löwenheim-Skolem theorems. In Jaquette, D., editor, *Philosophy of Logic*. Elsevier-North Holland.

Educational Development Center (2009). *CME Algebra I*. Pearson. EDC: Educational Development Center.

Ehrenfeucht, A. and Mostowski, A. (1956). Models of axiomatic theories admitting automophisms. *Fundamenta Mathematicae*, 43:50–68.

Ehrlich, P., editor (1994). *Real Numbers, Generalizations of the Reals, and Theories of Continuaa*. Kluwer Academic Publishers.

Ehrlich, P. (1995). Hahn's *Über die nichtarchimedischen Grössensysteme* and the development of the modern theory of magnitudes and numbers to measure them. In Hintikka, J., editor, *Essays on the Development of the Foundations of Mathematics*, pages 165–213. Kluwer Academic Publishers.

(1997). From completeness to Archimedean completeness. *Synthese*, 110:57–76.

(2001). Number systems with simplicity hierarchies: A generalization of Conway's theory of surreal numbers. *Journal of Symbolic Logic*, 66:1231–1258.

(2006). The rise of non-Archimedean mathematics and the roots of a misconception I: The emergence of non-Archimedean systems of magnitudes 1,2. *Archive for the History of the Exact Sciences*, 60:1–121.

(2012). The absolute arithmetic continuum and the unification of all numbers great and small. *Bulletin of Symbolic Logic*, 18:1–45.

Einstein, A. (2002). Geometry and experience. In *Collected Papers of Albert Einstein*, pages 383–485. Princeton University Press. 1921 German original in collected works; English: www-groups.dcs.st-and.ac.uk/history/Extras/Einstein_geometry.html.

Eklof, P. (1973). Lefschetz's principle and local functors. *Proceedings of the American Mathematical Society*, 37:333–339.

(1976). Whitehead's problem is undecidable. *American Mathematical Monthly*, 83:775–788.

Eklof, P. and Mekler, A. (2002). *Almost Free Modules: Set Theoretic Methods*. North Holland. 2nd edition.

Enderton, H. (2007). Second order and higher order logic. *The Stanford Encyclopedia of Philosophy (Fall 2015 Edition)*, accessed 2017 ⟨https://plato.stanford.edu/entries/logic-higher-order/⟩.

Euclid (1956). *Euclid's Elements*. Dover. In 3 volumes, translated by T. L. Heath; first edition 1908; online at http://aleph0.clarku.edu/~djoyce/java/elements/.

Evans, D. (1986). Homogeneous geometries. *Proceedings of the London Mathematical Society*, 52:305–327.

Fagin, R. (1976). Probabilities on finite models. *Journal of Symbolic Logic*, 41:50–58.

Farah, I., Hart, B., and Sherman, D. (2013). Model theory of operator algebras I: Stability. *Bulletin of the London Mathematical Society*, 45:825–838.

(2014). Model theory of operator algebras II: Model theory. *Israel Journal of Mathematics*, 201:477–505.

Feferman, S. (1978). The logic of mathematical discovery vs. the logical structure of mathematics. In *PSA: Proceedings of the Biennial Meeting of the Philosophy of Science Association, 1978.* Oxford University Press. Reprinted in collected papers, *In the Light of Logic*, Oxford University Press, 1998.

(2008a). Conceptions of the continuum. Expanded version of a lecture at the Workshop on Philosophical Reflections on Set Theory held at the Centre de Cultura Contemporánia de Barcelona on October 7, 2008; http://math .stanford.edu/~feferman/papers/ConceptContin.pdf.

(2008b). Tarski's conceptual analysis of semantic notions. In Patterson, D., editor, *New Essays on Tarski and Philosophy*. Oxford University Press.

(2012). The continuum hypothesis is neither a definite mathematical problem nor a definite logical problem. Revised version of a lecture (Harvard University, Oct. 5, 2011) in the series, Exploring the Frontiers of Incompleteness (EFI), organized by Peter Koellner.; https://math.stanford.edu/~feferman/ papers/CH_is_Indefinite.pdf.

Felgner, U. (1971). *Models of ZF Set Theory*. Springer-Verlag, LNM 223.

Ferreirós, J. and Gray, J. J. (2008). Introduction. In Ferreirós, J. and Gray, J. J., editors, *Architecture of Modern Mathematics: Essays in History and Philosophy*, pages 1–43. Oxford University Press.

Foreman, M. (2010). Chang's conjecture, generic elementary embeddings and inner models for huge cardinals. www2.kobe-u.ac.jp/~hsakai/RIMS2010/ slides/Foreman.pdf.

Fraenkel, A. (1928). *Einleitung in die Mengenlehre*. Springer. 3rd, revised edition.

Fraïssé, R. (1954). Sur quelques classifications des systèmes de relations. *Publ. Sci. Univ. Algeria Sèr. A*, 1:35–182.

(1985). Deux relations dénombrables, logiquement équivalentes pour le second ordre, sont isomorphes (modulo un axiome de constructibilité). In *Mathematical Logic and Formal Systems*, volume 94 of *Lecture Notes in Pure and Applied Mathematics*, pages 161–182. Dekker.

Franks, C. (2010a). *The Autonomy of Mathematical Knowledge: Hilbert's Program Revisited*. Cambridge University Press.

(2010b). Cut as consequence. *History and Philosophy of Logic*, 31:349–379.

(2014). Logical completeness, form, and content: An archaeology. In Kennedy, J., editor, *Interpreting Gödel: Critical Essays*, pages 78–106. Cambridge University Press.

Frege, G. and Hilbert, D. (1980). Frege-Hilbert. In Gabriel, G., Hermes, H., Kambartel, F., et al., editors, *Frege, Gottlob: Philosophical and Mathematical Correspondence*. Oxford.

Freitag, J. (2014). Isogeny in superstable groups. *Archive for Mathematical Logic*, 54:449–461.

(2015). Indecomposability for differential algebraic groups. *Journal of Pure and Applied Algebra*, 219:3009–3029.

Freitag, J. and Scanlon, T. (2015). Strong minimality and the j-function. *Journal of the European Mathematical Society*.

Frenkel, E. (2013). *Love and Math*. Basic Books, Perseus Group.

Freudenthal, H. (1957). Zur Geschichte der Grundlagen der Geometrie – zugleich eine Besprechung der 8. Aufl. von Hilberts Grundlagen der Geometrie. *Nieuw Archiefvoor Wiskunde*, 3:105–142.

Friedman, H. (1971). Higher set theory and mathematical practice. *Annals of Mathematical Logic*, 2:325–357.

   (1975). One hundred and two problems in mathematical logic. *Journal of Symbolic Logic*, pages 113–129.

   (1999). A consistency proof for elementary algebra and geometry. www.personal .psu.edu/t20/fom/postings/9908/msg00067.html.

Friedman, H., Simpson, S., and Smith, R. (1983). Countable algebra and set existence axioms. *Annals of Pure and Applied Logic*, 25:141–181.

Gao, S. (1996). On automorphism groups of countable structures. *Journal of Symbolic Logic*, 63:891–896.

Gehret, A. (2017). The asymptotic couple of the field of logarithmic transseries. *Journal of Symbolic Logic*, 82:35–61.

Giaquinto, M. (2008). Cognition of structures. In Mancosu, P., editor, *The Philosophy of Mathematical Practice*, pages 198–256. Oxford University Press.

Gillies, D., editor (2008a). *Revolutions in Mathematics*. Oxford University Press.

Gillies, D. (2008b). The Fregean revolution in logic. In Gillies, D., editor, *Revolutions in Mathematics*, pages 265–305. Oxford University Press.

Giovannini, E. (2013). Completitud y continuidad en fundamentos de la geometrìa de Hilbert: Acera de Vollständigkeitsaxiom. *Theoria*, 76:139–163.

   (2016). Bridging the gap between analytic and synthetic geometry. *Synthese*, 193:31–70.

Givant, S. and Tarski, A. (1999). Tarski's system of geometry. *Bulletin of Symbolic Logic*, 5:175–214.

Glebski, Y., Kogan, V., Liogon'kii, M., and Taimanov, V. (1969). The extent and degree of satisfiability of formulas of the restricted predicate calculus. *Kiberneticka*, 2:17–27.

Göbel, R. and Trlifaj, J. (2006). *Approximations and Endomorphism Algebras of Modules*. Walter De Gruyter.

Gödel, K. (1929). Über die Vollständigkeit des Logikkalküls. In Feferman, S. et al., editors, *Kurt Gödel: Collected Works, Volume I*, pages 60–101. Oxford University Press. 1929 PhD thesis reprinted.

   (1930). The completeness of the axioms of the functional calculus of logic. In Feferman, S. et al., editors, *Kurt Gödel: Collected Works, Volume I*, pages 103–123. Oxford University Press. Under the auspices of Association for Symbolic Logic 1930c in collected works; first appeared: Monatshefte für Mathematik und Physik.

   (1946). Remarks before the Princeton bicentennial conference on problems in mathematics, 1946. In Feferman, S. et al., editors, *Kurt Gödel: Collected Works, Volume I*. Oxford University Press.

Goldbring, I. and Sinclair, T. (2015). On Kirchberg's embedding problem. *Journal of Functional Analysis*, 269:155–198.

Goldfarb, W. (1999). On Gödel's way in: The influence of Rudolph Carnap. *Bulletin of Symbolic Logic*, 11:185–193.

Goncharov, S., Harizanov, V., Laskowski, M., Lempp, S., and McCoy, C. (2003). Trivial strongly minimal sets are model complete after naming constants. *Proceedings of the American Mathematical Society*, 131:3901–3912.

Gowers, T. (2000). The two cultures of mathematics. In Arnold, V. et al., editors, *Mathematics: Frontiers and Perspectives*, pages 65–78. American Mathematical Society.

Gowers, T., editor (2008). *The Princeton Companion to Mathematics*. Princeton University Press.

Grattan-Guinness, I. (2009). Numbers, magnitudes, ratios, and proportions in Euclid's elements: How did he handle them? In *Routes of Learning*, pages 171–195. Johns Hopkins University Press. (First appeared *Historia Mathematica*, 23 1996:355–375).

Gray, J. (2004). *Non-Euclidean Geometry and the Nature of Space*. Burndy Publications. Contains Latin and English translation of Bolyai.

(2011). *Worlds out of Nothing: A Course in the History of Geometry in the 19th Century*. Springer Undergraduate Mathematics Series. Springer Verlag. 2nd edition.

Greenberg, M. (2010). Old and new results in the foundations of elementary plane Euclidean geometry. *American Mathematical Monthly*, 117:198–219.

Grosholz, E. (1985). Two episodes in the unification of logic and topology. *British Journal of the Philosophy of Science*, 36:147–157.

Grossberg, R. and Hart, B. (1989). The classification theory of excellent classes. *Journal of Symbolic Logic*, 54:1359–1381.

Grossberg, R. and Shelah, S. (1983). On universal locally finite groups. *Israel Journal of Mathematics*, 44:289–302. Sh index 174.

(1986). On the number of non isomorphic models of an infinitary theory which has the order property part A. *Journal of Symbolic Logic*, 51:302–322. Sh index 222.

Grossberg, R. and VanDieren, M. (2006a). Galois stability for tame abstract elementary classes. *Journal of Mathematical Logic*, 6:1–24.

(2006b). Shelah's categoricity conjecture from a successor for tame abstract elementary classes. *Journal of Symbolic Logic*, 71:553–568.

Grothendieck, A. (1966). Elements de géometrie algébriques (rédigés avec la collaboration de J. Dieudonné), IV Etudes Locale des Schémas et des Morphismes de Schémas. *Publications Mathematiques de l'IHES*, 28:5–255.

Guicciardini, N. (2006). Method versus calculus in Newton's criticisms of Descartes and Leibniz. In Gillies, D., editor, *Proceedings of the International Congress of Mathematicians, Madrid, Spain*, pages 1799–1823. European Mathematical Society.

Hafner, J. and Mancosu, P. (2008). Beyond unification. In Mancosu, P., editor, *The Philosophy of Mathematical Practice*, pages 151–178. Oxford University Press.

Hájek, P. (1998). *Metamathematics of Fuzzy Logic*. Trends in Logic–Studia Logica Library. Kluwer Academic Publishers.

Hallett, M. (2008). Reflections on the purity of method in Hilbert's Grundlagen der Geometrie. In Mancosu, P., editor, *The Philosophy of Mathematical Practice*, pages 198–256. Oxford University Press.

Hallett, M. and Majer, U., editors (2004). *David Hilbert's Lectures on the Foundations of Geometry 1891–1902*. Springer.

Hanf, W. (1960). Models of languages with infinitely long expressions. In *Abstracts of Contributed Papers from the First Logic, Methodology and Philosophy of Science Congress, Vol. 1*, page 24. Stanford University.

Harnik, V. (2011). Model theory vs. categorical logic; two approaches to pretopos completion (a.k.a. T^{eq}). In Hart, B. et al., editors, *Models, Logics, and Higher-dimensional Categories: A Tribute to the Work of Mihály Makkai*, volume 53 of *CRM, Proceedings and Lecture Notes*, pages 79–107. American Mathematical Society.

Harrington, L. and Makkai, M. (1985). The main gap: Counting uncountable models of ω-stable and superstable theories. *Notre Dame Journal of Formal Logic*.

Harrington, L., Morley, M., Ščedrov, A., and Simpson, S. (1985). Introduction. In *Harvey Friedman's Research on the Foundations of Mathematics*, Studies in Logic and the Foundations of Mathematics, pages xvi + 408. North-Holland.

Harris, M. (2015). *Mathematics without Apologies: Portrait of a Problematic Vocation*. Princeton University Press.

Hart, B. (1989). A proof of Morley's conjecture. *Journal of Symbolic Logic*, 54: 1346–1358.

Hart, B., Hrushovski, E., and Laskowski, C. (2000). The uncountable spectra of countable theories. *Annals of Mathematics*, 152:207–257.

Hart, B. and Laskowski, C. (1997). A survey of the uncountable spectra of countable theories. In Hart, B. et al., editors, *Algebraic Model Theory*, pages 107–118. Kluwer Academic Publishers.

Hart, B. and Shelah, S. (1990). Categoricity over P for first order T or categoricity for $\phi \in l_{\omega_1\omega}$ can stop at $\aleph_k$ while holding for $\aleph_0, \ldots, \aleph_{k-1}$. *Israel Journal of Mathematics*, 70:219–235.

Hartshorne, R. (1967). *Geometry: Foundations of Projective Geometry*. W. A. Benjamin.

(1977). *Algebraic Geometry*. Springer Verlag.

(2000). *Geometry: Euclid and Beyond*. Springer Verlag.

Haskell, D., Hrushovski, E., and MacPherson, H. (2007). *Stable Domination and Independence in Algebraically Closed Valued Fields*. Lecture Notes in Logic. Association for Symbolic Logic.

Heath, T. (1921). *A History of Greek Mathematics*. Clarendon Press.

Henderson, D. and Taimina, D. (2005). How to use history to clarify common confusions in geometry. In Shell-Gellasch, A. and Jardine, D., editors, *From Calculus to Computers*, volume 68 of *MAA Math Notes*, pages 57–74. Mathematical Association of America.

Henkin, L. (1949). The completeness of the first-order functional calculus. *Journal of Symbolic Logic*, 14:159–166.

(1953). Some interconnections between modern algebra and mathematical logic. *Transactions of the American Mathematical Society*, 74:410–427.

(1954). A generalization of the notion of ω-consistent. *Journal of Symbolic Logic*, 19:183–196.

(1955). The representation theorem for cylindrical algebras. In *Mathematical Interpretation of Formal Systems*, page 85–97. North-Holland.

(1996). The discovery of my completeness proofs. *Bulletin of Symbolic Logic*, 2:127–158.

Henson, W. and Keisler, H. (1986). On the strength of nonstandard analysis. *Journal of Symbolic Logic*, 51:377–386.

Hessenberg, G. (1905). Beweis des Desarguesschen Satzes aus dem Pascalschen. *Mathematische Annalen*, 61:161–172.

Heyting, A. (1963). *Axiomatic Projective Geometry*. John Wiley & Sons, North Holland.

Hilbert, D. (1918a). Axiomatic thought. In Ewald, W., editor, *From Kant to Hilbert, volume 2*, pages 1105–1115. Oxford University Press. Collection published 2005.

(1918b). Axiomatisches Denken. *Mathematische Annalen*, 78:405–15.

(1962). *Foundations of Geometry*. Open Court Publishers. Original German publication 1899: reprinted with additions in Townsend translation (with additions) 1902: Gutenberg e-book #17384, www.gutenberg.org/ebooks/17384.

(1971). *Foundations of Geometry*. Open Court Publishers. Translation from 10th German edition, Bernays 1968.

Hilbert, D. and Ackermann, W. (1938). *Grundzüge der Theoretischen Logik, 2nd edition*. Springer. First edition, 1928.

Hirvonen, Å. (2015). Classification theory: Crash course to Shelah's main gap. slides from Scandinavian logic school. www.helsinki.fi/sls2015/program.html.

Hirvonen, Å. and Hyttinen, T. (2009). Categoricity in homogeneous complete metric spaces. *Archive for Mathematical Logic*, 48:269–322.

Hodges, W. (1987). What is a structure theory? *Bulletin of the London Mathematics Society*, 19:209–237.

(1993). *Model Theory*. Cambridge University Press.

Howard, W. (2013). Comments on the relations of Bourbaki and logicians. http://homepages.math.uic.edu/~jbaldwin/pub/howonbour.pdf.

Hrushovski, E. (1986). Contributions to stable model theory. PhD thesis, University of California, Berkeley.

(1989). Almost orthogonal regular types. *Annals of Pure and Applied Logic*, 45:139–155.

(1993). A new strongly minimal set. *Annals of Pure and Applied Logic*, 62: 147–166.

(1996). The Mordell-Lang conjecture over function fields. *Journal of the American Mathematical Society*, 9:667–690.

(1997). Stability and its uses. In *Current Developments in Mathematics, 1996 (Cambridge, MA)*, pages 61–103. International Press.

(1998). Geometric model theory. In *Proceedings of the International Congress of Mathematicians, Volume I (Berlin, 1998)*, pages 281–302 (electronic).

(2002). Computing the Galois group of a linear differential equation. In *Differential Galois Theory (Bedlewo 2001)*, volume 58 of *Banach Center Publications*, pages 97–138. Polish Academy of Sciences.

(2012). Stable group theory and approximate subgroups. *Journal of the American Mathematical Society*, 25:189–243.

Hrushovski, E. and Itai, M. (2003). Strongly minimal sets in differentially closed fields. *Transactions of the American Mathematical Society*, 355:4267–4296.

Hrushovski, E. and Loeser, F. (2016). *Non-Archimedean Tame Topology and Stably Dominated Types*, volume 192 of *Annals of Mathematical Studies*. Princeton University Press.

Hrushovski, E. and Pillay, A. (2011). On nip and invariant measures. *Journal of the European Mathematical Society*, 13:1005–1061.

Hrushovski, E. and Sokolović (1993). Minimal subsets of differentially closed fields. Submitted approximately 1993; accepted but did not appear.

Hrushovski, E. and Zilber, B. (1993). Zariski geometries. *Bulletin of the American Mathematical Society*, 28:315–324.

Hughes, D. and Piper, F. (1973). *Projective Planes*. Springer Verlag.

Huntington, E. (1911). The fundamental propositions of algebra. In Young, J., editor, *Lectures on the Fundamental Concepts of Algebra and Geometry*, pages 151–210. Macmillan.

Hyttinen, T., Kangas, K., and Väänänen, J. (2013). On second order characterizability. *Logic Journal of the IGPL*, 21:767–787.

Hyttinen, T., Lessmann, O., and Shelah, S. (2005). Interpreting groups and fields in some nonelementary classes. *Journal of Mathematical Logic*, 5(1):1–47.

Iovino, J. (1999). Stable Banach spaces and Banach space structures II. In *Models, Algebras, and Proofs*, volume 203 of *Lecture Notes in Pure and Applied Math.*, pages 97–117. Dekker.

Jacobson, N. (1964). *Lectures in Abstract Algebra III*. University Series in Higher Mathematics. Van Nostrand.

Jarden, A. and Shelah, S. (2013). Non-forking frames in abstract elementary classes. *Annals Pure and Applied Logic*, 164:135–191. Shelah index 875.

Jech, T. (1978). *Set Theory*, volume 79 of *Pure and Applied Mathematics*. Academic Press.

Johnson, W. (2014). On the proof of elimination of imaginaries in algebraically closed valued fields. math arXiv:1406.3654v1.

Johnstone, P. (1982). *Stone Spaces*. Cambridge University Press.

Jónsson, B. (1956). Universal relational systems. *Mathematica Scandinavica*, 4: 193–208.

  (1960). Homogeneous universal relational systems. *Mathematica Scandinavica*, 8:137–142.

Julien, V. (1964). *Philosophie Naturelle et géométrie au XVIIe siècle*. Honoré Champion.

Kang, M. (1993). Injective morphisms of affine varieties. *Proceedings of the American Mathematical Society*, 119:1–4.

Kantor, W., Liebeck, M., and Macpherson, H. (1989). $\aleph_0$-categorical structures smoothly approximable by finite substructures. *Proceedings of the London Mathematical Society*, 59:439–463.

Karp, C. (1964). *Languages with Expressions of Infinite Length*. North Holland.

Kazhdan, D. (2006). Lecture notes in motivic integration: Logic. www.ma.huji.ac.il/~kazhdan/Notes/motivic/b.pdf.

Keisler, H. (1971). *Model Theory for Infinitary Logic*. North-Holland.

  (1976). Six classes of theories. *Journal of the Australian Mathematical Society Series A*, 21:257266.

  (2017). Review: Three papers of Maryanthe Malliaris and Saharon Shelah. *Bulletin of Symbolic Logic*, 23:117–121.

Kennedy, J. (2011). Gödel's thesis: An appreciation. In Baez, M. et al., editors, *Kurt Gödel and the Foundations of Mathematics: Horizons of Truth*, pages 95–109. Cambridge University Press.

  (2013). On formalism freeness: Implementing Gödel's 1946 Princeton bicentennial lecture. *Bulletin of Symbolic Logic*, 19:351–393.

  (2015). On the 'Logic without borders' point of view. In Hirvonen, A., Kontinen, J., Kossak, R., and Villaceces, A., editors, *Logic without Borders*, pages 1–14. DeGruyter.

Kennedy, J., Magidor, M., and Väänänen, J. (2016). Inner models from extended logics. Preprint.

Kim, B. (1998). Forking in simple theories. *Journal of the London Mathematical Society*, 57:257–267.

Kim, B. and Pillay, A. (1997). Simple theories. *Annals of Pure and Applied Logic*, 88:149–164.

Kirby, J. (2008). Abstract elementary categories. www.uea.ac.uk/~ccf09tku/pdf/aecats.pdf.

  (2010). On quasiminimal excellent classes. *Journal of Symbolic Logic*, 75:551–564.

Klee, V. and Wagon, S. (1991). *Old and New Unsolved Problems in Plane Geometry and Number Theory*. Dolciani Mathematical expositions. Mathematical Association of America.

Klein, J. (1968). *Greek Mathematical Thought and the Origin of Algebra*. Dover. Original German edition 1934.

Knight, J., Montalban, A., and Schweber, N. (2016). Computable structures in generic extensions. *Journal of Symbolic Logic*, 81:814–832.

Kochen, S. (1961). Ultraproducts in the theory of models. *Annals of Mathematics*, 74:221–261.

Kojman, M. and Shelah, S. (1992). Non-existence of universal orders in many cardinals. *Journal of Symbolic Logic*, 57:875–891. Sh index 409.

Kolchin, E. (1973). *Differential Algebra and Algebraic Groups*. Academic Press.

Koponen, V. (2006). The finite submodel property and ω-categorical expansions of pregeometries. *Annals of Pure and Applied Logic*, 139:201–229.

Kossak, R. (2011). What is a resplendent structure? *Acta Math Hungarica*, 58: 812–814.

Kreisel, G. (1984). Logical foundations, a lingering malaise. Zitiert mit Genehmigung des Philosphischen Archivs der Universität Konstanz. Alle Recht vorbehalten.

Kunen, K. (1980). *Set Theory: An Introduction to Independence Proofs*. North-Holland.

Lachlan, A. (1972). A property of stable theories. *Fundamenta Mathematicae*, 77: 9–20.

(1975). Theories with a finite number of models in an uncountable power are categorical. *Pacific Journal of Mathematics*, 61:698–711.

Lakatos, D. (1976). *Proofs and Refutations*. Cambridge University Press.

Lang, S. (1964). *Algebraic Geometry*. Interscience.

Langford, C. (1926-27). Theorems on deducibility. *Annals of Mathematics*, 28: 459–471.

Larson, P. (2017). Scott processes. In Iovino, J., editor, *Beyond First Order Model Theory*. Chapman Hall.

Lascar, D. (1985). Why some people are excited by Vaught's conjecture. *Journal of Symbolic Logic*, 50:973–982.

Lascar, D. and Poizat, B. (1979). An introduction to forking. *Journal of Symbolic Logic*, 44:330–350.

Laskowski, M. C. (1988). Uncountable theories that are categorical in a higher power. *Journal of Symbolic Logic*, 53:512–530.

(1992). Vapnik-Chervonenkis classes of definable sets. *Journal of the London Mathematical Society*, 45:377–384.

Laskowski, M. C., Rast, R., and Ulrich, D. (2017). Borel completeness and potential canonical Scott sentences. *Fundamenta Mathematicae*.

Levi, F. (1939). On a fundamental theorem of geometry. *Journal of the Indian Mathematical Society*, 3–4:82–92.

Levi-Civita, T. (1892). Sugli infiniti ed infinitesimi attuali quali elementi analitici. *Atti del R. Istituto Veneto di Scienze Lettre ed Arti, Venezia (Serie 7)*, pages 1765–1815. Reprinted in *Opera Matematiche. Memorie e Note [Collected mathematical works. Memoirs and Notes]* Volume Primo (1893–1900) Pubblicate a cura dell'Accademia Nationale dei Lincei, Roma, Editor Zanichelli (1954).

Lieberman, M. (2013). Category-theoretic aspects of abstract elementary classes. *Annals of Pure and Applied Logic*, 162:903–915.

Lieberman, M. and Rosický, J. (2016). Classification theory for accessible categories. *Journal of Symbolic Logic*, 81:151–165.

——— (2017). Metric abstract elementary classes as accessible categories. *Journal of Symbolic Logic*. arXiv:1504.02660; doi:10.1017/jsl.2016.39.

Lindström, P. (1964). On model completeness. *Theoria*, 30:183–196.

Łos, J. (1954). On the categoricity in power of elementary deductive systems and related problems. *Colloquium Mathematicum*, 3:58–62.

Löwenheim, L. (1967). On possibilities in the calculus of relatives. In Van Heijenoort, J., editor, *From Frege to Gödel: A Sourcebook in Mathematical Logic, 1879–1931*. Harvard University Press. German original published in 1915.

Lyndon, R. (1967). *Notes on Logic*. Van Nostrand.

Macintyre, A. J. (1976). On definable subsets of p-adic fields. *Journal of Symbolic Logic*, 41:605–10.

——— (2003a). A history of interactions between logic and number theory. Notes from Arizona Winter School: http://swc.math.arizona.edu/aws/2003/03MacintyreNotes.pdf.

——— (2003b). Model theory: Geometrical and set-theoretic aspects and prospects. *Bulletin of Symbolic Logic*, 9:197–212.

——— (2011). The impact of Gödel's incompleteness theorems on mathematics. In Baez, M. et al., editors, *Kurt Gödel and the Foundations of Mathematics*, pages 3–26. Cambridge University Press.

MacLane, S. (1936). Some interpretations of abstract linear dependence in terms of projective geometry. *American Journal of Mathematics*, pages 236–240.

——— (1986). *Mathematics: Form and Function*. Springer Verlag.

Macpherson, D. and Steinhorn, C. (2011). Definability in classes of finite structures. In Esparza, J., Michaux, C., and Steinhorn, C., editors, *Finite and Algorithmic Model Theory*, London Mathematical Society Lecture Note Series (No. 379), pages 140–176. Cambridge University Press.

Maddy, P. (2007). *Second Philosophy: A Naturalistic Method*. Oxford University Press.

——— (2011). *Defending the Axioms*. Oxford University Press.

Magidor, M. (2013). Inner models constructed from generalized logics. Slides from Mostowski centennial. http://mostowski100.mimuw.edu.pl/lib/exe/fetch.php?media=magidor_mostowski100.pdf.

——— (2015). Inner models constructed from generalized logics. Slides from Logic Colloquium, 2015 Helsinki.

Makkai, M. and Paré, R. (1989). *Accessible Categories: The Foundations of Categorical Model Theory*, volume 104 of *Memoirs of the American Mathematical Society*. American Mathematical Society.

Makkai, M. and Reyes, G. (1977). *First Order Categorical Logic*, volume 111 of *Springer Lecture Notes*. Springer Verlag.

Makkai, M. and Shelah, S. (1990). Categoricity of theories in $L_{\kappa,\omega}$, with κ a compact cardinal. *Annals of Pure and Applied Logic*, 47:41–97.

Makowsky, J. A. (2013). Lecture 7: The decidability of elementary geometry. Topics in Automated Theorem Proving; www.cs.technion.ac.il/~janos/COURSES/THPR/2013-14/lect-7.pdf.

Malcev, A. (1971a). A general method for obtaining local theorems in group theory. In Wells, B., editor, *The Metamathematics of Algebraic Systems, Collected Papers: 1936–1967*, pages 15–21. North-Holland. 1941 Russian original in Uceny Zapiski Ivanov Ped. Inst.

(1971b). Investigations in the area of mathematical logic. In Wells, B., editor, *The Metamathematics of Algebraic Systems, Collected Papers: 1936–1967*, pages 1–14. North-Holland. 1936 German original in Mat. Sbornik.

Malliaris, M. (2009). Realization of ϕ-types and Keisler's order. *Annals of Pure and Applied Logic*, 157:220–224.

Malliaris, M. and Shelah, S. (2013). General topology meets model theory, on p and t. *Proceedings of the National Academy of Sciences of the USA*, 110:13300–13305.

(2014). Regularity lemmas for stable graphs. *Transactions of the American Mathematical Society*, 366:1551–1585.

(2015). Keisler's order has infinitely many classes. *Israel Journal of Mathematics*. arXiv:1503.08341.

(2016). Cofinality spectrum theorems in model theory, set theory, and general topology. *Journal of the American Mathematical Society*, 29:237–297.

Mancosu, P. (2008a). Mathematical explanation: Why it matters. In Mancosu, P., editor, *The Philosophy of Mathematical Practice*, pages 134–150. Oxford University Press.

Mancosu, P., editor (2008b). *The Philosophy of Mathematical Practice*. Oxford University Press.

Mancosu, P. and Zach, R. (2015). Howard Behman's 1921 lecture on the decision problem and the algebra of logic. *The Bulletin of Symbolic Logic*, 21:164–187.

Manders, K. (1984). Interpretations and the model theory of the classical geometries. In Müller, G. and Richter, M., editors, *Models and Sets*, pages 297–330. Springer Lecture Notes in Mathematics 1103.

(1987). Logic and conceptual relationships in mathematics. In The Paris Logic Group, editors, *Logic Colloquium '85*, pages 194–211. Elsevier Science Publishers B.V. (North-Holland).

(1989). Domain extension and the philosophy of mathematics. *Journal of Philosophy*, 86:553–562.

(2008). Diagram-based geometric practice. In Mancosu, P., editor, *The Philosophy of Mathematical Practice*, pages 65–79. Oxford University Press.

Manzano, M. (1996). *Extensions of First-Order Logic*. Cambridge University Press.

Marek, W. (1973). Consistance d'une hypothèse de Fraïssé sur la définissabilité dans un langage du second ordre. *Comptes rendus de l'Académie des sciences, série A-B* 276:A1147–A1150.

Marker, D. (1996). Model theory and exponentiation. *Notices of the American Mathematical Society*, 43:753–760.

(2000). Review of *Tame Topology and O-minimal Structures*, by Lou van den Dries. *Bulletin (New Series) of the American Mathematical Society*, 37:351–357.

(2002). *Model Theory: An Introduction*. Springer Verlag.

(2007). The number of countable differentially closed fields. *Notre Dame Journal of Formal Logic*, 48:99–113.

(2011). Scott ranks of counterexamples to Vaught's conjecture. Notes from 2011; http://homepages.math.uic.edu/~marker/harrington-vaught.pdf.

(2016). *Lectures on Infinitary Model Theory*. Lecture Notes in Logic. Association of Symbolic Logic, Cambridge University Press.

Marker, D., Messmer, M., and Pillay, A. (1996). *Model Theory of Fields*. Springer Verlag.

Marquis, J.-P. (2008). A path to the epistemology of mathematics: Homotopy theory. In Ferreirós, J. and Gray, J. J., editors, *Architecture of Modern Mathematics: Essays in History and Philosophy*, pages 239–260. Oxford University Press.

Martin, D. A. (2001). Multiple universes of sets and indeterminate truth values. *Topoi*, 20:5–16.

Martin, K. (2015). Paradox lost and paradox regained: The Banach-Tarski paradox and ancient paradoxes of infinity. PhD thesis, Department of Philosophy, University of Illinois at Chicago.

Mathias, A. (1992). The ignorance of Bourbaki. *Mathematical Intelligencer*, 14:4–13. also in: Physis Riv. Internaz. Storia Sci (New Series) 28 (1991):887–904.

(2012). Hilbert, Bourbaki, and the scorning of logic. Preprint www.dpmms.cam .ac.uk/~ardm/logbanfinalmk.pdf.

Mazur, B. (2008). When is one thing equal to some other thing? In Gold, B. and Simons, R., editors, *Proof and Other Dilemmas: Mathematics and Philosophy*, pages 221–242. Mathematics Association of America.

McGee, V. (1997). How we learn mathematical language. *The Philosophical Review*, 106:35–68.

McLarty, C. (2010). What does it take to prove Fermat's last theorem? Grothendieck and the logic of number theory. *Bulletin of Symbolic Logic*, 16:359–377.

Meadows, T. (2013). What can a categoricity theorem tell us? *Review of Symbolic Logic*, 6:524–544.

Menn, S. (2018). Eudoxus' theory of proportion and his method of exhaustion. In Reck, E., editor, *Logic, Philosophy of Mathematics, and their History: Essays in Honor of W.W. Tait*. College Publications.

Miller, N. (2007). *Euclid and his Twentieth Century Rivals: Diagrams in the Logic of Euclidean Geometry*. CSLI Publications.

Millman, R. and Parker, G. (1981). *Geometry: A Metric Approach with Models*. Undergraduate texts in mathematics. Springer Verlag.

Molland, A. (1976). Shifting the foundations: Descartes's transformation of ancient geometry. *Historia Mathematica*, 3:21–49.

Montalban, A. (2013). A computability theoretic equivalent to Vaught's conjecture. *Advances in Mathematics*, 235:56–73.

Moore, G. (1988). The emergence of first order logic. In Aspray, W. and Kitcher, P., editors, *History and Philosophy of Modern Mathematics*, pages 95–135. University of Minnesota Press.

—— (1997). The prehistory of infinitary logic: 1885–1955. In Maria Luisa Dalla Chiara et al., editors, *Structures and Norms in Science: Volume Two of the Tenth International Congress of Logic, Methodology and Philosophy of Science, Florence, August 1995*, pages 105–123. Springer Netherlands.

Moosa, R. (2005). On saturation and the model theory of compact Kähler manifolds. *Journal für die reine und angewandte Mathematik*, 586:1–20.

Morley, M. (1965a). Categoricity in power. *Transactions of the American Mathematical Society*, 114:514–538.

—— (1965b). Omitting classes of elements. In Addison, J., Henkin, L., and Tarski, A., editors, *The Theory of Models*, pages 265–273. North-Holland.

—— (1970). The number of countable models. *Journal of Symbolic Logic*, 35:14–18.

Morley, M. and Vaught, R. (1962). Homogeneous universal models. *Mathematica Scandinavica*, 11:37–57.

Mostowski, A. (1937). Abzählbare Boolesch Körper und ihre Anwendug auf die allgemeine Metamathematik. *Fundamenta Mathematicae*, 29:34–53.

Mueller, I. (2006). *Philosophy of Mathematics and Deductive Structure in Euclid's Elements*. Dover Books in Mathematics. Dover Publications. First published by MIT press in 1981.

Murphy, G. and Medin, D. (1985). Concepts and conceptual structure. *Psychological Review*, 92:289–314.

Nagloo, J. and Pillay, A. (2016). On algebraic relations between solutions of a generic Painlevé equation. *Journal für die reine und angewandte Mathematik (Crelle's Journal)*. DOI:10.1515/crelle-2014-0082.

Newton, I. (1769). On the linear construction of equations; Universal arithmetic. In Whiteside, D. T., editor, *The Mathematical Works of Isaac Newton*. (1964–1967). Johnson Reprint Corp., 2 volumes.

Orey, S. (1956). On ω-consistency and related properties. *Journal of Symbolic Logic*, 21:246–252.

Pambuccian, V. (2001). A methodologically pure proof of a convex geometry problem. *Beiträge zur Algebra und Geometrie, Contributions to Algebra and Geometry*, 42:40–406.

—— (2005). Euclidean geometry problems rephrased in terms of midpoints and point-reflections. *Elemente der Mathematik*, 60:19–24.

—— (2009). A reverse analysis of the Sylvester-Gallai theorem. *Notre Dame Journal of Formal Logic*, 50:245–259.

(2014a). Review of 'completeness and continuity in Hilbert's foundations of geometry: On the Vollständigkeitsaxiom' by Eduardo Giovannini. *Zbl 1269.03002*.

(2014b). Review of 'Giuseppe Veronese and the foundations of geometry' by Paola Cantú. *Zbl 1158.51300*.

Panza, M. (2011). Rethinking geometrical exactness. *Historia Mathematica*, 38: 42–95.

Parsons, C. (1990a). The structuralist view of mathematical objects. *Synthese*, 384:303–346.

(1990b). The uniqueness of the natural numbers. *Iyyun*, 39:13–44.

(2013). Some consequences of the entanglement of logic and mathematics. In Frauchiger, M., editor, *Reference, Rationality, and Phenomenology: Themes from Føllesdal*, pages 153–178. Walter de Gruyter.

Pavelka, J. (1979). On fuzzy logic III. *Z.Math. Logik Grundlagen*, 25:447–464.

Peretyatkin, M. G. (1997). *Finitely Axiomatizable Theories*. Springer Verlag.

Peterzil, Y. and Starchenko, S. (1998). A trichotomy theorem for o-minimal theories. *Proceedings of the London Mathematical Society*, 77:481–523.

(2000). Geometry, calculus, and Zilber's conjecture. *Bulletin of Symbolic Logic*, 2:72–82.

(2010). Tame complex analysis and o-minimality. In *Proceedings of the International Congress of Mathematicians, Vol. II (New Delhi, 2010)*, pages 58–81.

Pierce, D. (2011). Numbers. https://arxiv.org/pdf/1104.5311v1.pdf.

Pierce, D. and Pillay, A. (1998). A note on the axioms for differentially closed fields of characteristic zero. *Journal of Algebra*, 204:108–115.

Pillay, A. (1988). On groups and fields definable in o-minimal structures. *Journal of Pure and Applied Algebra*, 55:239–255.

(1995). Model theory, differential algebra, and number theory. In Chatterji, S., editor, *Proceedings of the International Congress of Mathematicians, Zurich, 1994*, pages 277–287. Birkhauser Verlag.

(1996). Differential algebraic groups and the number of countable differentially closed fields. In *Model Theory of Fields*, pages 114–134. Springer.

(1999). The model theoretic content of Lang's conjecture. In Bouscaren, E., editor, *Model Theory and Algebraic Geometry : An Introduction to E. Hrushovski's Proof of the Geometric Mordell-Lang Conjecture*, pages 101–106. Springer Verlag.

(2000). Model theory. *Notices of the American Mathematical Society*, 47:1373–1381.

(2010). Model theory. *Logic and Philosophy Today, Part 1: Journal of Indian Council of Philosophical Research*, 28. Reprinted in Logic and Philosophy Today, Studies in Logic 29 (& 30), College Publications, 2011.

Pillay, A. and Ziegler, M. (2003). Jet spaces of varieties over differential and difference fields. *Selecta Math. New Series*, 9:579–599.

Plato (1999). *Statesman*. Perseus digital library. Fowler translation 1921.

(2016). *Statesman*. Google Books. Jowett translation, 1892.

Plotkin, J. M. (1990). Who put the back in 'back-and-forth'? *Journal of Symbolic Logic*, 55:444–5. Abstract presented at joint meeting with APA, 1989.

Poincaré, H. (1903). H. Poincaré's review of Hilbert's Foundations of Geometry. *Bull. Amer. Math. Soc.*, 10:1–23. E. J. Townsend translated from French.

(1952). *Science and Hypothesis*. Dover, first published in French, 1905.

Poizat, B. (1983). Une théorie de Galois imaginaire. *Journal of Symbolic Logic*, 1151–1170.

(1985). *Cours de théories des modèles*. Nur Al-mantiq Wal-ma'rifah.

(1987). *Groupes stables*. Nur Al-mantiq Wal-ma'rifah.

(1990). Review of Baldwin, John T., Classification theory: 1985, *Journal of Symbolic Logic*, 878–881. Reviewed by pseudonym John B. Goode.

(1991). Review of Baldwin, John T., The spectrum of resplendency, *J. Symbolic Logic* 55 (1990). *Math Reviews*. MR1056376.

(2001). *Stable Groups*. American Mathematical Society. Translation of Poizat (1987).

Post, E. (1967). Introduction to a general theory of elementary propositions. In Van Heijenoort, J., editor, *From Frege to Gödel: A Sourcebook in Mathematical Logic, 1879–1931*, pages 264–283. Harvard University Press.

Presburger, M. (1930). Über die Vollständigkeit eines gewissen Systems der Arithmetic ganzer Zahlen, in welchem die Addition als einzige Operation hervortritt. In *Sprawozdanie z I Kongresu Mat. Karjów Słowiańskich*, pages 92–101. See Stansifer 1984 for translation.

Putnam, H. (1982). Peirce the logician. *Historia Mathematica*, 9:290–301.

Quine, W. V. O. (1977). Natural kinds. In Schwartz, S. P., editor, *Naming, Necessity, and Natural Kinds*, pages 155–175. Cornell University Press.

Ramsey, F. (1930). On a problem of formal logic. *Proceedings of the London Mathematical Society*, s2-30:264–286.

Reineke, J. (1975). Minimale Gruppen. *Zeitshrift Math. Logik. and Grundlagen der Math.*, 21:357–359.

Ritt, J. (1950). *Differential Algebra*, volume 70 of *Colloquium Publications*. American Mathematical Society. Revision of Differential equations from the algebraic standpoint, 1932.

Robinson, A. (1951). *Introduction to Model Theory and to the Metamathematics of Algebra*. Studies in Logic and the Foundations of Mathematics. North-Holland. 1st edition 1951; 2nd edition 1965.

(1952). On the application of symbolic logic to algebra. In *Proceedings of the International Congress of Mathematicians, Cambridge, Massachusetts, U.S.A., August 30–September 6, 1950*, Volume 1, pages 686–694. American Mathematical Society.

(1954). On predicates in algebraically closed fields. *Journal of Symbolic Logic*, 19:103–114.

(1956). *Complete Theories*. North-Holland.

(1959a). On the concept of a differentially closed field. *Bulletin of the Research Council of Israel*, 8F:113–128.

Robinson, G. D. (1959b). *Foundations of Geometry*. Toronto Press.

Rodin, A. (2014). *Axiomatic Method and Category Theory*. Springer.

Rogers, H. (1967). *Theory of Recursive Functions and Effective Computability*. McGraw-Hill.

Roitman, J. (1990). *Introduction to Modern Set Theory*. Wiley.

Rosenstein, J. (1982). *Linear Orderings*. Academic Press.

Rosický, J. (1981). Concrete categories and infinitary languages. *Journal of Pure and Applied Algebra*, 22:309–339.

Rusnock, P. (2000). *Bolzano's Philosophy and the Emergence of Modern Mathematics*. Rodopi.

Russell, B. and Whitehead, A. N. (1910). *Principia Mathematica*. Oxford.

Ryll-Nardzewski, C. (1959). On categoricity in power $\leq \aleph_0$. *Bulletin of the Polish Academy of Sciences; Series: Mathematics, Astronomy, and Physics*, 7:545–548.

Sacks, G. E. (1972). *Saturated Model Theory*. Benjamin.

(1975). Remarks against foundational activity. *Historia Mathematica*, 2:523–528.

Salanskis, J.-M. and Sinaceur, H., editors (1992). *Le labyrinthe du continu*. Springer.

Sànchez, O. and Pillay, A. (2016). Some definable galois theory and examples. Preprint.

Scanlon, T. (2001). Diophantine geometry from model theory. *Bulletin of Symbolic Logic*, 7:37–57.

(2002). Model theory and differential algebra. In Guo, L., Cassidy, P. J., Keigher, W. F., and Sit, W. Y., editors, *Differential Algebra and Related Topics: Proceedings of the International Workshop (Newark, November 2000)*, pages 125–150. World Scientific.

(2009). Motivic integration: An outsider's tutorial. Lecture, Durham, England; https://math.berkeley.edu/~scanlon/papers/scanlon_durham_motivic_integration_outsiders_tutorial.pdf.

(2012). Counting special points: Logic, Diophantine geometry, and transcendence theory. *Bulletin of the American Mathematical Society, New Series*, 49: 51–71.

Schiemer, G. (2016). Hilbert, duality, and the geometrical roots of model theory. *Review of Symbolic Logic*, 433–472.

Schiemer, G. and Reck, E. (2013). Logic in the 1930's: type theory and model theory. *Bulletin of Symbolic Logic*, 19:37–92.

Schlimm, D. (1985). Bridging theories with axioms: Boole, Stone, Tarski. In Van Kerhove, B., editor, *New Perspectives on Mathematical Practices: Essays in Philosophy and History of Mathematics*, pages 222–235. World Scientific.

(2013). Axioms in mathematical practice. *Philosophia Mathematica*, 21:37–92.

Seidenberg, A. (1954). A new decision method for elementary algebra. *Annals of Mathematics*, 60:365–374.

(1958). Comments on Lefschetz's principle. *American American Monthly*, 65: 685–690.

Shapiro, S. (1991). *Foundations without Foundationalism: A Case for Second Order Logic*. Oxford University Press, paperback 2000.

(1997). *Philosophy of Mathematics: Structure and Ontology*. Oxford University Press.

(2005). Categories, structures, and the Frege-Hilbert controversy: The status of meta-mathematics. *Philosophica Mathematica (III)*, 13:61–77.

Shelah, S. (1969). Categoricity and classes of models (Hebrew). PhD thesis, Hebrew University of Jerusalem.

(1970). Stability, the f.c.p., and superstability; model theoretic properties of formulas in first-order theories. *Annals of Mathematical Logic*, 3:271–362.

(1972). A combinatorial problem; stability and order for models and theories in infinitary languages. *Pacific Journal of Mathematics*, 41:247–261.

(1975). Categoricity in $\aleph_1$ of sentences in $L_{\omega_1,\omega}(Q)$. *Israel Journal of Mathematics*, 20:127–148. Sh index 48.

(1978). *Classification Theory and the Number of Nonisomorphic Models*. North-Holland.

(1980). Simple unstable theories. *Annals of Mathematical Logic*, 19:177–203.

(1982). The spectrum problem I, $\aleph_\epsilon$-saturated models the main gap. *Israel Journal of Mathematics*, 43:324–356.

(1983a). Classification theory for nonelementary classes. I. The number of uncountable models of $\psi \in L_{\omega_1\omega}$ part A. *Israel Journal of Mathematics*, 46;3: 212–240. Sh index 87a.

(1983b). Classification theory for nonelementary classes. II. The number of uncountable models of $\psi \in L_{\omega_1\omega}$ part B. *Israel Journal of Mathematics*, 46;3: 241–271. Sh index 87b.

(1985). Classification of first order theories which have a structure theory. *Bulletin of the American Mathematical Society*, 12:227–232. Sh index 200.

(1990). *Classification Theory and the Number of Nonisomorphic Models*. North-Holland. Second edition.

(1996). Toward classifying unstable theories. *Annals of Pure and Applied Logic*, 80:229–255. Sh index 500.

(1999). The future of set theory. *Israel Mathematical Conference proceedings*, 6. http://shelah.logic.at/files/E16.pdf.

(2000a). On what I do not understand (and have something to say): I. *Fundamenta Mathematicae*, 166:1–82.

(2000b). On what I do not understand (and have something to say), model theory, II. *Mathematica Japonica*, 51:329–377.

(2004). Classification theory for elementary classes with the dependence property – a modest beginning. *Scientiae Math Japonicae*, 59:265–316.

(2009a). *Classification Theory for Abstract Elementary Classes*. Studies in Logic. College Publications.

(2009b). Model theory without choice: Categoricity. *Journal of Symbolic Logic*, 74:361–401.

(2013a). Maximal failures of sequence locality in a.e.c. Preprint on archive.

(2013b). Response to the award of the 2013 Steele prize for seminal research. *A.M.S. Prize Booklet, 2013*, page 50. www.ams.org/profession/prizebooklet-2013.pdf.

(2015). Dependent theories and the generic pair conjecture. *Communications in Contemporary Math*, 17:64 pp. Sh index 900: had circulated since early 2000s.

Shelah, S., Harrington, L., and Makkai, M. (1984). A proof of Vaught's conjecture for ω-stable theories. *Israel Journal of Mathematics*, 49:259–280.

Shoenfield, J. (1967). *Mathematical Logic*. Addison-Wesley.

Sieg, W. (1994). Mechanical procedures and mathematical experiences. In George, A., editor, *Mathematics and Mind*, pages 71–117. Oxford University Press.

(1999). Hilbert's programs: 1917–1922. *Bulletin of Symbolic Logic*, 5:1–44.

(2013). *Hilbert's Programs and Beyond*. Oxford University Press.

Simon, P. (2011). Ordre et stabilité en les théories NIP. PhD thesis, l'Université Paris-Sud.

Simpson, S. (2009). *Subsystems of Second Order Arithmetic*. Cambridge University Press.

Skolem, T. (1920/1967a). Logico-combinatorial investigations in the satisfiability or provability of mathematical propositions: A simplified proof of a theorem of L. Löwenheim and generalizations of the theorem. In Van Heijenoort, J., editor, *From Frege to Gödel: A Sourcebook in Mathematical Logic, 1879–1931*, pages 252–263. Harvard University Press. German original published in 1920.

(1923/1967b). Some remarks on axiomatic set theory. In Van Heijenoort, J., editor, *From Frege to Gödel: A Sourcebook in Mathematical Logic, 1879–1931*, pages 290–301. Harvard University Press. German original published in 1923; address delivered in 1922.

(1934). Über die nicht-charakterisierbarkeit der Zahlenreihe mittels endlich oder abzhlbar unendlich vieler Aussagen mit ausschliesslich Zahlenvariablen. *Fundamenta Mathematica*, 23:150–161.

Smith, J. T. (2010). Definitions and non-definability in geometry. *American Mathematical Monthly*, 117:475–489.

Smorynski, C. (2008). *History of Mathematics: A Supplement*. Springer Verlag.

Smullyan, R. M. (1961). *Theory of Formal Systems*. Princeton University Press.

Sobociński, B. (1955). On well-constructed axiom systems. *Polish Society of Arts and Sciences Abroad*, pages 1–12.

Spivak, M. (1980). *Calculus*. Publish or Perish Press.

Stansifer, R. (1984). Presburger's article on integer arthmetic: Remarks and translation. Technical Report TR84-639, Cornell University, Computer Science Department.

Stein, H. (1990). Eudoxos and Dedekind: On the ancient Greek theory of ratios and its relation to modern mathematics. *Synthese*, 222:163–211.

Steinitz, F. (1910). Algebraische theorie der Körper. *Crelle's Journal*, 167–309.

Stekeler-Weithofer, P. (1992). On the concept of proof in elementary geometry. In Detlefsen, M., editor, *Proof and Knowledge in Mathematics*, pages 81–94. Routledge.

Suppes, P. (1956). *Introduction to Logic*. Van Nostrand.

Sylla, E. (1984). Compounding ratios, Bradwardine, Oresme, and the first edition of Newton's Principia. In Mendelsohn, E., editor, *Transformations and Tradition in the Sciences: Essays in Honor of I. Bernard Cohen*, pages 11–44. Cambridge University Press.

Tao, T. (2009). Infinite fields, finite fields, and the Ax-Grothendieck theorem. http://terrytao.wordpress.com/2009/03/07/infinite-fields-finite-fields-and-the-ax-grothendieck-theorem/.

Tappendon, J. (2008a). Mathematical concepts and definitions. In Mancosu, P., editor, *The Philosophy of Mathematical Practice*, pages 251–275. Oxford University Press.

(2008b). Mathematical concepts: Fruitfulness and naturalness. In Mancosu, P., editor, *The Philosophy of Mathematical Practice*, pages 276–301. Oxford University Press.

Tarski, A. (1931). Sur les ensembles définissables de nombres réels I. *Fundamenta Mathematica*, 17:210–239.

(1946). A remark on functionally free algebras. *Annals of Mathematics*, 47: 163–165.

(1950). Some notions and methods on the borderline of algebra and metamathematics. In *Proceedings of the International Congress of Mathematicians, Cambridge, Massachusetts, U.S.A., August 30–September 6, 1950*, volume 1, pages 705–720. American Mathematical Society, 1952.

(1951). *A Decision Method for Elementary Algebra and Geometry*. University of California Press, 2nd edition. In Caviness, B. F. and Johnson, J. R., editors, *Quantifier Elimination and Cylindrical Algebraic Decomposition*, pages 24–84. Texts and Monographs in Symbolic Computation. Springer, 1998.

(1954). Contributions to the theory of models, I and II. *Indagationes Mathematicae*, 16:572,582.

(1956). On some fundamental concepts of metamathematics. In *Logic, Semantics and Metamathematics: Papers from 1923–38*. Clarendon Press. Translated by J. H. Woodger.

(1959). What is elementary geometry? In Henkin, L., Suppes, P., and Tarski, A., editors, *The Axiomatic Method*, pages 16–29. North-Holland.

(1965). *Introduction to Logic and to the Methodology of the Deductive Sciences*. Galaxy Book. First edition: Oxford 1941.

Tarski, A., Mostowski, A., and Robinson, R. (1968). *Undecidable Theories*. North-Holland. First edition: Oxford 1953.

Tarski, A. and Vaught, R. (1956). Arithmetical extensions of relational systems. *Compositio Mathematica*, 13:81–102.

Teissier, B. (1997). Tame and stratified objects. In Schneps, L. and Lochak, P., editors, *Geometric Galois Actions I: Around Grothendieck's Equisse d'un Programme*, pages 231–242. Cambridge University Press.

Tsementzis, D. (2016). Univalent foundations as structuralist foundations. *Synthese*. doi:10.1007/s11229-016-1109-x.

Tsuboi, A. (2014). Dividing and forking – a proof of the equivalence. In *Model Theoretic Aspects of the Notion of Independence and Dimension*, Volume 1888, pages 23–28. RIMS Kokyuroku.

The Univalent Foundations Program (2015). *Homotopy Type Theory: Univalent Foundations of Mathematics*. Institute for Advanced Study, Princeton.

Väänänen, J. (2012). Second order logic or set theory. *Bulletin of Symbolic Logic*, 18:91–121.

(2014). Sort logic and foundations of mathematics. *Lecture Note Series, IMS, NUS*, pages 171–186.

Väänänen, J. and Wong, T. (2015). Internal categoricity in arithmetic and set theory. *Notre Dame Journal of Formal Logic*, 56:121–134. doi: 10.1215/00294527-2835038.

Vahlen, K. T. (1907). Über nicht-archimedische Algebra. *Jahresber. Deutsch. Math. Verein*, 16:409–421.

van der Hoeven, J. (2006). *Transseries and Real Differential Algebra*. Self-published.

Van der Waerden, B. L. (1949). *Modern Algebra*. Frederick Ungar Publishing. First German edition 1930.

Vasey, S. (2016a). Infinitary stability theory. *Archive for Mathematical Logic*, 55: 562–592.

(2016b). Toward a stability theory of tame abstract elementary classes. Preprint.

(2017a). Indiscernible extraction and Morley sequences. *Notre Dame Journal of Formal Logic*, 58:127–132.

(2017b). Shelah's eventual categoricity conjecture in universal classes II. *Selecta Mathematica*, 23:1469–1506.

Vaught, R. (1961). Denumerable models of complete theories. In *Infinitistic Methods, Proceedings of the Symposium on the Foundations of Mathematics, Warsaw, 1959*, pages 303–321. Państwowe Wydawnictwo Naukowe.

(1965). A Löwenheim-Skolem theorem for cardinals far apart. In Addison, J., Henkin, L., and Tarski, A., editors, *The Theory of Models*, pages 81–89. North-Holland.

(1986). Alfred Tarski's work in model theory. *Journal of Symbolic Logic*, 51: 869–882.

Veblen, O. (1904). A system of axioms for geometry. *Transactions of the American Mathematical Society*, 5:343–384.

(1914). The Foundations of Geometry. In Young, J., editor, *Monographs of Modern Mathematics Relevant to the Elementary Field*, pages 1–51. Longman, Green, and Company.

Veronese, G. (1889). Il continuo rettlineo e lassioma v di Archimede. *Memorie della Reale Accademia dei Lincei. Classe di scienze naturali, fisiche e matematiche*, 4:603–624.

——— (1891). Fondamenti di geometria a pi dimensioni e a pi specie di unit rettilinee esposti in forma elementare. *Lezioni per la Scuola di magistero in Matematica, Padua.*

Vis, T. (2009). Desargues theorem. Unpublished notes http://math.ucdenver.edu/~tvis/Teaching/4220spring09/Notes/Desargues.pdf.

Wagner, F. (1994). Relational structures and dimensions. In Kaye, R. et al., editors, *Automorphisms of First Order Structures*, pages 153–180. Clarendon Press.

Waterhouse, W. (1979). *Affine Group Schemes.* Springer Verlag.

Weil, A. (1950). The future of mathematics. *American Mathematical Monthly*, page 297.

——— (1962). *Foundations of Algebraic Geometry.* American Mathematical Society. 1st edition 1946.

Werndl, C. (2009). Justifying definitions in mathematics going beyond Lakatos. *Philosophia Mathematica*, 313–340.

White, N. and Nicoletti, G. (1985). Axiom systems. In White, N., editor, *Theory of Matroids*, volume 1, pages xvi+ 316. Cambridge University Press.

Wilkie, A. (1996). Model completeness results for expansions of the real field by restricted Pfaffian functions and exponentiation. *Journal of the American Mathematical Society*, 1051–1094.

——— (2007). O-minimal structures. *Séminaire Bourbaki*, 985. http://eprints.ma.man.ac.uk/1745/01/covered/MIMS_ep2012_3.pdf.

Wilson, M. (2006). *Wandering Significance: An Essay on Conceptual Behaviour.* Oxford University Press.

Yaqub, J. (1967). The Lenz-Barlotti classification. In Sandler, R., editor, *Proceedings of the Projective Geometry Conference, 1967*, pages 129–163. University of Illinois at Chicago Circle.

Zagzebski, L. (1996). *Virtues of the Mind: An Inquiry into the Nature of Virtue and the Ethical Foundations of Knowledge.* Cambridge University Press.

Zalamea, F. (2012). *Synthethic Philosophy of Contemporary Mathematics.* Urbanomic/Sequence Press.

Ziegler, M. (1982). Einige unentscheidbare Körpertheorien. *Enseignement Mathématique*, 28:269–280. Michael Beeson has an English translation.

Zilber, B. (1980). Strongly minimal countably categorical theories. *Siberian Mathematics Journal*, 24:219–230.

——— (1984a). Strongly minimal countably categorical theories II. *Siberian Mathematics Journal*, 25:396–412.

——— (1984b). Strongly minimal countably categorical theories III. *Siberian Mathematics Journal*, 25:559–571.

(1984c). The structure of models of uncountably categorical theories. In *Proceedings of the International Congress of Mathematicians August 16-23, 1983, Warszawa*, pages 359–68. Polish Scientific Publishers.

(1991). *Uncountably Categorical Theories*. Translations of the American Mathematical Society. American Mathematical Society. Summary of earlier work.

(2004). Pseudo-exponentiation on algebraically closed fields of characteristic 0. *Annals of Pure and Applied Logic*, 132:67–95.

(2005a). Analytic and pseudo-analytic structures. In Cori, R., Razborov, A., Todorcevic, S., and Wood, C., editors, *Logic Colloquium 2000; Paris, France July 23–31, 2000*, number 19 in Lecture Notes in Logic. Association of Symbolic Logic.

(2005b). A categoricity theorem for quasiminimal excellent classes. In *Logic and its Applications*, volume 380 of *Contemporary Mathematics*, pages 297–306. American Mathematical Society.

(2010). *Zariski Geometries: Geometry from the Logician's Point of View*. Number 360 in London Math. Soc. Lecture Notes. London Mathematical Society, Cambridge University Press.

Index

Printed in the United States
By Bookmasters